Knowledge Spaces

Springer
Berlin
Heidelberg
New York
Barcelona
Hong Kong
London
Milan
Paris
Singapore
Tokyo

Jean-Paul Doignon • Jean-Claude Falmagne

Knowledge Spaces

With 30 Figures and 28 Tables

 Springer

Jean-Paul Doignon

Université Libre de Bruxelles
Département de Mathématiques, c.p. 216
Boulevard du Triomphe
B-1050 Bruxelles, Belgium

Jean-Claude Falmagne

Department of Cognitive Science
University of California
Irvine, CA 92697-5100
USA

Library of Congress Cataloging-in-Publication Data

Doignon, Jean-Paul.
    Knowledge spaces/Jean-Paul Doignon, Jean-Claude Falmagne.
        p. cm.
    Includes bibliographical references and index.
    ISBN 3-540-64501-2 (softcover: alk. paper)
    1. Natural language processing (Computer science)
    2. Computational linguistics. 3. Artificial intelligence.
    I. Falmagne, Jean-Claude. II. Title.
    QA76.9.N38D65   1999
    006.3'5–dc21   98-29442
                                                              CIP

ISBN 3-540-64501-2 Springer-Verlag Berlin Heidelberg New York

© Springer-Verlag Berlin Heidelberg 1999
Printed in Germany

Typesetting: Camera-ready by the authors
Cover Design: Künkel+Lopka, Heidelberg
Printed on acid-free paper   SPIN 10663038   33/3142 – 5 4 3 2 1 0

Preface

The research reported in this book started during the academic year 1982–83. One of us (JCF) was on sabbatical leave at the University of Regensburg. For various reasons, the time was ripe for a new joint research subject. Our long term collaboration was thriving, and we could envisage an ambitious commitment. We decided to build an efficient machine for the assessment of knowledge—for example, that of students learning a scholarly subject. We began at once to work out the theoretical components of such a machine. Until then, we had been engaged in topics dealing mostly with geometry, combinatorics, psychophysics, and especially measurement theory. These subjects have some bearing on the content of this book. A close look at the foundations of measurement, in the sense that this term has in the physical sciences, may go a long way toward convincing you of its limited applicability. It seemed to us that in many scientific areas, from chemistry to biology and especially the behavioral sciences, theories must often be built on a very different footing than that of classical physics. Evidently, the standard physical scales such as time, mass, or energy may always be used in measuring aspects of phenomena. But the substrate proper to these other sciences may very well be, in most cases[1], of a fundamentally different nature. In short, nineteenth century physics is a bad example. This is not always understood. There was in fact a belief, shared by many scientists in the last century, that for an academic endeavour to be called a 'science', it had to resemble classical physics in critical ways. In particular, its basic observations had to be quantified in terms of measurement scales in the exact sense of classical physics.

Prominent advocates of that view were Francis Galton, Karl Pearson and William Thomson Kelvin. Because that position is still influential today, with a detrimental effect on fields such as 'psychological measurement', which is relevant to our subject, it is worth quoting some opinions in detail. In Pearson's biography of Galton (Pearson, 1924, Vol. II, p. 345), we can

---

[1] In some particular areas of the behavioral sciences, there is room for models in which the scaling of some numerical variable(s) plays a useful role. One possible example may be decision making, where the measurement of utility may be justified. Such situations, however, may be the exceptions rather than the rule.

find the following definition:

> **Anthropometry**, *or the art of measuring the physical and mental faculties of human beings, enables a shorthand description of any individual by measuring a small sample of his dimensions and qualities. These will sufficiently define his bodily proportions, his massiveness, strength, agility, keenness of sense, energy, health, intellectual capacity and mental character, and will substitute concise and exact* **numerical**[2] *values for verbose and disputable estimates*[3]*."*

For scientists of that era, it was hard to imagine a non-numerical approach to precise study of an empirical phenomenon. Karl Pearson himself, for instance—commenting on a piece critical of Galton's methods by the editor of the *Spectator*[4]—, wrote (Pearson, 1924, Vol. II, p. 345)

> *"There might be difficulty in ranking Gladstone and Disraeli for "Candour", but few would question John Morley's position relative to both of them in this quality. It would require an intellect their equal to rank truly in scholarship Henry Bradshaw, Robertson Smith and Lord Acton, but most judges would place all three above Sir John Seeley, as they would place Seeley above Oscar Browning. After all, there are such things as brackets, which only makes the statistical theory of ranking slightly less simple in the handling."*

In other words, measuring a psychical attribute such as 'candour' only requires fudging a bit around the edges of the order relation of the real numbers, making it either (in current terminology) a 'weak order' (cf. 0.19 in this book) or perhaps a 'semiorder' (cf. Problems 6 and 7 in Chapter 2).

As for Kelvin, his position on the subject is wel(known, and often represented in the form: *"If you cannot measure it, then it is not science."* (In French: *"Il n'y a de science que du mesurable."*) The full quotation is on next page.

---

[2] Our emphasis.
[3] This excerpt is from an address on "Anthropometry at Schools" given in 1905 by Galton at the London Congress of the Royal Institute for Preventive Medicine. The text was published in the *Journal of Preventive Medicine*, Vol. XIV, pp. 93–98, London, 1906.
[4] The *Spectator*, May 23, 1874. The editor of the *Spectator* was taking Galton to task for his method of ranking applied to psychical character. He used 'candour' and 'power of repartee' as examples.

*"When you can measure what you are speaking about, and express
it in numbers, you know something about it; but when you cannot
measure it, when you cannot express it in numbers, your knowledge
is of a meager and unsatisfactory kind: it may be the beginning of
knowledge, but you are scarcely, in your thoughts, advanced to the
stage of* **science**, *whatever the matter may be."* (Kelvin, 1889.)

Such a position, which equates precision with the use of numbers, was not
on the whole beneficial to the development of mature sciences outside of
physics[5]. It certainly had a costly impact on the assessment of mental
traits. For instance, for the sake of scientific precision, the assessment of
mathematical knowledge was superseded in the US by the measurement
of mathematical aptitude using instruments directly inspired from Galton
via Alfred Binet in France. They are still used today in such forms as the
S.A.T.[6], the G.R.E. (*Graduate Record Examination*), and other similar
tests.

In the minds of those nineteenth century scientists and their followers, the
numerical measurement of mental traits was to be a prelude to the estab-
lishment of sound, predictive scientific theories in the spirit of those used so
successfully in classical physics. The planned constructions, however, never
went much beyond the measurement stage[7].

Of course, we are enjoying the benefits of hindsight. In all fairness, there
were important mitigating circumstances affecting those who uphold the
cause of numerical measurement as a prerequisite to science. For one thing,
the appropriate mathematical tools were not yet available for different conceptions. More importantly, the 'Analytical Engine' of Charles Babbage was
still a dream, and close to another century had to pass before the appear-

---

[5] Note in passing that the contemporary movement called 'Postmodernism' may be regarded as a delayed reaction against the narrow scientism of Kelvin and the like. Curi-
ously, both these nineteeth century scientists and their postmodernist critics would then
be guilty of exactly the same misapprehension of the boundaries of scientific theory. The
postmodernists are saying, in effect, that if these are the limits of scientific theory, then
it cannot deal with what they are interested in.
[6] Interestingly, the meaning of the acronym S.A.T. has recently been changed by Education Testing Service from '*Scholastic Aptitude Test*' to '*Scholastic Assessment Test*',
suggesting that a different philosophy on the part of the test makers may be under
development.
[7] Sophisticated theories can certainly be found in some areas of the behavioral sciences,
but they do not generally rely on 'psychological measurement.'

ance of computing machines capable of handling the symbolic manipulations that would be required.

The material of this book represents a sharp departure from other approaches to the assessment of knowledge. Its mathematics is in the spirit of current research in combinatorics. No attempt is made to obtain a numerical representation. We start from the concept of a possibly large but essentially discrete set of 'units of knowledge.' In the case of elementary algebra, for instance, one such unit might be a particular type of algebra problem. The full domain for high school algebra may contain several hundred such problems. Our two key concepts are: the 'knowledge state', a subset of problems that some individual is capable of solving correctly, and the 'knowledge structure', which is a distinguished collection of knowledge states. For high school algebra, a useful knowledge structure may contain several hundred thousand feasible knowledge states. The term 'knowledge space', which gave this book its title, refers to a special kind of knowledge structure playing an important role in our work.

The two concepts of knowledge state and knowledge structure give rise to various lattice-theoretical developments motivated by the empirical application intended for them. This material is presented in Chapters 1–6. (Chapter 0 contains a general, non technical introduction to our subject.) The behavioral nature of the typical empirical observations—the responses of human subjects to questions or problems—practically guarantees noisy data. Also, it is reasonable to suppose that all the knowledge states (in our sense) are not equally likely in a population of reference. This means that a probabilistic theory had to be forged to deal with these two kinds of uncertainties. This theory is expounded in Chapters 7 and 8. Chapters 9, 10 and 11 are devoted to various practical schemes for uncovering an individual's knowledge state by sophisticated questioning. Chapter 12 describes some applications and simulations of our concepts.

Many worthwhile developments could not be included here. There is much on-going research, and we had to limit our coverage. Further theoretical concepts and results can be found in chapters of two edited volumes, by Albert (1994) and Albert and Lukas (1998). The second one also contains some applications to various domains of knowledge. Current references on knowledge spaces can be obtained at

    http://wundt.kfunigraz.ac.at/hockemeyer/bibliography.html

thanks to Cord Hockemeyer, who maintains an evolutive and searchable database. For a real-life demonstration of a system based on the concepts of this book, the reader is directed to either

   http://www.aleks.uci.edu   or   http://www.aleks.com

where a full-scale arithmetic program involving both an assessment module and a learning module is available (see page 9 of this book for a more detailed description). Several other systems, in different stages of completion, are under development.

Our enterprise, from the first idea to the completion of this monograph, took 13 years, during which Falmagne benefited from major help in the form of several grants from the National Science Foundation of the US, first at New York University, and then at the University of California at Irvine. JCF also ackowledges a grant from the Army Research Institute (to New York University). He spent the academic year 1987-88 at the Center for Advanced Study in the Behavioral Sciences in Palo Alto. JPD, as a Fulbright grantee, was a visitor at the Center for several months, and substantial progress on our topic was made during that period. We thank all these institutions for their financial support.

Numerous colleagues, students and former students were helpful at various stages of our work. Their criticisms, comments and suggestions certainly improved this book. We thank especially Dietrich Albert, Biff Baker, Eric Cosyn, Charlie Chubb, Yung-Fong Hsu, Geoffrey Iverson, Mathieu Koppen, Kamakshi Lakshminarayan, Damien Lauly, Josef Lukas, Bernard Monjardet, Cornelia Müller-Dowling, Louis Narens, Misha Pavel, Michel Regenwetter, Ragnar Steingrimsson, Ching-Fan Seu and Nicolas Thiéry. We also benefited from the remarks of students from two Erasmus courses given by JPD (Leuven, 1989, and Graz, 1998).

Some special debts must be ackowledged separately. One is to Duncan Luce, for his detailed remarks on a preliminary draft, many of which led us to alter some aspects of our text. As mentioned at the beginning of this preface, JCF spent the all important gestation period of 1982-83 at the University of Regensburg, in the stimulating atmosphere of Professor Jan Drösler's team there. This stay was made possible by a Senior US Scientist Award to Falmagne from the von Humboldt Foundation. The role of Drösler and his colleague and of the von Humboldt Foundation is gratefully recognized here. In the last few years, Steve Franklin's friendly criticisms and cooperation have been invaluable to JCF.

The manuscript was prepared in $\mathcal{AMS}$-TEX with our own macros. We are grateful to Peggy Leroy for help in preparing the index and most thankful to Wei Deng for her uncomprimising chase for our typographical errors and solecisms. Because this is a human enterprise, the reader will surely uncover some remaining incongruities, for which we accept all responsibilities. We have also benefited from the kind efficiency of Mrs. Glaunsinger, Mrs. Stricker and Dr. Engesser, all from Springer-Verlag, who much facilitated the production phase of this work.

Finally, but not lastly, our puzzled gratitude goes to our respective spouses, Monique Doignon and Dina Falmagne, for indulging our fancy for so long, and for their faithful support. To all, our many thanks.

Jean-Paul Doignon   Jean-Claude Falmagne
Auderghem, Belgium   Irvine, CA

Table of Contents

Chapter 0

Overview and Mathematical Glossary

A student is facing a teacher, who is probing her[1] knowledge of high school mathematics. The student, a new recruit, is freshly arrived from a foreign country, and important questions must be answered: to which grade should the student be assigned? What are her strengths and weaknesses? Should the student take a remedial course in some subject? Which topics is she ready to learn? The teacher will ask a question and listen to the student's response. Other questions will then be asked. After a few questions, a picture of the student's state of knowledge will emerge, which will become increasingly sharper in the course of the examination.

However refined the questioning skills of the teacher may be, some important aspects of her task are not a priori beyond the capability of a clever machine. Imagine a student sitting in front of a computer terminal. The machine selects a problem and displays it on the monitor. The student's response is recorded, and the data base—which keeps track of the set of all feasible knowledge states consistent with the responses given so far—is updated. The next question is selected so as to maximize (in some appropriate sense) the expected information in the student's response. The goal is to focus as fast as possible on some knowledge state capable of explaining all the responses.

The main purpose of this book is to expound a mathematical theory for the construction of such an assessment routine. We shall also describe some related computer algorithms, and a number of applications already performed or in progress.

One reason why a machine might conceivably challenge a human examiner can be found in the poor memory of the latter. No matter how the concept of a 'knowledge state in high school mathematics' is defined, a comprehensive list of such states will contain many thousands of entries. The human mind is not especially suitable for fast and accurate scanning of such large data

---

[1] In most instances, we shall conform to current standards in that the fictitious characters appearing in our story will remain nameless and genderless. In those cases in which the appropriate linguistic detours have failed us, the female characters appear in even chapters, and the male characters in odd chapters.

bases. We forget, confuse and distort routinely. A fitting comparison is
chess. We also have there, before any typical move, a very large number
of possibilities to consider. Today, the best human players are still able
to overcome, on occasion, the immense superiority of the machine as a
scanning device. However, the recent victory of `Deep Blue` over Kasparov
should dispel any notion that the human edge could be long lasting.[2]

Our developments will be based on a few commonsensical concepts, which
we introduce informally in the next section.

Main Concepts

0.1. The questions and the domain. We envisage a field of knowledge
that can be parsed into a set of questions each of which has a correct response. An instance of a question in high school algebra is

[P1] *What are the roots of the equation* $3x^2 + \frac{11}{2}x - 1 = 0$?

We shall consider a basic set of such questions, called the 'domain', that
is large enough to give a fine-grained, representative coverage of the field.
In high school algebra, this means a set containing at least several hundred questions. Let us avoid any misunderstanding. Obviously, we are not
especially interested in a student's capability of solving [P1] with the particular values 3, $\frac{11}{2}$ and -1 indicated for the coefficients. Rather, we want
to assess the student's capability of solving all quadratic equations of that
kind. In this book, the label 'question' (we also say 'problem', or 'item') is
reserved for a class of queries differing from each other solely by the choice
of some numbers in specified classes. In that sense, [P1] is an instance of
the question

[P2] *Express the roots of the equation* $\alpha x^2 + \beta x + \gamma = 0$ *in terms of*
     α, β *and* γ.

When the machine tests a student on question [P2], the numbers α, β and γ
are selected in some specified manner. Practical considerations may enter in
such a selection. For example, one may wish to choose α, β and γ in such a

---

[2] Besides, the world class chess players capable of challenging the best chess programs
have gone through a long and punishing learning process, during which poor moves were
immediately sanctioned by a loss of a piece—and of some self-esteem. No systematic
effort is made toward preparing human examiners with anything resembling the care
taken in training a good chess player.

way that the roots of the equation can be expressed conveniently as simple fractions or decimal numbers. Nevertheless, a subject's response reveals her mastery of [P2]. Objections to our choice of fundamental concepts are sometimes made, which we shall address later in this chapter.

0.2. The knowledge states. The 'knowledge state' of an individual is represented in our approach by the set of questions in the domain that she is capable of answering in ideal conditions. This means that she is not working under time pressure, is not impaired by emotional turmoil of any kind, etc. In reality, careless errors arise. Also, the correct response to a question may occasionally be guessed by a subject lacking any real understanding of the question asked. (This is certainly the case when a 'multiple choice' format is used, but it may also happen in other situations.) In general, an individual's knowledge state is thus not directly observable, and has to be inferred from the responses to the questions. The connections between the knowledge state and the actual responses are explored in Chapters 7 and 8, where the probabilistic aspects of the theory are elaborated.

0.3. The knowledge structure. In our experience, for any non-trivial domain, the number of feasible knowledge states tends to be quite large. In an experiment reported in Chapter 12, for example, the number of knowledge states obtained for a domain containing 50 questions in high school mathematics ranged from about 900 to a few thousand. The lists of states were obtained by interviewing five educators using an automated questioning technique called QUERY (Koppen and Doignon, 1988; Koppen, 1993). This is described in Chapter 12 (see also 0.6).

Several thousand feasible states may seem to be an excessively large number of possibilities to sort out. However, it is but a minute fraction of the set of all 2^{50} subsets of the domain. The collection of all the knowledge states captures the organization of the knowledge. It plays a central rôle and will be referred to as the 'knowledge structure.' A miniature example of a knowledge structure, for a domain

$$Q = \{a, b, c, d, e\},$$

is given in Figure 0.1. In this case, the number of states is small enough that a graphic representation is possible. More complex examples are discussed later in this book (see e.g. Chapter 12).

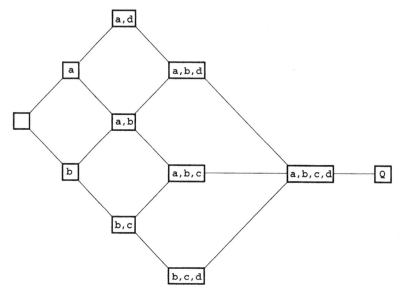

Fig. 0.1. The knowledge structure of Equation (1).

The graph in Figure 0.1 represents the knowledge structure

$$\mathcal{K} = \{\, \varnothing, \{a\}, \{b\}, \{a,b\}, \{a,d\}, \{b,c\}, \{a,b,c\},$$
$$\{a,b,d\}, \{b,c,d\}, \{a,b,c,d\}, Q \,\}. \qquad (1)$$

This knowledge structure contains 11 states. The domain Q and the empty set $\varnothing$, the latter symbolizing complete ignorance, are among them. The edges of the graph represent the *covering relation* of set inclusion: an edge linking a state K and a state K' located to its right in the graph means that $K \subset K'$ (where $\subset$ denotes the strict inclusion), and that there is no state K'' such that $K \subset K'' \subset K'$. This graphic representation is often used. When scanned from left to right, it suggests a learning process: at first, a subject knows nothing at all about the field, and is thus in state $\varnothing$, which is represented by the empty box on the left of the figure. She then gradually progresses from state to state, following one of the paths in Figure 0.1, until a complete mastery of the topic is achieved in state Q. This idea is systematically developed in Chapters 7 and 8. The knowledge structure of Figure 0.1 and Equation (1) was obtained empirically, and successfully tested, in the framework of a probabilistic model, on a large number of subjects. The questions $a, \ldots, e$ were problems in elementary Euclidean geometry. The text of these questions can be found in Figure 8.1. This work is due to Lakshminarayan (1995) and is discussed in Chapter 8.

0.4. Knowledge spaces. Examining Figure 0.1 reveals that the family of states satisfies an interesting property: if K and K' are any two states in $\mathcal{K}$, then $K \cup K'$ is also a state. In mathematical lingo: the family $\mathcal{K}$ is 'closed under union.' A knowledge structure satisfying this property is called a 'knowledge space.' This concept plays a central rôle in this book. It must be pointed out, however, that the closure under union, while useful for the developments of the theory described here, is not essential and can be dispensed with in many cases.

0.5. Surmise functions and clauses. Knowledge spaces can be regarded as a generalization of quasi orders (i.e. reflexive and transitive relations; see 0.17, 0.18). According to a classical result of Birkhoff (1937), any family of sets closed under both union and intersection can be recoded as a quasi order, without loss of information. A similar representation can be obtained for families of sets closed only under union, but the representing concept is not a quasi order, nor even a binary relation. Rather, it is a function σ associating a family $\sigma(q)$ of subsets of Q to each item q in Q, with the following interpretation: if a subject has mastered question q, that subject must also have mastered all the items in at least one of the members of $\sigma(q)$. One may think of a particular member of $\sigma(q)$ as a possible set of prerequisites for q. This is consistent with the view that there may be more than one way to achieve the mastery of any particular question q. In this book, the function σ (under certain conditions) will be referred to as a 'surmise function' and the elements of $\sigma(q)$ for a particular item q will be called the 'clauses' for q or the 'backgrounds' of q. In the example of Figure 0.1, a subject having mastered item d must also have mastered either at least item a, or at least items b and c. Item d has two clauses: we have

$$\sigma(d) = \{ \{a, d\}, \{b, c, d\} \}.$$

Notice that these two clauses for d contain d itself. By convention, any clause for an item contains that item. In other words, any item is a prerequisite for itself.

In the special case where the family of states is closed under both union and intersection, any question q has a unique clause. In this case, the surmise function σ is essentially a quasi order (i.e. becomes a quasi order after a trivial change of notation): the unique clause for an item q contains q itself, plus all the items equivalent to q or preceding q in the quasi order.

0.6. Entailments relations and human expertise. Another representation of a knowledge space will also be introduced, which plays an important rôle in practical applications. Human experts (such as experienced teachers) may possess critical information concerning the knowledge states which are feasible in some empirical situation. However, direct querying would not work. We cannot realistically ask a teacher to give a complete list of the feasible states, with the hope of getting a useful response. Fortunately, a recoding of the concept of knowledge space is possible, which leads to a more fruitful approach. Consider asking a teacher queries of the type

[Q1] *Suppose that a student has failed items q_1, q_2, ..., q_n. Do you believe this student would also fail item q_{n+1}? You may assume that chance factors, such as lucky guesses and careless errors, play no rôle in the student's performance, which reflects her actual mastery of the field.*

Positive responses to *all* queries of that kind, for a given domain Q, define a binary relation $\mathcal{P}$ pairing subsets of Q with elements of Q. Thus, a positive response to query [Q1] is coded as

$$\{q_1, \ldots, q_n\} \, \mathcal{P} \, q_{n+1}.$$

It can be shown that if the relation $\mathcal{P}$ satisfies certain natural conditions, then it uniquely specifies a particular knowledge space. Such a relation is then called an 'entailment.' Algorithms based on that relation have been written, which are helpful when querying an expert. In applying one of these algorithms, it is supposed that an expert relies on an implicit knowledge space to respond to queries of the type [Q1]. The output of the procedure is the personal knowledge space of an expert. In practice, only a very small fraction of the queries of the form [Q1] must be asked.

The three equivalent concepts of knowledge spaces, surmise functions and entailments form the core of the theory. As mentioned in 0.2, in realistic situations, subjects sometimes fail problems that they fully understand, or provide amazingly correct responses to problems that they do not understand at all. That is, the knowledge states are not directly observable. The usual way out of such difficulties is a probabilistic approach.

0.7. Probabilistic knowledge structures. Probabilities enter in two ways in the theory. We begin by supposing that to each state K of the knowledge structure is attached a number $P(K)$, which can be interpreted as the probability of finding, in the population of reference, a subject in state K. The triple $(Q, \mathcal{K}, P)$ will be called a 'probabilistic knowledge structure.' Next, we shall introduce, for any subset R of the domain and any state K, a conditional probability $r(R, K)$ that a subject in state K would provide correct responses to all the items in R, and only to those items. Any subset R of the domain Q is a 'response pattern.' The overall probability $\rho(R)$ of observing a response pattern R can thus be computed from the weighted sum

$$\rho(R) = \sum_{K \in \mathcal{K}} r(R, K) P(K).$$

Formal definitions of these probabilistic concepts will be found in Chapters 7 and 8, where specific forms of the functions P and r are also investigated. Empirical tests are also described.

0.8. Assessment procedures. The machinery of knowledge states and (probabilistic) structures and spaces provides the foundation for a number of assessment procedures. The goal of such an algorithm is to uncover, as efficiently as possible, the knowledge state of an individual by asking appropriate questions from the domain. The algorithm proceeds by setting up a plausibility function on the set of all states, which is updated after each response of the student. The choice of the next question is based on the current plausibility function. The output of the assessment algorithm is a state (or possibly, a handful of closely related states) which best represents the student's performance.

Since this output may be unwieldy, it is sometimes subjected to an elaborate analysis which results in a report summarizing the accomplishments of the student, and making recommendations. A practical implementation of an automated system built on the concepts of this book is briefly described later in this chapter.

Possible Limitations

The concept of a knowledge state, which is at the core of our work, is sometimes criticized on the grounds that it trivializes an important idea. In the minds of some critics, the concept of a 'knowledge state' should cover much more than a set of questions fully mastered by a student. It should contain many other features related to the student's current understanding of the material, such as the type of errors that she is likely to make. A reference to Van Lehn's work (1988) is often made in this connection.

For the most part, such criticisms come from a misconception of the exact status of the definition of 'knowledge state' in our work. An important aspect of this work has consisted in developing a formal language within which critical aspects of knowledge assessment could be discussed and manipulated. 'Knowledge state' is a defined concept in that formal language. We make no claim that the concept of a 'knowledge state' in our sense captures all the cognitive features that one might associate with such a word, any more than in topology, say, the concept of a compact set captures the full physical intuition that might be evoked by the adjective "compact." As far as we know, the concept of a knowledge state has never been formally defined in the literature pertaining to computerized learning. There was thus no harm, we believed, in appropriating the term, and giving it a formal status. On the other hand, it is certainly true that if some solid information is available regarding refined aspects of the students' performance, such information should be taken into account in the assessment. There are various ways this can be achieved.

As a first example, we take the type of error mechanisms discussed by Van Lehn (1988). Suppose that for some or all of our questions, such error mechanisms have been elucidated. In other words, erroneous responses are informative, and can be attributed to specific faulty algorithms. One possibility would be to analyze or redesign our domain (our basic set of questions) in such a manner that a knowledge state itself would involve the diagnosis of error mechanisms. After all, if a student is a victim of some faulty algorithm, this algorithm should be reflected in the pattern of responses to suitably chosen questions. A description of the knowledge state could then include such error mechanisms.

More generally, a sophisticated description of the knowledge states could be obtained by a precise and detailed tagging of the items. For example, it is possible to associate, to each item in the domain, a detailed list of informations featuring entries such as: the part of the field to which this item belongs (e.g.: calculus, derivatives; geometry, right triangles), the type of problem (e.g.: word problem, computation), the expected grade at which this item should be learned, the concepts used in the formulation or in the solution of this item, the most frequent type of misconceptions, etc.

When an assessment algorithm has uncovered the state of some individual, these tags can be used to prepare, via various manipulations, a comprehensive description of the state in everyday language. A brief description of a system implementing these ideas is given below.

A Practical Application: The Aleks System

The computer educational system Aleks provides on the Internet[3] a multilingual educational environment with two modules: an *assessment module* and a self paced *teaching module* with many tools. 'Aleks' is an acronym for 'Assessment and Learning in Knowledge Spaces.' The assessment module is based on the work described in this monograph. Its working version implements one of the continuous Markov procedures presented in Chapter 11. It relies on knowledge structures built with the techniques of Chapters 5 and 12. The system currently[4] covers basic arithmetic, algebra and geometry. The assessment is comprehensive in the sense that the set of all possible questions covers the whole curriculum in the topic. An 'answer editor' permits the student to enter the responses in a style imitating a paper and pencil method. All the questions have open responses. At the end of the assessment, the system delivers a detailed report describing the student's accomplishments, making recommendations for further learning, and giving immediate access to the teaching module.

While using the teaching module, the student may request a 'rephrasing' or an 'explanation' of any problem proposed by the system. A graphic calculator and a mathematics dictionary are available on-line. The dictio-

---

[3] www.aleks.com. This system was built with the financial support of the National Science Foundation.
[4] In the fall of 1997.

nary provides definition of all the technical terms and is accessed simply by clicking on any highlighted word.

Both the report and the teaching are consistent with programmable educational standards. The current default standards of the system are the California Standards. A user-friendly tool allows the teacher to modify these standards, if necessary. When such a modification is made, both the assessment module and the teaching module are automatically adapted to the new standards.

Other systems, also based on the concepts of knowledge structures and knowledge states in the sense of this book and similar in spirit to the Aleks system, are under construction at Graz, Austria (by Albert, Hockemeyer and Kaluscha; see e.g. Albert and Hockemeyer, 1997; Hockemeyer, 1997; Hockemeyer, Held and Albert, 1998) and at Braunschweig, Germany (by Dowling; see e.g. Dowling, Hockemeyer and Ludwig, 1996).

Potential Applications to Other Fields

Even though our theoretical developments have been primarily guided by a specific application in education, the basic concept of a knowledge structure (or space) is very general, and has potential uses in superficially quite different fields. A few examples are listed below.

0.9. Failure analysis. Consider a complex device, such as a telephone interchange (or a computer, or a nuclear plant). At some point, the device's behavior indicates a failure. The system's administrator (or a team of experts), will perform a sequence of tests to determine the particular malfunction responsible for the difficulty. Here, the domain is the set of observable signs. The states are the subsets of signs induced by all the possible malfunctions.

0.10. Medical diagnosis. A physician examines a patient. To determine the disease (if any), the physician will check which symptoms are present. As in the preceding example, a carefully designed sequence of verifications will take place. Thus, the system is the patient, and the state is a subset of symptoms specifying her medical condition. (For an example of a computerized medical diagnosis system, see Shortliffe, 1976, and Shortliffe and Buchanan, 1975.)

0.11. Pattern Recognition. A pattern recognition device analyzes a visual display to detect one of many possible patterns, each of which is defined by a set of specified features. Consider a case in which the presence of features is checked sequentially, until a pattern can be identified with an acceptable risk of error. In this example, the system is a visual display, and the possible patterns are its states. (For a first contact with the vast literature on pattern recognition, consult, for instance, Duda and Hart, 1973, and Fu, 1974.)

0.12. Axiomatic Systems. Let E be a collection of well-formed expressions in some formal language, and suppose that we also have a fixed set of derivation rules. Consider the relation $\mathcal{I}$ on the set of all subsets of E, with the following interpretation: we write $A \mathcal{I} B$ if all the expressions in B are derivable from the expressions in A by application of the derivation rules. We call any $K \subseteq E$ a *state* of $\mathcal{I}$ if $B \subseteq K$ whenever $A \subseteq K$ and $A\mathcal{I}B$. It is easily shown that the collection $\mathcal{L}$ of all states is closed under intersection; that is, $\cap \mathcal{F} \in \mathcal{L}$ for any $\mathcal{F} \subseteq \mathcal{L}$ (See Problem 5 in Chapter 1). Notice that this constraint is the dual of that defining a knowledge space, in the sense that the set

$$\overline{\mathcal{L}} = \{Z \in 2^E \mid \overline{Z} \in \mathcal{L}\}$$

is closed under union.

On the Content and Organization of this Book

A glossary of some basic mathematical concepts will be found in the next section of this chapter. This glossary contains entries such as 'binary relation', 'partial order', 'chains', 'Hausdorff maximal principle', etc. In writing this book, we had in mind a reader with a minimum mathematical background corresponding roughly to that of a mathematics major: e.g. a three semester sequence of calculus, a couple of courses in algebra, and a couple of courses in probability and statistics. However, a reader equipped with just that may find the book rough going, and will have to muster up a fair amount of patience and determination. To help all readers, exercises are provided at the end of each chapter.

Chapters 1–6 and 9 are devoted to algebraic aspects of the theory. They cover knowledge structures and spaces, surmise functions, entailments, and the concept of an 'assessment language' (Chapter 9). Chapters 7 and 8 deal with probabilistic knowledge structures. These two chapters develop a number of stochastic learning models describing the successive transitions, over time, between the knowledge states. Chapter 10 and 11 are devoted to various stochastic algorithms for assessing knowledge. Chapter 12 describes some applications of the theory to the construction of a knowledge structure in a practical situation.

Chapters are divided into sections and paragraphs. Only the paragraphs are numbered. The title "On the Content and Organization of this Book" above is that of a section. The paragraphs are numbered lexicographically. When a number "n.m" is used for a paragraph, the "n" denotes the chapter number, and the "m" refers to the paragraph number within that chapter. The most frequent titles for paragraphs will be "Examples", "Definition", "Theorem", and "Remark(s)." Usually, minor results will appear as remarks. Especially difficult chapters, sections, paragraphs and exercises are marked by a star. We also use a star to indicate a part of the book that can be omitted at first reading without loss of continuity.

Some printing conventions should be remembered. We put single quotation marks around a term used in a technical sense, but yet undefined. A word or phrase mentioned, but not used, is flanked by double quotation marks. (This convention does not apply to mathematical symbols, however.) A slanted font is used for the definition of technical terms and for stating theorems. Many applications of all these conventions can be found in the preceding pages.

Glossary of Mathematical Concepts

We review here some basic conventions regarding terminology and notation.

0.13. Set theory, relations, mappings. Standard set theoretical notation will be used throughout. We use $+$ for disjoint union. Set inclusion is denoted by $\subseteq$ and proper (or strict) inclusion by $\subset$. The *size*, or *cardinality*, or *cardinal number* of a set Q is denoted by $|Q|$. An element (x,y) of the Cartesian product $X \times Y$ is often abbreviated as xy. A *relation* $\mathcal{R}$ from a set X to a set Y is a subset of $X \times Y$. The *complement* of $\mathcal{R}$ (*with respect to* $X \times Y$) is the relation $\overline{\mathcal{R}} = (X \times Y) \setminus \mathcal{R}$, also from X to Y. The specification "with respect to $X \times Y$" is usually omitted when no ambiguity can arise. The phrase "with respect to" is occasionally abbreviated by the acronym "w.r.t." The *converse* of $\mathcal{R}$ is the relation

$$\mathcal{R}^{-1} = \{yx \in Y \times X \mid x\mathcal{R}y\}$$

from Y to X.

A *mapping* f from the set X to the set Y is a relation from X to Y such that for any $x \in X$, there exists exactly one $y \in Y$ with $x\,f\,y$; we then write $y = f(x)$. This mapping is *injective* if $f(x) = f(x')$ implies $x = x'$ (for all $x, x' \in X$). It is *surjective* if for any $y \in Y$ there is some $x \in X$ with $y = f(x)$. Finally, f is *bijective*, or *one-to-one*, if f is both injective and surjective.

A detailed treatment of relations can be found, for example, in Suppes (1960) or Roberts (1979). Some basic facts and concepts are recalled in the following few paragraphs.

0.14. Relative product. The *(relative) product* of two relations $\mathcal{R}$ and $\mathcal{S}$ is denoted by

$$\mathcal{RS} = \{xy \mid \exists z: \ x\mathcal{R}z \wedge z\mathcal{S}y\}.$$

When $\mathcal{R}$ and $\mathcal{S}$ are explicitly given as relations from X to Y and from Z to W respectively, the element z in the above formula is to be taken in $Y \cap Z$, and the relation $\mathcal{RS}$ is from X to W. It is easy to check that the relative product is an associative operation: for any three relations $\mathcal{R}$, $\mathcal{S}$ and $\mathcal{T}$, we always have $(\mathcal{RS})\mathcal{T} = \mathcal{R}(\mathcal{ST})$. Hence, the parentheses play no useful rôle. Accordingly, we shall write $\mathcal{R}_1\mathcal{R}_2 \cdots \mathcal{R}_k$ for the relative product of k relations $\mathcal{R}_1$, $\mathcal{R}_2$, ..., $\mathcal{R}_k$. For any relation $\mathcal{R}$ and any positive integer n, we write $\mathcal{R}^n$ for the nth *(relative) power* of the relation $\mathcal{R}$, that is

$$\mathcal{R}^n = \underbrace{\mathcal{R}\,\mathcal{R}\cdots\mathcal{R}}_{n\text{ times}}.$$

By convention, $\mathcal{R}^0$ is the *identity relation* on the ground set, which will always be specified by the context. For a relation $\mathcal{R}$ from X to Y, we set $\mathcal{R}^0 = \{xx \mid x \in X \cup Y\}$.

0.15. Properties of relations. A relation $\mathcal{R}$ on a set X is

reflexive (on X) when $x\mathcal{R}x$ for all $x \in X$;

symmetric (on X) when $x\mathcal{R}y$ implies $y\mathcal{R}x$ for all $x, y \in X$;

asymmetric (on X) when $x\mathcal{R}y$ implies $\neg(y\mathcal{R}x)$ for all $x,y \in X$ (the symbol $\neg$ denotes logical negation);

antisymmetric (on X) if ($x\mathcal{R}y$ and $y\mathcal{R}x$) implies $x = y$ for all $x,y \in X$.

0.16. Transitive closure. A relation $\mathcal{R}$ is *transitive* if whenever $x\mathcal{R}y$ and $y\mathcal{R}z$, we also have $x\mathcal{R}z$. More compactly, $\mathcal{R}$ is transitive if $\mathcal{R}^2 \subseteq \mathcal{R}$. The *transitive closure* (or *reflexo-transitive closure*) of a relation $\mathcal{R}$ is the relation $t(\mathcal{R})$ defined by

$$t(\mathcal{R}) = \mathcal{R}^0 \cup \mathcal{R}^1 \cup \cdots \cup \mathcal{R}^k \cup \cdots = \cup_{k=0}^{\infty} \mathcal{R}^k.$$

0.17. Equivalence relations and set partitions. An *equivalence relation* $\mathcal{R}$ on a set X is a reflexive, transitive, and symmetric relation on X. It corresponds biunivocally to a *partition* of X, that is, a family of nonempty subsets of X which are pairwise disjoint and whose union is X: these subsets, or *classes* of the equivalence relation $\mathcal{R}$, are all the subsets of X of the form $\{x \in X \,|\, x\mathcal{R}z\}$, with $z \in X$.

0.18. Quasi orders, partial orders. A *quasi order* on a set X is any relation which is transitive and reflexive on X, as for instance $\leq$ on the set of all real numbers. A set equipped with a quasi order is *quasi ordered*. An antisymmetric quasi order is a *partial order*.

Any quasi order $\mathcal{P}$ on X gives rise to the equivalence relation $\mathcal{P} \cap \mathcal{P}^{-1}$. A partial order $\mathcal{P}^*$ is obtained on the set X^* of all equivalence classes by setting $C\mathcal{P}^*C'$ if $c\mathcal{P}c'$ for some (and thus for all) c in C and c' in C'. The partially ordered set $(X^*, \mathcal{P}^*)$ is called the *reduction* of the quasi ordered set $(X, \mathcal{P})$. An element x in a quasi order $(X, \mathcal{P})$ is *maximal* when $x\mathcal{P}y$ implies $y\mathcal{P}x$, for all $y \in X$. It is a *maximum* if moreover $y\mathcal{P}x$ for all $y \in X$. *Minimal elements* and *minimum* are similarly defined. A partially ordered set can have at most one maximum and one minimum.

0.19. Weak orders, linear orders. A *weak order* $\mathcal{P}$ on the set X is a quasi order on X which is *complete*, in the sense that for all $x,y \in X$ we have $x\mathcal{P}y$ or $y\mathcal{P}x$. The reduction of a weak order is a *linear*, or *simple*, or *total order*.

0.20. Covering relation, Hasse diagram. In a partially ordered set $(X, \mathcal{P})$, the element x is *covered by* the element y when $x\mathcal{P}y$ with $x \neq y$

and moreover $x\mathcal{P}t\mathcal{P}y$ implies $x = t$ or $t = y$. The *covering relation* or *Hasse diagram* of $(X,\mathcal{P})$ is the relation consisting of all pairs xy with y covering x. When X is infinite, the Hasse diagram of $(X,\mathcal{P})$ may be empty even though $\mathcal{P}$ itself is not empty. When X is finite, the Hasse diagram of $(X,\mathcal{P})$ provides a comprehensive summary of $\mathcal{P}$ in the sense that the transitive closure of the Hasse diagram of $(X,\mathcal{P})$ is equal to $\mathcal{P}$. In fact, in this case, the Hasse diagram of $(X,\mathcal{P})$ is the smallest relation having its transitive closure equal to $\mathcal{P}$. When X is a small set, the Hasse diagram of $\mathcal{P}$ can be conveniently displayed by a graph drawn according to the following conventions: the elements of X are represented by points on a page, with an ascending edge from x to y when x is covered by y.

0.21. Chain, Hausdorff Maximality Principle. A *chain* in a partially ordered set $(X,\mathcal{P})$ is any subset C of X such that $c\mathcal{P}c'$ or $c'\mathcal{P}c$ for all $c, c' \in C$ (in other words, the order induced by $\mathcal{P}$ on C is linear). In several proofs belonging to starred material, Hausdorff maximality principle is invoked to establish the existence of a maximal element in a partially ordered set. This principle is equivalent to Zorn's Lemma and states that a quasi ordered set $(X,\mathcal{P})$ admits a maximal element whenever all its chains are bounded above, that is: for any chain C in X, there exists b in X with $c\mathcal{P}b$ for all $c \in C$. For details on Hausdorff Maximality Principle and related conditions, see Dugundji (1966) or Suppes (1960).

0.22. Real numbers. The following notation is used for the basic sets:

N	the set of natural numbers (excluding 0);
Z	the set of integer numbers;
Q	the set of rational numbers;
R	the set of real numbers;
R$^+$	the set of (strictly) positive real numbers.

Real open intervals are denoted by $]x, y[= \{z \in \mathbf{R} \mid x < z < y\}$. The corresponding notation for the half open and closed intervals is $]x, y]$, $[x, y[$ and $[x, y]$.

0.23. Metric Spaces. A *distance* d on a set X is any mapping $d : X \times X \to \mathbf{R}$ satisfying for $x, y, z \in X$:

(1) $d(x, y) \geq 0$, with $d(x, y) = 0$ iff $x = y$ (i.e., d is *definite positive*),

(2) $d(x, y) = d(y, x)$ (i.e. d is *symmetric*),

(3) $d(x, y) \leq d(x, z) + d(z, y)$ (the *triangular inequality*).

A *metric space* is a set equipped with a distance. As an example, let E be any finite set. The *symmetric-difference distance* on the *power set* 2^E (consisting of all subsets of E) is defined by setting, for $A, B \in 2^E$:

$$d(A, B) = |A \triangle B|.$$

Here, $A \triangle B = (A \setminus B) \cup (B \setminus A)$ denotes the *symmetric difference* of the sets A and B.

0.24. Probabilistic and statistical concepts. Standard techniques of probability theory, stochastic processes and statistics are used whenever needed. The symbol **P** denotes the probability measure of the sample space under consideration. Some statistical techniques dealing with goodness-of-fit tests are briefly reviewed in Chapter 7.

Original Sources and Related Works

A first pass at developing a mathematical theory for knowledge structures was made in Doignon and Falmagne (1985). This paper contains the algebraic core of the theory. Its content is covered and extended in Chapters 1 to 6 of this book. A rather technical follow up of this work was given in Doignon and Falmagne (1988). A comprehensive description of our program, intended for non mathematicians, is given in Falmagne, Koppen, Villano, Doignon and Johanessen (1990). Short introductions to knowledge space theory are contained in Doignon and Falmagne (1987) and Falmagne (1989b). A more leisurely paced text is Doignon (1994a). More specific references pertaining to particular aspects of our work are given in the last section of each chapter of this book.

As indicated in Paragraphs 0.9 to 0.12, our results are potentially applicable to other fields, such as computerized medical diagnosis, pattern recognition or the theory of feasible symbologies (for the latter, see Problem 15 of Chapter 1). The literature on computerized medical diagnosis is vast. We only mention here the well-known example of Shortliffe (1976) and Shortliffe and Buchanan (1975). For pattern recognition, the reader is referred to Duda and Hart (1973), and Fu (1974). For symbology theory, the reader may consult Jameson (1992).

There are obvious similarities between knowledge assessment in the framework of knowledge structures as developed in this book, and the technique known as 'tailor testing' in psychometric practice. In both situations, subjects are presented with a sequence of well-chosen questions, and the goal is to determine, as accurately and as efficiently as possible, their mastery of a given field. There is, however, an essential difference in the theoretical foundations of the two approaches. In psychometric theory, it is assumed that the responses to the items reflect primarily the subject's ability with respect to one or more intellectual trait. In constructing the test, the psychometrician chooses and formats the items so as to minimize the impact of other determinants of the subject's performance, such as schooling, culture, or knowledge in the sense of this book. In fact, the primary or sole aim of such psychometric tests is to measure the subject's abilities on some numerical scales. The ubiquitous I.Q. test is the exemplary case of this enterprise. Accordingly, the models underlying a 'tailor testing' procedure tend to be simple numerical structures, in which a subject's ability is represented as a real number or as a real vector with a small number of dimensions. A standard source for psychometric theory is still Lord and Novick (1974) (see also Wainer and Messick, 1983). For 'tailor testing' the reader could consult, for example, Lord (1974) or Weiss (1983). A few authors have proposed to escape the unidimensionality of classical 'tailor testing.' As an example, Durnin and Scandura (1973) developed tests under the assumption that the subject performs a well-defined, algorithmic treatment of tasks to be solved.

Paragraph 0.12 provides examples of families of subsets that are closed under intersection. In a sense which will be made precise in Definition 1.7, these families are dual to knowledge spaces. There is a vast mathematical literature concerned with the general study of such families, referred to as 'closure spaces', 'abstract convexities', and a few other terms. More detailed comments on relationships of our work with that of others can be found in the sources sections of Chapters 1, 2 and 3.

Chapter 1

Knowledge Structures and Spaces

Suppose that some complex system is assessed by an expert, who checks for the presence or absence of some revealing features. Ultimately, the state of the system is described by the subset of features (from a possibly large set) which are detected by the expert. This concept is very general, and becomes powerful only on the background of specific assumptions, in the context of some applications. We begin with the combinatoric underpinnings of the theory.

Fundamental Concepts

1.1. Example. *(Knowledge structures in education.)* A teacher is examining a student to determine, for instance, which mathematics courses would be appropriate at this stage of the student's career, or whether the student should be allowed to graduate. The teacher will ask one question, then another, chosen as a function of the student's response to the first one. After a few questions, a picture of the student's knowledge state will emerge, which will become increasingly more precise in the course of the examination. By 'knowledge state' we mean here the set of all problems that the student is capable of solving in ideal conditions. (We assume, for the time being, that careless errors and lucky guesses do not occur.) The next definition formalizes these ideas.

1.2. Definition. A *knowledge structure* is a pair $(Q, \mathcal{K})$ in which Q is a nonempty set, and $\mathcal{K}$ is a family of subsets of Q, containing at least Q and the empty set $\varnothing$. The set Q is called the *domain* of the knowledge structure. Its elements are referred to as *questions* or *items* and the subsets in the family $\mathcal{K}$ are labeled *(knowledge) states*. Occasionally, we shall say that $\mathcal{K}$ is a *knowledge structure on* a set Q to mean that $(Q, \mathcal{K})$ is a knowledge structure. The specification of the domain can be omitted without ambiguity since we have $\cup \mathcal{K} = Q$.

1.3. Example. Consider the set $U = \{a, b, c, d, e, f\}$ equipped with the knowledge structure

$$\mathcal{H} = \{\, \varnothing, \{d\}, \{a, c\}, \{e, f\}, \{a, b, c\}, \{a, c, d\}, \{d, e, f\},$$
$$\{a, b, c, d\}, \{a, c, e, f\}, \{a, c, d, e, f\}, U \,\}.$$

As illustrated by this example, we do not assume that all subsets of the domain are states. The knowledge structure $\mathcal{H}$ contains eleven states out of sixty-four possible ones.

1.4. Definition. Let $\mathcal{K}$ be a knowledge structure. We shall denote by $\mathcal{K}_q$ the collection of all states containing item q. In Example 1.3, we have, for instance

$$\mathcal{H}_a = \{\, \{a, c\}, \{a, b, c\}, \{a, c, d\}, \{a, b, c, d\}, \{a, c, e, f\}, \{a, c, d, e, f\}, U \,\},$$
$$\mathcal{H}_e = \{\, \{e, f\}, \{d, e, f\}, \{a, c, e, f\}, \{a, c, d, e, f\}, U \,\}.$$

Notice that items a and c carry the same information relative to $\mathcal{H}$, in the sense that they are contained in the same states: any state containing a also contains c, and vice versa. In other terms, we have $\mathcal{H}_a = \mathcal{H}_c$. From a practical viewpoint, any individual having mastered item a has necessarily mastered item c, and vice versa. Thus, in testing the acquired knowledge of a subject, only one of these two questions must be asked. Similarly, we also have $\mathcal{H}_e = \mathcal{H}_f$.

In general, the set of all items contained in the same states as item q will be denoted by q^* and will be called a *notion*. Thus, for a knowledge structure $\mathcal{K}$ and any item q in its domain Q, we have the notion

$$q^* = \{r \in Q \,|\, \mathcal{K}_q = \mathcal{K}_r\}.$$

The collection Q^* of all notions is a partition of the set Q of items. When two items belong to the same notion, we shall sometimes say that they are *equally informative*. In such a case, the two items form a pair in the equivalence relation on Q associated to the partition Q^*.

In Example 1.3, we have the four notions

$$a^* = \{a, c\}, \quad b^* = \{b\}, \quad d^* = \{d\}, \quad e^* = \{e, f\},$$

forming the partition

$$U^* = \big\{ \{a,c\}, \{b\}, \{d\}, \{e,f\} \big\}.$$

A knowledge structure in which each notion contains a single item is called *discriminative*. A discriminative knowledge structure can always be manufactured from an arbitrary knowledge structure $(Q, \mathcal{K})$ by forming the notions, and constructing the knowledge structure $\mathcal{K}^*$ induced by $\mathcal{K}$ on Q^* through the definition

$$\mathcal{K}^* = \{K^* \mid K \in \mathcal{K}\},$$

where, for any K in $\mathcal{K}$, we have $K^* = \{q^* \mid q \in K\}$.

The knowledge structure $(Q^*, \mathcal{K}^*)$ is called the *discriminative reduction* of $(Q, \mathcal{K})$. Since this construction is straightforward, we shall often simplify matters and suppose that a particular knowledge structure under consideration is discriminative.

1.5. Example. Starting from the knowledge structure $(U, \mathcal{H})$ from Example 1.3, we describe its discriminative reduction by setting

$$a^* = \{a,c\}, \quad b^* = \{b\}, \quad d^* = \{d\}, \quad e^* = \{e,f\};$$
$$U^* = \{a^*, b^*, d^*, e^*\};$$
$$\mathcal{H}^* = \{\varnothing, \{d^*\}, \{a^*\}, \{e^*\}, \{a^*, b^*\}, \{a^*, d^*\}, \{d^*, e^*\},$$
$$\{a^*, b^*, d^*\}, \{a^*, e^*\}, \{a^*, d^*, e^*\}, U^*\}.$$

Thus, $(U^*, \mathcal{H}^*)$ is formed by aggregating equally informative items from U.

1.6. Definition. A knowledge structure $(Q, \mathcal{K})$ will be called *finite* (respectively, *essentially finite*), if Q (respectively $\mathcal{K}$) is finite. A similar definition holds for *countable* (respectively, *essentially countable*) knowledge structures.

Typically, knowledge structures encountered in education are essentially finite. They may not be finite however: at least conceptually, some notions may contain a potentially infinite number of equally informative questions. Problems 2 and 3 require the reader to show that an arbitrary knowledge structure $\mathcal{K}$ and its discriminative reduction $\mathcal{K}^*$ have the same cardinality, and that the knowledge structure $(Q, \mathcal{K})$ is essentially finite if and only if Q^* is finite.

As suggested by these first few definitions, our choice of terminology is primarily guided by Example 1.1, which has also guided many of our theoretical developments. Our results are also applicable to very different fields, however (cf. Examples 0.9 to 0.12).

An important special case of a knowledge structure arises when the family of states is closed under union. This is the topic of our next section.

Knowledge Spaces

1.7. Definition. When the family $\mathcal{K}$ of a knowledge structure $(Q, \mathcal{K})$ is *closed under union* — that is, when $\cup \mathcal{F} \in \mathcal{K}$ whenever $\mathcal{F} \subseteq \mathcal{K}$ — we shall say that $(Q, \mathcal{K})$ is a *(knowledge) space*, or equivalently, that $\mathcal{K}$ is a *(knowledge) space on* Q. The *dual* of a knowledge structure $\mathcal{K}$ on Q is the knowledge structure $\overline{\mathcal{K}}$ containing all the *complements* of the states of $\mathcal{K}$; that is

$$\overline{\mathcal{K}} = \{K \in 2^Q \mid Q \setminus K \in \mathcal{K}\}.$$

Thus $\mathcal{K}$ and $\overline{\mathcal{K}}$ have the same domain.

1.8. Remarks. Obviously, any mathematical result concerning knowledge spaces can readily be transported to their duals, and vice versa. This interplay is one reason for our interest in knowledge spaces: the duals of knowledge spaces belong to an important family of mathematical structures called 'closure spaces.' Much is known regarding 'closure spaces' that has a useful translation in our context. These structures are introduced in the next section.

Another reason comes from the empirical interpretation of knowledge spaces which has inspired our work (and our terminology). Consider the case of two students engaged in extensive interactions for a long time, and suppose that their initial knowledge states with respect to a particular body of information are K and K'. At some point, one of these students could conceivably have acquired the joint knowledge of both. The knowledge state of this student would then be $K \cup K'$. Obviously, there is no certainty that this will happen. However, requiring the existence of a state in the structure to cover this case is reasonable.

One could object to this argument that a student having acquired all the items in two states K and K' may actually end up in a state containing not only all the items in $K \cup K'$, but also some other items, which would have been derived spontaneously, so to speak, from the knowledge of the items in $K \cup K'$. Moreover, it is conceivable that such derivations are 'automatic'; that is, they always occur, no matter who the students are. This would mean, of course, that the state $K \cup K'$ never occurs empirically.[1] Adopting the axiom of closure under union may thus result in the addition of a number of nonfeasible states in the structure.

However, the axiom of closure under union may not be as wasteful as this objection suggests. One reason is that the number of such 'useless' states may be negligible. More importantly, this axiom considerably simplifies the structure, and makes it easy to summarize and to store in a computer's memory (see the Section on Bases and Atoms and Theorem 1.17).

Other advantages of this axiom are less obvious and involve a change in the primitives. For instance, it will be shown in Chapter 3 (see Theorem 3.14) that knowledge spaces are equivalent to a class of AND/OR graphs encountered in artificial intelligence. A prominent example of a mathematical concept equivalent to knowledge spaces was introduced in 0.6 under the name of 'entailment.' We mentioned there that entailments can be used to construct a knowledge space by querying an expert without actually asking him or her to provide an explicit list of all the knowledge states. Imagine that an experienced teacher is asked, in a systematic way, questions of the following type:

> *Suppose that a student under examination has just provided wrong responses to all the items q_1, q_2, ..., q_n. Is it practically certain that this student will also fail item q_{n+1}? We assume that careless errors and lucky guesses are excluded.*[2]

The responses to all such items define a relation $\mathcal{R}$ on $2^Q \setminus \{\varnothing\}$, with the interpretation: for any two nonempty sets A, B of items

$$A\mathcal{R}B \quad \text{if and only if} \quad \left\{ \begin{array}{l} \text{from the failure of all the items in } A \\ \text{we infer the failure of all the items in } B. \end{array} \right.$$

---

[1] An axiom generalizing the closure under union and covering this case is discussed in Problem 11.

[2] In practice, we suppose also that n is small, say $n \leq 5$. We shall see is Chapter 12 that this assumption is empirically justified.

It turns out that any relation $\mathcal{R}$ on $2^Q \backslash \{\varnothing\}$ specifies a unique knowledge space. The definition of the states is given in the next theorem, which extends the discussion to sets of arbitrary cardinality.

1.9. Theorem. *Let Q be a non empty set, and let $\mathcal{R}$ be a relation on $2^Q \backslash \{\varnothing\}$. Let $\mathcal{S}$ be the collection of all subsets K of Q satisfying the following condition:*

$$K \in \mathcal{S} \quad \Longleftrightarrow \quad \big(\forall (A, B) \in \mathcal{R} : \quad A \cap K = \varnothing \Rightarrow B \cap K = \varnothing\big). \quad (1)$$

Then $\mathcal{S}$ contains $\varnothing$ and Q, and is closed under union.

PROOF. Take $\mathcal{F} \subseteq \mathcal{S}$ and suppose that $A\mathcal{R}B$, with $A \cap (\cup \mathcal{F}) = \varnothing$. We obtain: $A \cap K = \varnothing$, for all K in $\mathcal{F}$. Using (1), we derive that we must also have $B \cap K = \varnothing$ for all K in $\mathcal{F}$. We conclude that $B \cap (\cup \mathcal{F}) = \varnothing$ and thus $\cup \mathcal{F} \in \mathcal{S}$. Since the right member of (1) is trivially satisfied for $\varnothing$ and for Q, both must be in $\mathcal{S}$. □

Notice that the knowledge space $(Q, \mathcal{S})$ in Theorem 1.9 needs not be closed under intersection (see next example). This relation $\mathcal{R}$ provides a powerful method for constructing knowledge spaces in practice, whenever one is willing to rely on human expertise. Chapters 5 and 12 are devoted to this topic. The equivalence between entailments and knowledge spaces is established in Theorem 5.5.

1.10. Example. With $Q = \{a, b, c\}$ and $\mathcal{R}$ containing the single pair $(\{a, b\}, \{c\})$, both $\{a, c\}$ and $\{b, c\}$ are knowledge states in $(Q, \mathcal{S})$; however, their intersection $\{c\}$ is *not* a knowledge state.

Closure Spaces

1.11. Definition. By a *collection* on Q we mean a collection $\mathcal{K}$ of subsets of the *domain* Q. We often then write $(Q, \mathcal{K})$ to denote the collection. Thus, a knowledge structure $(Q, \mathcal{K})$ is a collection $\mathcal{K}$ which contains both $\varnothing$ and Q. Notice that a collection may be empty. A collection $(Q, \mathcal{L})$ is a *closure space* when the family $\mathcal{L}$ contains Q and is closed under intersection. This closure space is *simple* when $\varnothing$ belongs to $\mathcal{L}$. Thus, a collection $\mathcal{K}$ of subsets of a domain Q forms a knowledge space on Q iff the dual structure $\overline{\mathcal{K}}$ is a simple closure space.

Examples of closure spaces abound in mathematics.

1.12. Example. Let $\mathbf{R}^3$ be the set of all points of a 3-dimensional Euclidean space, and let $\mathcal{L}$ be the family of all affine subspaces (that is: the empty set, all singletons, lines, planes and $\mathbf{R}^3$ itself). Then $\mathcal{L}$ is closed under intersection. Another example is obtained by taking the family of all convex subsets of $\mathbf{R}^3$.

These two examples of closure spaces are only instances of general classes (replace 3-dimensional Euclidean space by any affine space over an (ordered) skew field). Moreover, other classes of examples can be found in almost all branches of mathematics (e.g. by taking subspaces of a vector space, subgroups of a group, ideals in a ring, closed subsets of a topological space). Our next example comes from another discipline.

1.13. Example. In Example 0.12, we considered the collection L of all the well-formed expressions in some formal language, together with a fixed set of derivation rules, and a relation $\mathcal{I}$ on the set of all subsets of L, defined by: $A\,\mathcal{I}\,B$ if all the expressions in B are derivable from the expressions in A by application of the derivation rules. A knowledge structure can be obtained by calling any $K \subseteq L$ a *state* of $\mathcal{I}$ if $B \subseteq K$ whenever $A \subseteq K$ and $A\mathcal{I}B$. It is easily shown that the collection $\mathcal{L}$ of all states is closed under intersection; that is, $\cap \mathcal{F} \in \mathcal{L}$ for any $\mathcal{F} \subseteq \mathcal{L}$ (see Problem 5).

Closure spaces are sometimes called 'convexity structures.' References are given in the source section. We record below an obvious construction.

1.14. Theorem. *Let $(Q, \mathcal{L})$ be a closure space. Then any subset A of Q is contained in a unique, smallest element of $\mathcal{L}$, denoted as A'; we have for $A, B \in 2^Q$,*

   (i) $A \subseteq A'$;
   (ii) $A' \subseteq B'$ *when* $A \subseteq B$;
   (iii) $A'' = A'$.

Conversely, any mapping $2^Q \to 2^Q : A \mapsto A'$ which satisfies Conditions (i) to (iii) is obtained from a unique closure space on Q; this establishes a one-to-one correspondence between those mappings and the closure spaces on Q. Moreover, $\varnothing' = \varnothing$ iff $\varnothing \in \mathcal{L}$.

PROOF. Given $A \in 2^Q$, the intersection of all elements of $\mathcal{L}$ that contain A is an element of $\mathcal{L}$ again, and this intersection is the smallest possible element of $\mathcal{L}$ that contains A (remember that Q itself is an element of $\mathcal{L}$, and contains A). Proving Conditions (i) to (iii) is easy, and left to the reader. Conversely, if a mapping $2^Q \to 2^Q : A \mapsto A'$ which satisfies Conditions (i) to (iii) is given, we set $\mathcal{L} = \{A \in 2^Q \mid A' = A\}$. Then $\mathcal{L}$ is closed under intersection (as easily verified). Moreover, the mapping $A \mapsto A'$ is obtained from $\mathcal{L}$ through the construction just introduced. It is now easy to establish the one-to-one correspondence mentioned in the statement; we leave the rest of the proof to the reader.   □

1.15. Definition. In the notation of Theorem 1.14, A' is called the *closure* of A (in the closure space $(Q, \mathcal{L})$).

Substructures

It may happen that only a part of the domain of a knowledge structure is of interest for some particular purpose. For example, the domain covers the full curriculum in high school mathematics, but the researcher is only concerned with those questions relevant to algebra. It is natural to ask which properties of the original knowledge structure will be preserved if only those algebra questions are retained. The next theorem introduces the key concepts.

1.16. Theorem. *Let $(Q, \mathcal{K})$ be a knowledge structure, and let A be any non empty subset of its domain Q. Then, the family*

$$\mathcal{H} = \{H \in 2^A \mid H = A \cap K \text{ for some } K \in \mathcal{K}\} \tag{2}$$

is a knowledge structure. Moreover, if $\mathcal{K}$ is a space (resp. is discriminative), then $\mathcal{H}$ is also a space (resp. is discriminative). The reverse implications do not hold.

PROOF. Since $A = A \cap Q$ and $\varnothing = A \cap \varnothing$, and both Q and $\varnothing$ are states of $\mathcal{K}$, both A and $\varnothing$ must be states of $\mathcal{H}$. For any family $\mathcal{F} \subseteq \mathcal{H}$, consider the family $\mathcal{F}' = \{F' \in \mathcal{K} \mid F = A \cap F' \text{ for some } F \in \mathcal{F}\}$. If $\mathcal{K}$ is a space, then $\cup \mathcal{F}' = S \in \mathcal{K}$. We have $A \cap S \in \mathcal{H}$, with successively

$$A \cap S = A \cap (\cup \mathcal{F}') = \cup_{F' \in \mathcal{F}'} (A \cap F') = \cup \mathcal{F}.$$

Thus $\mathcal{H}$ is a space. Example 1.18 shows that the reverse implication does not hold.

Suppose that $\mathcal{K}$ is discriminative, and take any two distinct items q, q' in A. By assumption, there exists a state K in $\mathcal{K}$ with $q \in K$ and $q' \notin K$ (or the contrary). Then also $q \in A \cap K$ and $q' \notin A \cap K$. As $A \cap K \in \mathcal{H}$, it follows that $\mathcal{L}$ is discriminative. A proof that the reverse implication does not hold is contained in Example 1.18.   □

1.17. Definition. Let $(Q, \mathcal{K})$ be a knowledge structure, let A be some nonempty subset of Q, and let $\mathcal{H}$ be defined by Equation (2). Then $(A, \mathcal{H})$ is called a *substructure* of the *parent structure* $(Q, \mathcal{K})$. A substructure $(A, \mathcal{H})$ is said to be a *proper* substructure if A is a proper subset of Q. If $(Q, \mathcal{K})$ is a space, then $(A, \mathcal{H})$ is a *subspace* of $(Q, \mathcal{K})$. When $H = A \cap K$ for two states H of $\mathcal{H}$ and K in $\mathcal{K}$, we say that H is the *trace* of K *on* A. In keeping with this terminology, we shall sometimes say that a knowledge structure $\mathcal{H}$ is the *trace of a knowledge structure* $\mathcal{K}$ on a set A to state that $(A, \mathcal{H})$ is a substructure of $(Q, \mathcal{K})$. A property of a knowledge structure is called *hereditary* if its validity for a knowledge structure implies its validity for any of its substructures. (Thus, Theorem 1.16 states that the properties of being a space and being discriminative are hereditary.)

We shall often denote by $\mathcal{K}|_A$ a substructure of the knowledge structure $\mathcal{K}$ having domain A included in $\cup \mathcal{K}$.

1.18. Example. Consider the two knowledge structures

$$\mathcal{K} = \big\{ \varnothing, \{c, e\}, \{d\}, \{a, c, d, e\}, \{b, c, d, e\}, \{a, b, c, d, e\} \big\},$$
$$\mathcal{K}|_{\{a,b\}} = \big\{ \varnothing, \{a\}, \{b\}, \{a, b\} \big\}.$$

Then $\mathcal{K}|_{\{a,b\}}$ is a discriminative knowledge space which is the trace of $\mathcal{K}$ on the set $\{a, b\}$, but $\mathcal{K}$ is neither discriminative, nor is it a space.

Bases and Atoms

As mentioned earlier, knowledge structures encountered in practical applications may have a large number of states, which raises the problem of describing such structures economically (for instance, in order to store the information in a computer's memory). In the case of a finite knowledge space, only some of the states must be specified, the remaining ones being generated by taking unions.

1.19. Definition. The *span* of a family $\mathcal{F}$ of sets is the family $\mathcal{F}'$ of all sets which are unions of some members of $\mathcal{F}$. In such a case, we also say that $\mathcal{F}$ *spans* $\mathcal{F}'$. A *base* for a knowledge structure $(Q, \mathcal{K})$ is a minimal family $\mathcal{B}$ of states spanning $\mathcal{K}$ (where "minimal" means "minimal w.r.t. set inclusion", that is: if $\mathcal{F} \subseteq \mathcal{B}$ is any family of states spanning $\mathcal{K}$, then necessarily $\mathcal{F} = \mathcal{B}$). By a standard convention, the empty set is the union of the empty subfamily of $\mathcal{B}$. Thus, the empty set never belongs to a base.

Clearly, a knowledge structure has a base only if it is a knowledge space. Moreover, a state K belonging to some base $\mathcal{B}$ of $\mathcal{K}$ cannot be the union of other elements of $\mathcal{B}$; thus, K cannot be the union of other states of $\mathcal{K}$.

1.20. Theorem. *Let $\mathcal{B}$ be a base for a knowledge space $(Q, \mathcal{K})$. Then $\mathcal{B} \subseteq \mathcal{F}$ for any subfamily $\mathcal{F}$ of states spanning $\mathcal{K}$. Consequently, a knowledge space admits at most one base.*

PROOF. Let $\mathcal{B}$ and $\mathcal{F}$ be as in the hypotheses of the theorem, and suppose that $K \in \mathcal{B} \setminus \mathcal{F}$. Then, $K = \cup \mathcal{H}$ for some $\mathcal{H} \subseteq \mathcal{F}$. Since $\mathcal{B}$ is a base, any member of $\mathcal{H}$ is a union of some elements from $\mathcal{B}$. This implies that K is a union of sets in $\mathcal{B} \setminus \{K\}$, contradicting the minimality property of a base. The uniqueness of a base is now obvious. □

1.21. Example. We show that some infinite knowledge space has no base. Consider the collection $\mathcal{O}$ of all open sets of $\mathbf{R}$. This space $\mathcal{O}$ is spanned by the family $\mathcal{I}_1$ of all open intervals with rational endpoints, as well as by the family $\mathcal{I}_2$ of all open intervals with irrational endpoints. If $\mathcal{O}$ had a base $\mathcal{B}$, we would have by Theorem 1.20 $\mathcal{B} \subseteq \mathcal{I}_1 \cap \mathcal{I}_2 = \varnothing$, which is absurd. Thus, $\mathcal{O}$ has no base (in the sense of Definition 1.19). In the finite case, however, a base always exists.

1.22. Theorem. *If a knowledge space is essentially finite, it necessarily has a base.*

Indeed, since the number of states is finite, there must be a minimal spanning subfamily of states, that is, a base.

The next definition will be useful for the specification of the base when it exists. Notice that we do not restrict ourselves to the essentially finite case.

1.23. Definition. For any item q, an *atom* at q is a minimal knowledge state containing q. A state K is called an *atom* if it is an atom at q for some item q.

Notice that we use the term "atom" with a meaning different from the one it has in lattice theory (see e.g. Birkhoff, 1967, or Davey and Priestley, 1990).

1.24. Example. For the space $\mathcal{K} = \{\varnothing, \{a\}, \{a,b\}, \{b,c\}, \{a,b,c\}\}$, the state $\{b,c\}$ is an atom at b and also an atom at c. There are two atoms at b, namely $\{a,b\}$ and $\{b,c\}$. There is only one atom at a, which is $\{a\}$. (However, a also belongs to the atom $\{a,b\}$, but the state $\{a,b\}$ is not an atom at a.)

The knowledge space from Example 1.21 has no atoms. On the other hand, in an essentially finite knowledge structure, there is at least one atom at every item.

Another characterization of the atoms in a knowledge space is given below.

1.25. Theorem. *A state K in a knowledge space $(Q, \mathcal{K})$ is an atom iff $K \in \mathcal{F}$ for any subfamily of states $\mathcal{F}$ satisfying $K = \cup \mathcal{F}$.*

PROOF. (Necessity) Suppose that K is an atom at q, and that $K = \cup \mathcal{F}$ for some subfamily $\mathcal{F}$ of states. Thus, q must belong to some $K' \in \mathcal{F}$, with necessarily $K' \subseteq K$. We must have $K = K'$, since K is a minimal state containing q. Thus, $K \in \mathcal{F}$.

(Sufficiency.) If K is not an atom, for each $q \in K$, there must be some state $K'(q)$ with $q \in K'(q) \subset K$. With $\mathcal{F} = \{K'(q) \mid q \in K\}$, we have thus $K = \cup \mathcal{F}$, and $K \notin \mathcal{F}$.   □

1.26. Theorem. *Suppose a knowledge space has a base. Then this base is formed by the collection of all the atoms.*

PROOF. Let $\mathcal{B}$ be the base of a knowledge space $(Q, \mathcal{K})$, and let $\mathcal{A}$ be the collection of all the atoms. (We do not assume that there is an atom at every item.) We have to show that $\mathcal{A} = \mathcal{B}$. If some $K \in \mathcal{B}$ is not an atom, then, for every $q \in K$, there exists a state $K'(q)$ with $q \in K'(q) \subset K$. But then $K = \cup_{q \in K} K'(q)$ and we cannot have $K \in \mathcal{B}$ (since each $K'(q)$ is a union of elements in $\mathcal{B}$, we see that K is a union of other elements from $\mathcal{B}$). Thus, K must be an atom for at least one item. Every element of the base is thus an atom, and we have $\mathcal{B} \subseteq \mathcal{A}$. Conversely, take any $K \in \mathcal{A}$. Then, $K = \cup \mathcal{F}$ for some $\mathcal{F} \subseteq \mathcal{B}$. By Theorem 1.25, we have $K \in \mathcal{F} \subseteq \mathcal{B}$. Thus, $\mathcal{A} = \mathcal{B}$. □

Even when the base exists, there may not be an atom at every item.

1.27. Example. Define

$$\mathcal{G} = \{[0, \frac{1}{n}] \mid n \in \mathbb{N}\} \cup \{\varnothing\}.$$

Then $([0,1], \mathcal{G})$ is a knowledge space, with a base consisting of all states except $\varnothing$; every item has an atom, except 0. Note that $([0,1], \mathcal{G})$ is not discriminative. However, its discriminative reduction $([0,1]^*, \mathcal{G}^*)$ (cf. Definition 1.4) provides a similar counterexample. (It has a base but no atom at 0^*.)

As the base $\mathcal{B}$ of a knowledge space $(Q, \mathcal{K})$ provides a compact encoding of this space, it is important to have efficient algorithms for constructing the base of a given space, and also for generating all the states from the base. Two such algorithms are sketched in the next two sections.

An Algorithm for Constructing the Base

We assume here that the domain Q is finite, with say $|Q| = m$ and $|\mathcal{K}| = n$. By Theorem 1.26, the base of a knowledge space is formed by all the atoms. Recall from Definition 1.23 that an atom at some item is a minimal knowledge state containing this item. A simple algorithm for building the base is grounded on this concept of an atom. Dowling (1993b) has nicely formulated this algorithm as follows.

1.28. Sketch of Algorithm. We arbitrary list the items as q_1, q_2, ..., q_m. We also list the states as K_1, K_2, ..., K_n in such a way that $K_i \subset K_j$ implies $i < j$ for $i, j \in \{1, 2, ..., n\}$. (This is achieved by listing the states according to nondecreasing size, and arbitrarily for states of the same size.) Form a $n \times m$ array T with the rows and the columns representing the states and the items respectively; i.e. the rows are indexed from 1 to n and the columns from 1 to m. At any step of the algorithm, a cell of T contains one of the symbols '$*$', '$+$' or '$-$'. Initially, set T_{ij} to $*$ if state K_i contains item q_j; otherwise, set T_{ij} equal to $-$. The algorithm inspects rows $i = 1, 2, \ldots, n$ and transforms into $+$ each value $*$ contained in a cell T_{ij} satisfying the following condition: there exists an index p such that $1 \leq p < i$, state K_p contains item q_j and $K_p \subset K_i$. When this is done, the atoms are the states K_i for which row i still contains at least one $*$.

1.29. Example. Take the space $\mathcal{K} = \{\varnothing, \{a\}, \{a, b\}, \{b, c\}, \{a, b, c\}\}$ from Example 1.24. The initial array T is shown on the left of Table 1.1. From the final value of T on the right, we conclude that the base is $\{\{a\}, \{a, b\}, \{b, c\}\}$.

	a	b	c			a	b	c
$\varnothing$	−	−	−		$\varnothing$	−	−	−
$\{a\}$	*	−	−		$\{a\}$	*	−	−
$\{a, b\}$	*	*	−		$\{a, b\}$	+	*	−
$\{b, c\}$	−	*	*		$\{b, c\}$	−	*	*
$\{a, b, c\}$	*	*	*		$\{a, b, c\}$	+	+	+

Table 1.1. The initial and final values of the array T in Example 1.29.

It is easily checked that Algorithm 1.28 also works when provided with a spanning family instead of the space itself (see Problem 8).

The description given for this algorithm is intended for an initial encoding of a space or of a spanning family $\mathcal{F}$ as an $n \times m$ array with each cell (i, j) indicating whether state K_i contains item q_j. It is not difficult to redesign the search for atoms in the case of a different encoding of the space (for instance, as a list of states with each state being a list of items).

An Algorithm for Generating a Space from its Base

The solution described below owes much to that of Dowling (1993b). However, by clarifying some of the underlying ideas, we improve both the principle and the efficiency of the algorithm.

Keeping the same notation, we suppose that the base $\mathcal{B}$ contains p states, with $\mathcal{B} = \{B_1, B_2, \ldots, B_p\}$. The states are obtained by a sequential procedure, based on considering increasingly larger subfamilies of the base. We set $\mathcal{G}_0 = \{\varnothing\}$, and for $i = 1, 2, \ldots, p$, we define $\mathcal{G}_i$ as the space spanned by $\mathcal{G}_{i-1} \cup \{B_i\}$. This is the general scheme. Nevertheless, some care must be taken to ensure efficiency. Clearly, at any step i of the algorithm, the new states created by taking the span of $\mathcal{G}_{i-1} \cup \{B_i\}$ are all of the form $G \cup B_i$ with $G \in \mathcal{G}_{i-1}$. However, some states formed by taking the union of B_i with some states in $\mathcal{G}_{i-1}$ may already exist in $\mathcal{G}_i$. A straightforward application of this scheme would require verifying for each newly generated state whether it was obtained before. As in general the number n of states can grow exponentially as a function of p, such verifications may be prohibitive. Accordingly, we want to form $G \cup B_i$ only when this union delivers a state not encountered before (whether at a previous step, or at the current one). Here is the crucial remark: among all states G from $\mathcal{G}_{i-1}$ producing $K = G \cup B_i$, there is a unique maximum one that we denote by M. We have thus $K = M \cup B_i$, and moreover $K = G \cup B_i$ for $G \in \mathcal{G}_{i-1}$ implies $G \subseteq M$. The existence and uniqueness of M follow at once from the fact that $\mathcal{G}_{i-1}$ is closed under union. The clue to the algorithm is Condition (ii) in the result below, which provides a manageable characterization of M. In this theorem, we consider the situation in which an arbitrary subset B of a domain Q is added to the base $\mathcal{D}$ of a knowledge space $\mathcal{G}$ on Q.

1.30. Theorem. *Let $(Q, \mathcal{G})$ be a knowledge space with base $\mathcal{D}$, and take $M \in \mathcal{G}$ and $B \in 2^Q$. The following two conditions are equivalent:*

  (i) $\forall G \in \mathcal{G}: \quad M \cup B = G \cup B \Rightarrow G \subseteq M$;
  (ii) $\forall D \in \mathcal{D}: \quad D \subseteq M \cup B \Rightarrow D \subseteq M$.

PROOF. (i) $\Rightarrow$ (ii). If $D \subseteq M \cup B$ for some $D \in \mathcal{D}$, we get $M \cup B = (M \cup D) \cup B$. As $M \cup D \in \mathcal{G}$, our assumption implies $M \cup D \subseteq M$, that is $D \subseteq M$.

  (ii) $\Rightarrow$ (i). If $M \cup B = G \cup B$ with $G \in \mathcal{G}$, there exists a subfamily $\mathcal{E}$ of $\mathcal{D}$ such that $G = \cup \mathcal{E}$. For $D \in \mathcal{E}$, we have $D \subseteq M \cup B$, hence by our assumption $D \subseteq M$. We conclude that $G \subseteq M$. $\square$

Returning to our discussion, we now have a way to generate, at the main stage i, only distinct elements $G \cup B_i$: it suffices to take such a union exactly when G from $\mathcal{G}_{i-1}$ satisfies

$$\forall D \in \{B_1, B_2, \ldots, B_{i-1}\}: \quad D \subseteq G \cup B_i \Longrightarrow D \subseteq G. \tag{3}$$

We must also avoid generating a state $G \cup B_i$ belonging to $\mathcal{G}_{i-1}$ (i.e., a state that was generated at some earlier main stage). To this effect, notice that for $G \in \mathcal{G}_{i-1}$ satisfying (3), we have $G \cup B_i \in \mathcal{G}_{i-1}$ iff $B_i \subseteq G$.

1.31. Sketch of the Algorithm. Let $\mathcal{B} = \{B_1,\ B_2,\ \ldots,\ B_p\}$ be the base of some knowledge space $\mathcal{K}$ on Q to be generated by the algorithm. Initialize $\mathcal{G}$ to $\{\varnothing\}$. At each step $i = 1, 2, \ldots, p$, perform the following:

(1) Initialize $\mathcal{H}$ to $\varnothing$.

(2) For each $G \in \mathcal{G}$, check whether
$$B_i \not\subseteq G \text{ and } \forall D \in \{B_1, B_2, \ldots, B_{i-1}\} : \ D \subseteq G \cup B_i \Rightarrow D \subseteq G.$$
If the condition holds, add $G \cup B_i$ to $\mathcal{H}$.

(3) When all G's from $\mathcal{G}$ have been considered, replace $\mathcal{G}$ with $\mathcal{G} \cup \mathcal{H}$. (This terminates step i.)

The family $\mathcal{G}$ obtained after step p is the desired space $\mathcal{K}$.

1.32. Example. For the base $\mathcal{B} = \{\{a\}, \{a, b\}, \{b, c\}\}$, Table 1.2 displays the successive values of $\mathcal{G}$.

Main stage	base element	states in $\mathcal{G}$
initialization		$\varnothing$
1	$\{a\}$	$\varnothing, \{a\}$
2	$\{a, b\}$	$\varnothing, \{a\}, \{a, b\}$
3	$\{b, c\}$	$\varnothing, \{a\}, \{a, b\}, \{b, c\}, \{a, b, c\}$

Table 1.2. The successive values of $\mathcal{G}$ in Example 1.32.

1.33. Example. Here is another example, with $\mathcal{B} = \{\{a\}, \{b, d\}, \{a, b, c\}, \{b, c, e\}\}$. In Table 1.3, we only show on each line the base element considered at this main stage together with the additional state(s) produced.

Base element	states in $\mathcal{H}$
initialization	$\varnothing$
$\{a\}$	$\{a\}$
$\{b, d\}$	$\{b, d\}, \{a, b, d\}$
$\{a, b, c\}$	$\{a, b, c\}, \{a, b, c, d\}$
$\{b, c, e\}$	$\{b, c, e\}, \{b, c, d, e\}, \{a, b, c, e\}, \{a, b, c, d, e\}$

Table 1.3. The successive values of $\mathcal{H}$ in Example 1.33.

1.34. Remarks. a) In principle, Algorithm 1.31 can be applied to any spanning subfamily $\mathcal{F}$ of a knowledge space $\mathcal{K}$ to produce that space (see Problem 9). In the case of a spanning subfamily $\mathcal{F}$ which is not the base, we recommend however to first construct the base $\mathcal{B}$ of $\mathcal{K}$ by Algorithm 1.28 to $\mathcal{F}$. Algorithm 1.31 can then be applied to $\mathcal{B}$ to produce $\mathcal{K}$.

b) A few words about the efficiency of Algorithm 1.31 are in order. Experiments show that the execution time of a `Pascal` program implementing this algorithm depends on the order in which the basis elements (or spanning elements) are listed. No 'best' rule seems to emerge about a plausible, optimal way of encoding the basis elements. On the other hand, an improvement is obtained with many data sets by adding the following extension to Algorithm 1.31. At each step i, build the union U of all B_j's with $0 < j < i$ and $B_j \subset B_i$. When a $G \in \mathcal{G}$ is taken into consideration, first verify whether $U \subseteq G$. If $U \nsubseteq G$, the verification of step (2) in Algorithm 1.31 can be skipped, because the condition cannot hold. Dowling's original algorithm uses set U heavily. This algorithm is also sensitive to the selected ordering. Our extended version performs generally faster by 10% to 30%.

c) On the theoretical side, the complexity of Algorithm 1.31 (in the sense of Garey and Johnson, 1979) is good. Because the cardinality n of the family $\mathcal{K}$ of sets spanned by a basis $\mathcal{B}$ with p elements in a domain Q of m items can grow exponentially with p, we analyze the complexity in terms of m, p and n together. Algorithm 1.31 has execution time in $O(n \cdot p^2 \cdot m)$, in other words there exist natural numbers m_0, p_0 and n_0 and a positive real number c such that execution on a domain of size $m \geq m_0$, with a basis of size $p \geq p_0$ producing a space of size $n \geq n_0$ will always take less than $c \cdot n \cdot p^2 \cdot m$ steps (see Problem 17).

Bases and Atoms: the Infinite Case*

The results on bases and atoms are straightforward in the case of essentially finite knowledge structures. In the infinite case, however, there is no guarantee that atoms exist (see Example 1.21). These difficulties vanish under the 'finitary' case defined below. The term 'finitary' comes from the theory of closure spaces (see the Sources Section at the end of this chapter.)

1.35. Definition. A knowledge structure $\mathcal{K}$ is *finitary* when the intersection of any chain of states in $\mathcal{K}$ is again a state. We shall also say that $\mathcal{K}$ is *granular* if for any state K containing some item q, there is an atom at q that is included in K. Obviously, any essentially finite knowledge structure is both finitary and granular. Another example is given below.

1.36. Example. Let $(V, \mathcal{S})$ be a knowledge structure, where V is a real vector space V and $\mathcal{S}$ is the family of all subspaces of V. Then its dual knowledge structure $(V, \overline{\mathcal{S}})$ is a finitary and granular knowledge space.

1.37. Theorem. *Any finitary knowledge structure is granular.*

PROOF. Consider the collection $\mathcal{F}$ of states that are included in a state K and contain an item q, ordered by inclusion. By Hausdorff's Maximal Principle, $\mathcal{F}$ must include at least one maximal chain $\mathcal{C}$ (cf. 0.21). If the knowledge structure is finitary, $\cap \mathcal{C}$ is a state and an atom at q.   □

This implies that the span of the family $\mathcal{A}$ of all the atoms of a finitary knowledge structure $\mathcal{K}$ necessarily includes $\mathcal{K}$. The converse of Theorem 1.37 does not hold, even for spaces. The next example provides a granular knowledge space which is not finitary.

1.38. Example. Consider the following subsets of $[0, 2]$:

$$\{0\} \cup [\frac{1}{k}, \frac{2}{k}], \qquad \text{for } k \in \mathbb{N}.$$

Since none of these subsets includes any other one, their collection forms the base $\mathcal{B}$ of a granular knowledge space $\mathcal{K}$ (in this case, any state in the base is an atom at each of the items it contains). On the other hand, $\mathcal{K}$ is not finitary. Since $[0, \frac{2}{k}] = \{0\} \cup \bigcup_{j=k}^{\infty} [\frac{1}{j}, \frac{2}{j}]$, the family of intervals $[0, \frac{2}{k}]$, for $k \in \mathbb{N}$, constitutes a chain in $\mathcal{K}$ whose intersection $\{0\}$ is not in $\mathcal{K}$.

Together with Theorem 1.37, the following result shows that any finitary knowledge space has a base.

1.39. Theorem. *Any granular knowledge space has a base.*

PROOF. Consider the collection $\mathcal{B}$ of all the atoms in a granular knowledge space $(Q, \mathcal{K})$. By Definition 1.35, any knowledge state in $\mathcal{K}$ is the union of all the atoms it contains. Thus the family $\mathcal{B}$ spans $\mathcal{K}$, and it is clearly minimal with this property.   □

A knowledge space may have a base without being granular. This happens in Example 1.27: there is no atom at 0.

We now study the condition of closure under intersection from the point of view of the atoms of the knowledge space.

1.40. Theorem. *A knowledge space $\mathcal{K}$ closed under intersection has exactly one atom at each item q, which is specified by $\cap \mathcal{K}_q$. Moreover, a granular knowledge space $\mathcal{K}$ having exactly one atom at each item is necessarily closed under intersection.*

PROOF. The assertions in the first sentence are obvious. Suppose that the knowledge space $\mathcal{K}$ has exactly one atom at each item and let $\mathcal{F}$ be a subfamily of $\mathcal{K}$. If $\cap\mathcal{F} = \varnothing$, then $\cap\mathcal{F}$ is a state. Otherwise, take any $q \in \cap\mathcal{F}$, and let $K(q)$ be the unique atom at q. For all $K \in \mathcal{F}$, we must have $K(q) \subseteq K$ (by granularity there is an atom at q contained in K). Hence $K(q) \subseteq \cap\mathcal{F}$, and thus

$$\cap\mathcal{F} = \bigcup_{q \in \cap\mathcal{F}} K(q) \in \mathcal{K},$$

since $\mathcal{K}$ is a knowledge space.   □

The following Example shows that the granularity assumption cannot be dropped in the second statement of Theorem 1.40.

1.41. Example. Take the knowledge space $\mathcal{K}$ on $\mathbf{R}$ with base

$$\{ [0, \frac{1}{n}] \mid n \in \mathbf{N} \} \cup \{]-\infty, 0], \mathbf{R} \}.$$

Then for each $r \in \mathbf{R}$ there is a unique atom at r. However, the intersection of states $]-\infty, 0] \cap [0, 1] = \{0\}$ does not belong to $\mathcal{K}$.

1.42. Corollary. *A granular knowledge space $\mathcal{K}$ is closed under intersection iff there is exactly one atom at each item.*

A more systematic study of knowledge spaces closed under intersection is contained in the next section.

The Surmise Relation

An important part of this book concerns the analysis, within the framework
of knowledge structures, of the possible ways of acquiring the material in a
domain Q. Here, we only consider the concept of 'predecessor' of a question.
Intuitively, an item r is a predecessor of an item q if r is never mastered
after q, either for logical or historical reasons. The next definition formalizes
the compelling idea that the predecessors of some item q are the items
contained in all the states containing q.

1.43. Definition. Let $(Q, \mathcal{K})$ be a knowledge structure, and let $\precsim$ be a
relation on Q defined by

$$r \precsim q \Longleftrightarrow r \in \cap \mathcal{K}_q. \tag{4}$$

The relation $\precsim$ will be called the *surmise relation* or sometimes the *prece-*
dence relation of the knowledge structure (the usage of the two terminologies
is discussed in Remark 1.45). When $r \precsim q$ holds, we say that r is *surmis-*
able from q, or that r *precedes* q. If moreover $q \precsim r$ does not hold, then
we write $r \prec q$ and say that r *strictly precedes* q.

Notice the equivalence:

$$r \precsim q \Longleftrightarrow \mathcal{K}_r \supseteq \mathcal{K}_q \tag{5}$$

which holds for any knowledge structure $\mathcal{K}$ and any items q, r in its domain.
(We leave the verification of this fact to the reader; see Problem 12). This
immediately implies the following result.

1.44. Theorem. *The surmise relation of a knowledge structure is a quasi*
order. When the knowledge structure is discriminative, this quasi order is
a partial order.

By abuse of language, we shall occasionally talk about the Hasse diagram
of a knowledge structure $\mathcal{K}$, to mean the Hasse diagram of the surmise
relation of the discriminative reduction $\mathcal{K}^* = \{K^* \mid K \in \mathcal{K}\}$ (cf. 1.4).

1.45. Remark. Two viewpoints can be taken with regard to the relation
$\precsim$. One is that of inference: if $r \precsim q$, then the mastery of r can be surmised
from that of q. The other one is that of learning: $r \precsim q$ means that r is
always mastered before or at the same time as q, either for logical reasons or

because this is the custom in the population of reference. As an illustration, consider the following two questions in European history:

Question q: *Who was the prime minister of Great Britain*
                just before World War II?
Question r: *Who was the prime minister of Great Britain*
                during World War II?

Today, anybody knowing that the answer to question q is 'Neville Chamberlain' would also know that the next prime minister was Winston Churchill. In our terms, this means that any state containing q would also contain r, that is, $r \precsim q$. Obviously, logic plays no rôle in this dependency, which only relies on the structure of the collection of states, which itself is a reflection of the population of subjects under consideration. In many cases, however, especially in mathematics or science, the formula $r \precsim q$ will mean that, for logical reasons, r must be mastered before or at the same time as q.

In Chapter 3, we shall discuss a generalization of the concept of a surmise relation formalizing the following natural idea: to any item q in the structure is attached a collection of possible learning histories leading to q.

We examine an example of a surmise relation.

1.46. Example. Consider the knowledge structure

$$\mathcal{G} = \big\{ \varnothing, \{a\}, \{b\}, \{a,b\}, \{b,c\}, \{a,b,c\}, \{b,c,e\}, $$
$$\{a,b,c,e\}, \{a,b,c,d\}, \{a,b,c,d,e\} \big\}. \tag{6}$$

It is easy to verify that $\mathcal{G}$ is a discriminative knowledge space (cf. 1.4 and 1.7). The surmise relation, which we denote by $\precsim$, is thus a partial order. The Hasse diagram of $\precsim$ is given in Figure 1.1. We shall not give all the details of the construction of $\precsim$ from the knowledge structure $\mathcal{G}$. As an illustration, notice that

$$c \in \{a,b,c,d\} \cap \{a,b,c,d,e\} = \cap \mathcal{G}_d.$$

We obtain $c \precsim d$. On the other hand, $a \notin \{b,c,e\}$. Thus $a \notin \mathcal{G}_e$, yielding $\neg(a \precsim e)$: in the Hasse diagram, there is no broken line descending from e to a. The surmise relation offers a compact summary of the information contained in a knowledge structure, especially when the domain is finite,

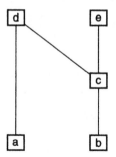

Fig. 1.1. Hasse diagram of the surmise relation of the knowledge structure $\mathcal{G}$ specified by Equation (6).

with a small number of elements. It must be realized, however, that some information might be missing: different knowledge structures may have the same surmise relation.

For instance, the knowledge structure $\mathcal{G}' = \mathcal{G} \setminus \{\{b,c\}\}$ has the same surmise relation as $\mathcal{G}$. This raises the question: When is a knowledge structure fully described by its surmise relation? Theorem 1.49 in the next section answers the question.

Quasi Ordinal Spaces

1.47. Definition. A knowledge space closed under intersection is called a *quasi ordinal space*. The chief reason for this terminology is that, in such a case, the knowledge structure is characterized by a quasi order, namely, its surmise relation (see Theorem 1.49). A discriminative and quasi ordinal space is a *(partially) ordinal space*. The surmise relation of such a space is a partial order. Clearly, a quasi ordinal space is finitary (cf. 1.35).

1.48. Theorem. *Let* $\mathcal{K}$, $\mathcal{K}'$ *be two quasi ordinal spaces on the same domain* Q. *Then,*

$$(\forall q, s \in Q: \quad \mathcal{K}_q \subseteq \mathcal{K}_s \Leftrightarrow \mathcal{K}'_q \subseteq \mathcal{K}'_s) \quad \Longleftrightarrow \quad \mathcal{K} = \mathcal{K}'. \qquad (7)$$

PROOF. The implication from right to left is trivial. To establish the converse implication, suppose that $K \in \mathcal{K}$. We have then

$$K \subseteq \bigcup_{q \in K} (\cap \mathcal{K}'_q) = K' \tag{8}$$

for some state $K' \in \mathcal{K}'$. We show that, in fact, $K = K'$. Take any $s \in K'$. There must be some $q \in K$ such that $s \in \cap\mathcal{K}'_q$. Thus, $\mathcal{K}'_q \subseteq \mathcal{K}'_s$, which implies $\mathcal{K}_q \subseteq \mathcal{K}_s$, by the assumed left side of '$\Longleftrightarrow$' in (7). We obtain $s \in \cap\mathcal{K}_q$, yielding $s \in K$. This gives $K' \subseteq K$, and by (8), $K = K'$. We conclude that $\mathcal{K} \subseteq \mathcal{K}'$, and by symmetry, $\mathcal{K} = \mathcal{K}'$.   □

1.49. Theorem. (Birkhoff, 1937.) *There exists a one-to-one correspondence between the collection of all quasi ordinal spaces $\mathcal{K}$ on a domain Q, and the collection of all quasi orders $\mathcal{Q}$ on Q. Such a correspondence is defined through the two equivalences*

$$p\mathcal{Q}q \iff (\forall K \in \mathcal{K}: q \in K \Rightarrow p \in K) \tag{9}$$
$$K \in \mathcal{K} \iff (\forall(p,q) \in \mathcal{Q}: q \in K \Rightarrow p \in K). \tag{10}$$

Moreover, under this correspondence, ordinal spaces are mapped onto partial orders.

Notice that Equation (9) can be written more compactly as

$$p\mathcal{Q}q \iff \mathcal{K}_p \supseteq \mathcal{K}_q. \tag{11}$$

Thus, the quasi order $\mathcal{Q}$ defined by (11) from the knowledge structure $\mathcal{K}$ is nothing else than the surmise relation of $\mathcal{K}$.

PROOF. Equation (9) clearly defines a quasi order on Q (cf. Theorem 1.44). Conversely, for any quasi order $\mathcal{Q}$ on Q, Equation (10) defines a family $\mathcal{K}$ of subsets of Q. We establish properties of this last family. First, $\mathcal{K}$ necessarily contains Q. The family $\mathcal{K}$ also contains $\varnothing$, since $q \in \varnothing \Rightarrow p \in \varnothing$ always holds for any $(p,q) \in \mathcal{Q}$. Thus, $\mathcal{K}$ is a knowledge structure. We show that $\mathcal{K}$ is closed under intersection. Take any $K, K' \in \mathcal{K}$ and suppose that $p\mathcal{Q}q$, with $q \in K \cap K'$. We obtain $q \in K$, $q \in K'$, which implies, by (10), $p \in K$, $p \in K'$, yielding $p \in K \cap K'$. Thus, $K \cap K' \in \mathcal{K}$. Similarly, the intersection of any subfamily of $\mathcal{K}$ belongs to $\mathcal{K}$. The proof that $\mathcal{K}$ is closed under union is similar.

We write $\tilde{\mathbf{K}}^{\mathbf{O}}$ for the collection of all quasi ordinal spaces $\mathcal{K}$ on a domain Q, and $\tilde{\mathbf{R}}^{\mathbf{O}}$ for the collection of all quasi orders $\mathcal{Q}$ on Q. The result obtains if the two mappings

$$f : \quad \tilde{\mathbf{K}}^{\mathbf{O}} \to \tilde{\mathbf{R}}^{\mathbf{O}} : \quad \mathcal{K} \mapsto f(\mathcal{K}) = \mathcal{Q},$$
$$g : \quad \tilde{\mathbf{R}}^{\mathbf{O}} \to \tilde{\mathbf{K}}^{\mathbf{O}} : \quad \mathcal{Q} \mapsto g(\mathcal{Q}) = \mathcal{K}$$

respectively defined by (9) and (10) are mutual inverses. By Equation (11) and Theorem 1.48, f is an injective function. Let $\mathcal{Q}$ be any quasi order on Q, with $\mathcal{K} = g(\mathcal{Q})$ and $f(\mathcal{K}) = \mathcal{Q}'$. Using (10), $p\mathcal{Q}q$ implies that for all $K \in \mathcal{K}$, $q \in K \Rightarrow p \in K$, yielding $p\mathcal{Q}'q$ by (9). Moreover, if $p\mathcal{Q}'q$, we take $K = \{x \in Q \mid x\mathcal{Q}q\}$: since $K \in \mathcal{K}$ and $q \in K$, we have $p \in K$, hence $p\mathcal{Q}q$. Thus, $\mathcal{Q} = \mathcal{Q}'$. Hence, any quasi order $\mathcal{Q}$ is in the range of the function f. We conclude that f and g are mutually inverse functions.

The last assertion (about ordinal spaces) is obvious.   □

1.50. Definition. Referring to the correspondence described in Theorem 1.49, we say that the quasi ordinal space $g(\mathcal{Q})$ is *derived* from the quasi order $\mathcal{Q}$, and similarly that the quasi order $f(\mathcal{K})$ is *derived* from the quasi ordinal knowledge structure $\mathcal{K}$.

For later reference, we notice that Equation (10) can be used with any relation $\mathcal{Q}$ to produce a knowledge space. The proof of the next theorem is left as Problem 15.

1.51. Theorem. *Let $\mathcal{Q}$ be any relation on a domain Q, and define a collection $\mathcal{K}$ of subsets of Q using Equation (10):*

$$K \in \mathcal{K} \quad \Longleftrightarrow \quad (\forall (p, q) \in \mathcal{Q} : q \in K \Rightarrow p \in K). \qquad (10)$$

Then $\mathcal{K}$ is a quasi ordinal knowledge space on Q.

1.52. Definition. Referring to Theorem 1.51, we say that the quasi ordinal space $\mathcal{K}$ is *derived* from the relation $\mathcal{Q}$.

Original Sources and Related Works

As indicated in Chapter 0, the theory of knowledge spaces was initiated by Doignon and Falmagne (1985). Most of that paper was restricted to the finite case. Additional references can be found in the Sources Section of Chapter 0. The results presented here for the infinite case are original, as for instance, those concerning the concept of a granular knowledge structure.

Dowling (1993b) contains two algorithms, one to construct the base of a finite knowledge space, the other to generate the space spanned by a finite family of sets. For the second task, we provide another algorithm which, although similar in spirit, is much easier to grasp and slightly more efficient on the average. A different algorithm, due to Ganter (1984, 1987; see also Ganter and Wille, 1996) in the setting of concept lattices, can also be used for the second task; it has the same overall theoretical efficiency but avoids to store the previously generated states.

Birkhoff's Theorem (1937), in our terminology, concerns quasi ordinal spaces. A variant for knowledge spaces in general is given in Chapter 3.

We mentioned in Example 1.12 a few of the mathematical examples of families of sets closed under intersection. A "closure space" (Definition 1.11) is often also called a "convexity space." The first term is used e.g. by Birkhoff (1967) or Buekenhout (1967), while the second can be found e.g. in Sierksma (1981). Also, Birkhoff (1967) calls "Moore family" any family of subsets closed under intersection. The excellent book by Van de Vel (1993) concerns "convex structures" (also called "aligned spaces" after Jamison, 1982), which are dual to finitary knowledge spaces (cf. Definition 1.35). The word "finitary" was used to qualify such closure spaces (which Buekenhout, 1967, calls "espace à fermeture finie"). It is motivated by the following result (often taken as a definition of "finitary closure space"): The closure space $(Q, \mathcal{L})$ is finitary iff the closure of any subset A of Q is the union of all closures of finite subsets of A (see Problem 18). Chapter 6 extends the concept of a closure to the general context of a quasi order.

Problems

1. Construct the discriminative reduction of the knowledge structure

$$\mathcal{K} = \big\{ \varnothing, \{a,c,d\}, \{b,e,f\}, \{a,c,d,e,f\}, \{a,b,c,d,e,f\} \big\}.$$

2. Do we have $|\mathcal{K}| = |\mathcal{K}^*|$ for any knowledge structure $\mathcal{K}$? Prove your answer.

3. Show that a knowledge structure $(Q, \mathcal{K})$ is essentially finite if and only if Q^* is finite.

4. How many states are contained in the dual of the knowledge structure $\mathcal{H}$ of Example 1.3? Specify some of these states.

5. Prove that the collection $\mathcal{L}$ of states in Example 1.13 is closed under intersection. Explain how this result is related to Theorem 1.9.

6. Does Theorem 1.9 still hold if $\mathcal{R}$ is a relation on 2^Q? Provide a counterexample if your response is negative.

7. Suppose that all the proper substructures of a knowledge structure $\mathcal{K}$ are spaces (resp. discriminative structures, closure spaces). Is it necessarily true, then, that $\mathcal{K}$ itself is a space (resp. discriminative structure, closure space)?

8. Show that Algorithm 1.28 also correctly builds the base when provided with a spanning family instead of the space itself.

9. Show that Algorithm 1.31 for producing the space from the base can also be used to produce the space from any spanning family.

10. Construct the Hasse diagram of the structure $\mathcal{H}$ of Example 1.3.

11. Consider the following axiom generalizing the closure under union.

[JS] *For any subfamily of states $\mathcal{F}$ in a knowledge structure $(Q, \mathcal{K})$, there exists a unique minimal state $K \in \mathcal{K}$ such that $\cup \mathcal{F} \subseteq K$.*

(Under this axiom, $\mathcal{K}$ is thus a 'join semi lattice' with respect to inclusion.) Construct a finite example in which this axiom is not satisfied.

12. Prove Equation (5) in Definition 1.43.

13. For each of the following properties of a knowledge structure, check whether it implies the same property for the dual structure: (a) being a space; (b) quasi ordinal; (c) ordinal.

14. It was proved that the properties of being a space and discriminativity were hereditary (in the sense of Def. 1.17). Are quasi ordinality and ordinality hereditary?

15. Prove Theorem 1.51.

16. (*Feasible Symbologies.*) Not all arbitrarily chosen set of symbols constitutes a symbology (or alphabet) that is appropriate for communication purposes. Conflicting considerations enter into the construction of an acceptable symbology S. On the one hand, any symbol in S must, in principle, be readily recognized as such, which means that these symbols must be easy to distinguish from other symbols available in some larger set. On the other hand, these symbols must also be discriminable from each other (see Jameson, 1992). For example, it is plausible that the set

$$\{\spadesuit, \heartsuit, \diamondsuit, \clubsuit, 0, 1, 2, \ldots, 9\} \tag{12}$$

would not be considered to form an appropriate symbology, while its two subsets $\{\spadesuit, \heartsuit, \diamondsuit, \clubsuit\}$ and $\{0, 1, 2, \ldots, 9\}$ would no doubt be acceptable. From a formal viewpoint, this situation is actually similar to that of Example 1.10. Consider a set C of symbols forming the universe of discourse. That is, the symbols in C are the only ones under consideration. (The set specified in Equation (12) is an instance of such a set S.) It is conceivable that several subsets of C could form acceptable symbologies. Discuss this example in the style of Example 1.13. (Try to adapt Theorem 1.9.)

17. Establish the assertions about execution time of Algorithm 1.31 (see Remarks 1.34).

18. (*Finitary closure spaces.*) As in Definitions 1.11 and 1.15, let $(Q, \mathcal{L})$ be a closure space, with A' denoting the closure of a subset A of Q. Dually to Definition 1.19, we say that $(Q, \mathcal{L})$ is $\cap$-*finitary* when the union of any chain of states is again a state. Show that $(Q, \mathcal{L})$ is $\cap$-finitary if it satisfies: for any $p \in Q$ and $A \subseteq Q$, we have $p \in A'$ iff $p \in F'$ for some *finite* subset F of A. The converse also holds, but its proof is probably less obvious (cf. e.g. Cohn, 1965, or van de Vel, 1993).

Chapter 2

Well-Graded Knowledge Structures

The knowledge state of an individual may vary over time. For example, the following learning scheme is reasonable. A novice student is in the empty state and thus knows nothing at all. Then, one or a few items are mastered; next, another batch is absorbed, etc., up to the eventual mastery of the full domain of the knowledge structure. There may be many possible learning sequences, however. Forgetting may also take place. More generally, there may be many ways of traversing a knowledge structure, evolving at each step from one state to another closely resembling one, and various reasons for doing so.

In this Chapter, we study knowledge structures from the viewpoint of the relationship between the states, based on the items they contain. The results will have application elsewhere in this book. For example, they will provide the combinatoric skeleton for the stochastic learning theories developed in Chapters 8 and 9 and for the assessment procedures described in Chapters 10 and 11. As one example will show, applications also extend beyond education. We first consider the finite case.

Essentially Finite Structures

2.1. Definition. A *learning path* in a knowledge structure $(Q, \mathcal{K})$ (finite or infinite) is a maximal chain $\mathcal{C}$ in the partially ordered set $(Q, \subseteq)$. According to the definition of 'chain' in 0.21, we have thus $C \subseteq C'$ or $C' \subseteq C$ for all $C, C' \in \mathcal{C}$. Moreover, the chain $\mathcal{C}$ is *maximal* if whenever $\mathcal{C} \subseteq \mathcal{C}'$ for some chain of states $\mathcal{C}'$, then $\mathcal{C} = \mathcal{C}'$. Thus, a maximal chain necessarily contains $\varnothing$ and Q. In some situations, the student could learn the items one at a time. For example, in the case of a finite domain Q containing m elements, a learning path could take the form

$$\varnothing \subset \{q_1\} \subset \{q_1, q_2\} \subset \cdots \subset \{q_1, q_2, \ldots, q_m\} = Q \qquad (1)$$

for some particular order $q_1, q_2, \ldots, q_m$ of the elements of Q. Such a learning path is called a *gradation*. Note that a gradation in the sense of

Equation (1) exists in a knowledge structure only if it is discriminative. Indeed, for any two items, there must be a state in the gradation from Equation (1) containing one item and not the other. Hence, these two items cannot be equally informative (cf. Definition 1.4). This means that each notion contains a single item. In other words, the knowledge structure is discriminative. On the other hand, a learning path in a discriminative structure is not necessarily a gradation. In fact, some discriminative structures have no gradations.

2.2. Example. Take a domain Q containing more than two elements, and let $\mathcal{F}$ be the family containing $\varnothing$ and Q, and all the subsets of Q containing exactly two elements. Then $\mathcal{F}$ is a discriminative knowledge structure, and all the learning paths are of the form: $\varnothing \subset \{q_1, q_2\} \subset Q$.

For general knowledge structures (i.e. possibly infinite or non discriminative), the concept of 'gradual learning' still makes sense, but the above definition of 'gradation' must be enlarged. The next example (which was encountered earlier, cf. 1.3) paves the way to a more comprehensive definition for the case of essentially finite structures.

2.3. Example. Let $U = \{a, b, c, d, e, f\}$, and

$$\mathcal{H} = \{\varnothing, \{d\}, \{a, c\}, \{e, f\}, \{a, b, c\}, \{a, c, d\}, \{d, e, f\}, \\ \{a, b, c, d\}, \{a, c, e, f\}, \{a, c, d, e, f\}, U\}. \qquad (3)$$

Notice that a and c are equally informative (cf. 1.4). This is reflected in the learning path:

$$\varnothing \subset \{d\} \subset \{d, a, c\} \subset \{d, a, c, b\} \subset U.$$

A student progressing along this learning path would first master item d, and then a and c jointly, and finally e and f jointly. However, the discriminative reduction $(U^*, \mathcal{H}^*)$ (cf. 1.4) has the learning path

$$\varnothing^* \subset \{d^*\} \subset \{d^*, a^*\} \subset \{d^*, a^*, b^*\} \subset U^*,$$

in which two successive states differ by exactly one element of U^*. In other words, the empty state is connected to the domain by successively adding notions one at a time, never leaving the knowledge structure along the way.

This observation readily suggests a more general type of linkage of a knowledge structure, in which any two states can be connected by successively adding or removing notions. For example, consider the two incomparable states $\{a, b, c\}$ and $\{a, c, e, f\}$ in the knowledge structure $\mathcal{H}$ of Equation (3). These two states can be connected by the sequence of states

$$\{a, b, c\}, \quad \{a, c\}, \quad \{a, c, e, f\},$$

generated as follows: first, $\{a, c\}$ is obtained by removing from $\{a, b, c\}$ the notion $b^* = \{b\}$; then $\{a, c, e, f\}$ is manufactured by adding the notion $e^* = \{e, f\}$ to $\{a, c\}$. Such generalized 'paths' are defined below.

We recall (cf. 0.23) that the canonical distance d between two finite sets A and B is defined by counting the number of elements in their symmetric difference $A \triangle B$:

$$d(A, B) = |A \triangle B| = |(A \setminus B) \cup (B \setminus A)|.$$

A similar concept introduced in the next definition relies on counting the number of notions by which two states of a knowledge structure differ. Remember from 1.4 that for any state K, we set $K^* = \{q^* \mid q \in K\}$.

2.4. Definition. Suppose that $(Q, \mathcal{K})$ is an essentially finite knowledge structure. (The general situation is considered in Definitions 2.17 and 2.19). Let K and L be two states in $(Q, \mathcal{K})$. The *essential distance* between K and L is defined by

$$e(K, L) = |K^* \triangle L^*|,$$

where K^* and L^* are defined as in 1.4. The function $e : \mathcal{K} \times \mathcal{K} \to \mathbf{R}$ is a distance in the usual sense (cf. 0.23).

A learning path $\mathcal{C}$ in $\mathcal{K}$ is called a *gradation* if for any $K \in \mathcal{C} \setminus \{Q\}$, there exists $q \in Q \setminus K$ such that $K \cup q^* \in \mathcal{C}$. (Or equivalently: for any $K \in \mathcal{C} \setminus \{\varnothing\}$, there exists $q \in K$ such that $K \setminus q^* \in \mathcal{C}$.)

A knowledge structure is *well-graded* if for any two distinct states K, L there exist some integer $h \geq 0$ and a sequence of states

$$K = K_0, \quad K_1, \quad \ldots, \quad K_h = L \tag{4}$$

such that for $j = 0, 1, \ldots, h - 1$, the two following conditions hold:

$$e(K_j, K_{j+1}) = 1, \tag{5}$$
$$e(K, L) > e(K_{j+1}, L). \tag{6}$$

A sequence of states $(K_j)_{0 \le j \le h}$ satisfying (4) and (5) is called a *(stepwise) path connecting* K and L. If, in addition, the sequence $(K_j)_{0 \le j \le h}$ verifies (6), the path is said to be *bounded*. Thus, a knowledge structure is well-graded if any two of its states are connected by a bounded path. We say that $(Q, \mathcal{K})$ is *1-connected* if any two of its states are connected by a path, not necessarily bounded. Finally, a path $(K_j)_{0 \le j \le h}$ connecting K and L is called *tight* if $e(K, L) = h$ (which is equivalent to $e(K, K_j) = j$ for all $0 \le j \le h$). Thus, a tight path is always bounded.

Note that, even for a discriminative space, 1-connectedness is not equivalent to wellgradedness.

2.5. Example. Let $\mathcal{K}$ be the knowledge space with base

$$\{ \{c\},\ \{a, b\},\ \{b, c\},\ \{c, d\},\ \{d, e\} \}.$$

Then $\mathcal{K}$ is 1-connected since there is a stepwise path from any state to $\mathsf{U}\mathcal{K} = \{a, b, c, d, e\}$. However, $\mathcal{K}$ is not well-graded: there is no bounded path connecting the two states $\{a, b\}$ and $\{d, e\}$.

Theorem 2.7 contains various characterizations of essentially finite well-graded knowledge structures. We need a few additional concepts, that we motivate on the knowledge structure of Example 2.3.

2.6. Example. Let $U = \{a, b, c, d, e, f\}$, and

$$\mathcal{H} = \{\ \varnothing,\ \{d\},\ \{a, c\},\ \{e, f\},\ \{a, b, c\},\ \{a, c, d\},\ \{d, e, f\},$$
$$\{a, b, c, d\},\ \{a, c, e, f\},\ \{a, c, d, e, f\},\ U \ \}.$$

Consider the state $\{a, b, c, d\}$ in the knowledge structure $(U, \mathcal{H})$. The states at essential distance 1 from $\{a, b, c, d\}$ are the following three states:

$$\{a, c, d\} = \{a, b, c, d\} \setminus \{b\},$$
$$\{a, b, c\} = \{a, b, c, d\} \setminus \{d\},$$
$$U = \{a, b, c, d\} \cup \{e, f\}.$$

The first two of these states result from the deletion of either the notion $\{b\}$ or the notion $\{d\}$ from $\{a, b, c, d\}$. We say that b and d form the 'inner fringe' of $\{a, b, c, d\}$. Similarly, the state U results from the addition of the notion $\{e, f\}$. We say that e and f form the 'outer fringe' of $\{a, b, c, d\}$.

2.7. Definition. The *inner fringe* of a state K in a knowledge structure $(Q, \mathcal{K})$, is the set of items

$$K^{\mathcal{I}} = \{\, q \in K \mid K \setminus q^* \in \mathcal{K} \,\}.$$

The *outer fringe* of a state K is the set

$$K^{\mathcal{O}} = \{\, q \in Q \setminus K \mid K \cup q^* \in \mathcal{K} \,\}.$$

The *fringe* of K is the union of the inner and outer fringes. We write

$$K^{\mathcal{F}} = K^{\mathcal{I}} \cup K^{\mathcal{O}}.$$

Let $N(K, 1)$ be the set of all states whose essential distance from K is at most one:

$$N(K, 1) = \{ L \in \mathcal{K} \mid e(K, L) \leq 1 \}.$$

Then we also have (cf. Problem 12)

$$K^{\mathcal{F}} = \big(\cup N(K, 1) \big) \setminus \big(\cap N(K, 1) \big).$$

We recall (cf. 1.7) that the complement of a state K in a knowledge structure $(Q, \mathcal{K})$ is denoted by $\bar{K} = Q \setminus K$.

2.8. Theorem. *For any essentially finite knowledge structure* $(Q, \mathcal{K})$, *the following six conditions are equivalent:*

(i) $(Q, \mathcal{K})$ *is well-graded;*

(ii) *for any two states* K, L *with* $e(K, L) = h$, *there exists a tight path* $(K_j)_{0 \leq j \leq h}$ *connecting* K *and* L;

(iii) *for any two states* K *and* L *there exists a path* $(K_j)_{0 \leq j \leq h}$ *connecting* K *and* L *and such that, for all* $0 \leq j \leq h - 1$:

$$K_j \cap L \subseteq K_{j+1} \subseteq K_j \cup L;$$

(iv) *for any two distinct states* K, L,

$$(K \triangle L) \cap K^{\mathcal{F}} \neq \emptyset; \tag{7}$$

(v) *any two states* K *and* L *in* $\mathcal{K}$ *which satisfy* $K^{\mathcal{I}} \subseteq L$ *and* $K^{\mathcal{O}} \subseteq \bar{L}$ *must be equal;*

(vi) *any two states* K *and* L *in* $\mathcal{K}$ *which satisfy* $K^{\mathcal{I}} \subseteq L$, $K^{\mathcal{O}} \subseteq \bar{L}$, $L^{\mathcal{I}} \subseteq K$, $L^{\mathcal{O}} \subseteq \bar{K}$ *must be equal.*

PROOF. We prove (i) $\Rightarrow$ (ii) $\Rightarrow$ (iii) $\Rightarrow$ (iv) $\Rightarrow$ (v) $\Rightarrow$ (vi) $\Rightarrow$ (i).

(i) $\Rightarrow$ (ii). Take any two states K and L and suppose $e(K, L) = \ell$. Property (ii) clearly holds if $\ell = 0$ or $\ell = 1$. We proceed by induction on ℓ, assuming $\ell > 1$. From the definition of wellgradedness (see 2.4), we know that there is a bounded path $(K_j)_{0 \leq j \leq h}$ connecting K and L (thus, with $K = K_0$ and $L = K_h$). Setting $\ell' = e(K_1, L)$, we must have $\ell' < \ell$. Moreover, using the triangular inequality, we obtain

$$\ell = e(K, L) \leq e(K, K_1) + e(K_1, L) \leq 1 + \ell',$$

and thus $\ell' = \ell - 1$. By the induction hypothesis, there is a tight path $(K'_j)_{1 \leq j \leq \ell}$ connecting K_1 and L. Then, $K = K_0$, K_1, K'_2, K'_3, ..., $K'_\ell = L$ is a tight path connecting K and L.

(ii) $\Rightarrow$ (iii). We leave to the reader to verify that any tight path connecting K and L fulfills the requirements of Condition (iii).

(iii) $\Rightarrow$ (iv). Take any two states K and L, and let $(K_j)_{0 \leq j \leq h}$ be the path described in Condition (iii). Then K and K_1 differ by exactly one notion q^*, and we have moreover $K \cap L \subseteq K_1 \subseteq K \cup L$. Either the item q belongs to K, or it belongs to L, but not both. Hence q belongs to $(K \bigtriangleup L) \cap K^{\mathcal{F}}$.

(iv) $\Rightarrow$ (v). We use contradiction. Let K and L be two distinct states satisfying $K^{\mathcal{I}} \subseteq L$ and $K^{\mathcal{O}} \subseteq \bar{L}$. Take any $q \in (K \bigtriangleup L) \cap K^{\mathcal{F}}$. If $q \in K$, then $q \in K^{\mathcal{I}} \subseteq L$, contradicting $q \in K \bigtriangleup L$. Hence $q \notin K$, but then $q \in L \cap K^{\mathcal{O}}$, and we obtain $q \in L$ and $q \in K^{\mathcal{O}} \subseteq \bar{L}$, a contradiction.

(v) $\Rightarrow$ (vi). Obvious.

(vi) $\Rightarrow$ (i). Let K and L be two distinct states in $\mathcal{K}$ with $e(K, L) = h > 0$. We construct a tight path $(K_i)_{0 \leq j \leq h}$ connecting K and L. (This establish the wellgradedness of $\mathcal{K}$, since any tight path is bounded.) Since $K \neq L$, Condition (vi) implies that for some item q, we must have:

$$q \in (K^{\mathcal{I}} \setminus L) \cup (K^{\mathcal{O}} \cap L) \cup (L^{\mathcal{I}} \setminus K) \cup (L^{\mathcal{O}} \cap K).$$

If $q \in K^{\mathcal{I}} \setminus L$, we set $K_1 = K \setminus q^*$. In the other three cases, we set $K_1 = K \cup q^*$ or $K_{h-1} = L \setminus q^*$ or $K_{h-1} = L \cup q^*$. We obtain either $e(K_1, L) = h - 1$ (in the first two cases), or $e(K, K_{h-1}) = h - 1$ (in the last two cases). The result follows by induction. $\qquad\square$

2.9. Remarks. a) The implication (i) $\Rightarrow$ (v) in Theorem 2.8 may have important practical applications. In a well-graded structure $\mathcal{K}$, a state is fully specified by its two fringes in the sense that

$$\forall K, L \in \mathcal{K}: \quad (K^{\mathcal{I}} = L^{\mathcal{I}} \text{ and } K^{\mathcal{O}} = L^{\mathcal{O}}) \iff K = L. \qquad (8)$$

In other words, in a well-graded structure, the state of a student is accurately described by listing all those notions that she may have recently acquired (i.e. the notions in the inner fringe of her state) and the notions that she may be ready to master (i.e. the notions in the outer fringe of her state).

b) With regard to the path $(K_j)_{0 \le j \le h}$ in Theorem 2.8(iii), we point out that K_{i+1} is derived either by deleting from K_i a notion included in $K_i \setminus L$, or by adding to K_i a notion included in $L \setminus K_i$. This construction will be used in 2.9 for the generalization to infinite knowledge structures.

c) Note that a knowledge structure in which all the learning paths are gradations is not necessarily well-graded. As an example, consider the knowledge structure

$$\{ \emptyset, \{a\}, \{c\}, \{a,b\}, \{b,c\}, \{a,b,d\}, \{b,c,d\}, \{a,b,c,d\} \}.$$

This structure has two learning paths, both of which are gradations, but is not well-graded: the two states $\{a,b\}$ and $\{b,c\}$ have the same inner fringe $\{b\}$ and outer fringe $\{d\}$ but are different (cf. Condition (v) of Theorem 2.8). Moreover, there is no bounded path connecting those states; in fact,

$$(\{a,b\} \bigtriangleup \{b,c,\}) \cap \{a,b\}^{\mathcal{F}} = \{a,c\} \cap \{b,d\} = \emptyset.$$

Such a situation cannot arise in an essentially finite knowledge space.

2.10. Theorem. *For any essentially finite knowledge space* $(Q, \mathcal{K})$, *the following three conditions are equivalent:*

    (i) $(Q, \mathcal{K})$ *is well-graded;*

    (ii) *all the learning paths in* $(Q, \mathcal{K})$ *are gradations;*

    (iii) *whenever* $K \subset L$ *for any two states* K, L, *there exist a positive integer* h *and a chain of states* $K = K_0 \subset K_1 \subset \cdots \subset K_h = L$ *such that* $|K_{j+1}^*| = |K_j^*| + 1$ *for* $0 \le j \le h - 1$.

PROOF. (i) $\Rightarrow$ (ii). Let $\mathcal{C}$ be any learning path, and take any K in $\mathcal{C}\backslash\{Q\}$. Denote by L the state that follows K in $\mathcal{C}$. By Theorem 2.8(iv) there exists some q in $(K \triangle L) \cap K^{\mathcal{F}}$. Thus, $K + q^*$ is a state included in L. Since $\mathcal{C}$ is maximal, we must have $K + q^* = L \in \mathcal{C}$. This implies that $\mathcal{C}$ is a gradation.

(ii) $\Rightarrow$ (iii). We leave this part of the proof to the reader (Problem 4).

(iii) $\Rightarrow$ (i) We prove Condition (iv) in Theorem 2.8(iv), denoting by V the left member of Equation (7). Suppose first that $L \setminus K \neq \varnothing$. Since $K \subset K \cup L$ and $K \cup L$ is also a state, there exists by Condition (iii) a state $K_1 = K + q^*$ for some item $q \in L \setminus K$. Thus, $q \in V$. On the other hand, if $L \subset K$, there is by Condition (iii) a state of the form $K \setminus q^*$ for some q in K. Again, $q \in V$. $\qquad\square$

Two other characterizations of finite, discriminative well-graded knowledge spaces will be given in Chapter 7 (Theorems 7.20 and 7.21).

2.11. Theorem. *Any essentially finite, quasi ordinal space $(Q, \mathcal{K})$ is well-graded.*

In Theorem 2.22, we extend this result to all quasi ordinal spaces (having defined 'wellgradedness' for infinite structures in 2.19).

PROOF. We establish (iii) in Theorem 2.10. Take any pair of states $K \subset L$. Suppose that there are $p, q \in L \setminus K$, with p, q not equally informative. (Otherwise, we set $h = 1$.) We may assume, without loss of generality, that there is a state M with $p \in M$ and $q \notin M$. Hence $K' = K \cup (M \cap L)$ is a state containing p but not q, and such that $K \subset K' \subset L$. The result follows by induction. $\qquad\square$

A Well-Graded Family of Relations: the Biorders*

An interesting example of a well-graded structure arises in the theory of order relations, in the guise of the 'biorders.' In fact, if we slightly relax the conditions defining a knowledge structure and drop the requirement that the domain itself be a state, then several families of order relations—regarded as sets of pairs—can be shown to be well-graded in the sense of 2.4. We indulge in this detour into the theory of order relations to illustrate potential applications of our results beyond the main focus of this monograph. Note that we restrict considerations to finite structures.

2.12. Definition. Let X and Y be two basic finite, nonempty sets, with Y not necessarily disjoint or distinct from X. We abbreviate the pair $(x, y) \in X \times Y$ as xy. A relation R from X to Y, that is $R \subseteq X \times Y$, is called a *biorder* if for all $x, x' \in X$ and $y, y' \in Y$, we have

[B] $(xRy, \neg(x'Ry)$ and $x'Ry') \Rightarrow xRy'$.

Using the compact notation for the (relative) product introduced in 0.13 and 0.14, Condition [B] can also be presented by the formula

[B$'$] $R\bar{R}^{-1}R \subseteq R$.

It is easy to check that the complement $\bar{R}$ of a biorder is itself a biorder. Accordingly, [B$'$] is equivalent to

[B$''$] $\bar{R}R^{-1}\bar{R} \subseteq \bar{R}$.

The interest for biorders stems in part from their numerical representation. Ducamp and Falmagne (1969) have shown that for finite sets X and Y, Condition [B] was necessary and sufficient to ensure the existence of two functions $f : X \to \mathbf{R}$ and $g : Y \to \mathbf{R}$ satisfying

$$xRy \quad \Longleftrightarrow \quad f(x) > g(y). \tag{9}$$

The term "biorder" was coined by Doignon, Ducamp and Falmagne (1984), who extended this representation to infinite sets X and Y. The concept is used in psychometrics, where X and Y represent, respectively, a set of subjects and a set of questions of a test of some ability. The notation xRy formalizes the fact that subject x has solved question y. The right member of the equivalence (9) is then interpreted as meaning: the ability $f(x)$ of subject x exceeds the difficulty $g(y)$ of question y. In this context, the relation R is coded as a rectangular 0-1 array and referred to as a Guttman's scale (Guttman, 1944). Condition [B] means that such a 0-1 array never contains a sub-array of the form shown in Table 2.1.

	y	y'
x	1	0
x'	0	1

Table 2.1. Forbidden pattern in a 0-1 array representing a biorder.

Condition [B] enters as one of the defining conditions of other standard order relations in encountered in measurement theory and utility theory such as the interval orders and the semiorders. The semiorders were introduced by Luce (1956; see also Scott and Suppes, 1959). The interval orders are due to Fishburn (1970). For background and references, the reader is referred to the Source section at the end of the chapter.

In this section, we consider the full family of all the biorders from X to Y as a family of subsets of a basic set $Q = X \times Y$. Thus, each of the biorders is regarded as a set of pairs. We shall prove the following result, which is due to Doignon and Falmagne (1997).

2.13. Theorem. *The family $\mathcal{B}$ of all biorders between two finite sets X and Y is a well-graded discriminative knowledge structure. Moreover, the inner and outer fringes (cf. 2.7) of any relation $R \in \mathcal{B}$ are specified by the two equations:*

$$R^{\mathcal{I}} = R \setminus R\bar{R}^{-1}R, \qquad R^{\mathcal{O}} = \bar{R} \setminus \bar{R}R^{-1}\bar{R}.$$

Clearly, both $\varnothing$ and $X \times Y$ are biorders. Hence, $\mathcal{B}$ is a knowledge structure, which is also discriminative because $\{xy\}$ is a biorder for all $x \in X$ and $y \in Y$. Checking that expressions for the inner and outer fringes are indeed those specified by the Theorem is easy, and is left to the reader. To establish that the family $\mathcal{B}$ of all biorders between two finite sets X and Y is well-graded, we shall prove Condition (v) in Theorem 2.8. The proof given in 2.15 relies on some auxiliary results.

We recall that for any relation R, the products $R\bar{R}^{-1}$ and $\bar{R}^{-1}R$ are irreflexive. Moreover, if R is a biorder, then for any positive integer n, the nth power $(R\bar{R}^{-1})^n$ of the product $R\bar{R}^{-1}$ is also irreflexive. We shall use the following fact:

2.14. Lemma. *Let R be a biorder from a finite set X to a finite set Y. Then we necessarily have*

$$R = \bigcup_{k=0}^{\infty}(R\bar{R}^{-1})^k R = \bigcup_{k=0}^{\infty}(R^{\mathcal{I}}(R^{\mathcal{O}})^{-1})^k R^{\mathcal{I}}.$$

PROOF. We show that

$$R \subseteq \cup_{k=0}^{\infty}(R\bar{R}^{-1})^k R \subseteq \cup_{k=0}^{\infty}(R^{\mathcal{I}}(R^{\mathcal{O}})^{-1})^k R^{\mathcal{I}} \subseteq R.$$

The first inclusion is obvious: take $k = 0$ and use a convention from 0.14: as $R\bar{R}^{-1}$ is a relation on X, we have $(R\bar{R}^{-1})^0$ equal to the identity relation on X. To establish the second inclusion, suppose that $xy \in (R\bar{R}^{-1})^k R$ for some $k \geq 0$. Because $(R\bar{R}^{-1})^n$ is irreflexive for any positive integer n and X is finite, we can assume without loss of generality that k is maximal. This implies that each of the $k + 1$ factors R in the formula $(R\bar{R}^{-1})^k R$ can be replaced with $R^{\mathcal{I}}$ while keeping xy in the full product. Indeed, if this were not the case, such a factor R could be replaced with $R\bar{R}^{-1}R$ and we would find $xy \in (R\bar{R}^{-1})^{k+1}R$, contradicting the maximality of k. The fact that each of the k factors $\bar{R}^{-1}$ in the formula $(R\bar{R}^{-1})^k R$ can be replaced with $(R^{\mathcal{O}})^{-1}$ is proved by similar arguments. We conclude that the second inclusion holds.

The third inclusion results from the biorder inclusion $R\bar{R}^{-1}R \subseteq R$ together with $R^{\mathcal{I}} \subseteq R$ and $R^{\mathcal{O}} \subseteq \bar{R}$. □

2.15. Theorem. *Let R and S be two biorders from X to Y. Then*

$$(R^{\mathcal{I}} \subseteq S \text{ and } R^{\mathcal{O}} \subseteq \bar{S}) \implies R = S.$$

PROOF. The inclusion $R \subseteq S$ follows from

$$
\begin{aligned}
xy \in R \quad &\Rightarrow \quad x\big(R^{\mathcal{I}}(R^{\mathcal{O}})^{-1}\big)^k R^{\mathcal{I}} y, \quad \text{for some } k \geq 0 \text{ (by Proposition 2.14)}\\
&\Rightarrow \quad x(S\bar{S}^{-1})^k Sy \qquad \text{(by hypothesis, } R^{\mathcal{I}} \subseteq S \text{ and } R^{\mathcal{O}} \subseteq \bar{S})\\
&\Rightarrow \quad xy \in S \qquad\qquad \text{(by Proposition 2.14).}
\end{aligned}
$$

To prove the converse inclusion, notice that $\bar{R}$ and $\bar{S}$ are themselves biorders. Moreover, $(\bar{R})^{\mathcal{I}} = R^{\mathcal{O}}$ and $(\bar{R})^{\mathcal{O}} = R^{\mathcal{I}}$. This means that our hypothesis translates as $(\bar{R})^{\mathcal{I}} \subseteq \bar{S}$ and $(\bar{R})^{\mathcal{O}} \subseteq \overline{(\bar{S})}$. The argument used above gives thus $\bar{R} \subseteq \bar{S}$, that is $S \subseteq R$. □

2.16. Proof of Theorem 2.13. From the discussion after the statement, it only remains to show that $\mathcal{B}$ is well-graded. This results from Theorem 2.15 which establishes Condition (v) of Theorem 2.8. □

2.17. Remarks. a) The family $\mathcal{B}$ of biorders of Theorem 2.13 is neither a knowledge space nor a closure space (in the sense of Definitions 1.7 and 1.11). Indeed, each of the four relations $\{ab\}, \{a'b'\}, \{ab, a'b, a'b'\}, \{ab, ab', a'b'\}$ is a biorder from $\{a, b\}$ to $\{a', b'\}$ with $a \neq b$ and $a' \neq b'$, but

$$\{ab\} \cup \{a'b'\} = \{ab, a'b, a'b'\} \cap \{ab, ab', a'b'\} = \{ab, a'b'\} \notin \mathcal{B}.$$

(In fact, $\{ab, a'b'\}$ is the forbidden subrelation represented by the 0-1 array of Table 2.1.)

b) As mentioned earlier, similar result have been obtained regarding the wellgradedness of other families of order relations under a slightly more general definition of a knowledge structure which does not require that the domain of the knowledge structure be a state. Examples are the partial orders, the interval orders and the semiorders (Doignon and Falmagne, 1997; see Problems 5-8).

The Infinite Case*

We now turn to the wellgradedness of general (that is, possibly infinite) knowledge structures. In this case, a state in a gradation may be obtained as the 'limit' of the states that it includes in this gradation.

2.18. Definition. A *gradation* in a knowledge structure $(Q, \mathcal{K})$ is a learning path $\mathcal{C}$ such that for any $K \in \mathcal{C} \setminus \{\varnothing\}$, one of the following holds:

$$K = K' \cup q^*, \text{ for some } q \in K \text{ and } K' \in \mathcal{C} \setminus \{K\}, \tag{10}$$
$$\text{or} \quad K = \bigcup \{L \in \mathcal{C} \mid L \subset K\}. \tag{11}$$

When the knowledge structure $(Q, \mathcal{K})$ is essentially finite, the 'limit' situation described by Equation (11) does not occur, and this definition coincides with that given in 2.4.

2.19. Example. As in 1.21, let $\mathcal{O}$ be the knowledge space formed by the collection of all open subsets of $\mathbf{R}$. Then any two states O, O' in $\mathcal{O}$ with $O \subset O'$ belong to a gradation. In fact, any maximal chain $\mathcal{C}$ of states containing O and O' must be a gradation (by Hausdorff maximality principle, cf. 0.21, there exists at least one such maximal chain). Indeed, suppose that some K in $\mathcal{C}$ does not satisfy Equation (11). Pick some item q in $K \setminus \bigcup \{L \in \mathcal{C} \mid L \subset K\}$. Then $K \setminus \{q\} = K' \in \mathcal{C}$ since $\bigcup \{L \in \mathcal{C} \mid L \subset K\} \subset K' \subset K$, with $K = K' \cup q^*$, and K satisfies Equation (10).

Defining well-graded structures in the infinite case requires more powerful tools than the finite sequences used in Equation (4). A suitable device is suggested by the formulation in Theorem 2.8(iii) (cf. Remark 2.9(b)). We begin by generalizing the concept of a path connecting two states.

2.20. Definition. A family $\mathcal{D}$ of states in a knowledge structure $(Q, \mathcal{K})$ is a *bounded path* connecting a state K and a state L if it contains K and L and the following three conditions hold: for all distinct D and E in $\mathcal{D}$,

(1) $\qquad\qquad\qquad K \cap L \subseteq D \subseteq K \cup L;$

(2) either $\qquad D \setminus L \subseteq E \setminus L \quad$ and $\quad D \setminus K \supseteq E \setminus K,$

     or $\qquad\quad D \setminus L \supseteq E \setminus L \quad$ and $\quad D \setminus K \subseteq E \setminus K;$

(3) either (a) $\quad \exists F \in \mathcal{D} \setminus \{D\}, \exists q \in D \setminus F : F \cup q^* = D;$

$$\text{or} \qquad \text{(b)} \left\{ \begin{array}{l} D \setminus K = \cup\{G \setminus K \mid G \in \mathcal{D}, \, G \setminus K \subset D \setminus K\}, \\ \text{and} \\ D \setminus L = \cup\{G \setminus L \mid G \in \mathcal{D}, \, G \setminus L \subset D \setminus L\}. \end{array} \right.$$

The knowledge structure is *well-graded* if any two of its states are connected by a bounded path. When the knowledge structure is essentially finite, this definition of wellgradedness becomes identical to that in 2.4: Case (3)(b) does not arise, and Theorem 2.8(iii) applies.

2.21. Example. Examples of bounded paths are easy to manufacture. Consider the two states $]a, b[$ and $]c, d[$, with $a < c < b < d$, in the knowledge space formed by the open subsets of $\mathbf{R}$ (cf. 2.19). Define a bounded path $\mathcal{A}$ connecting these two states by

$$\mathcal{A} = \{]a, b[, \,]c, d[\} \cup \{A(x) \mid x \in \mathbf{R}\}$$

containing all the open intervals

$$A(x) =]g(x)(c - a) + a, g(x)(d - b) + b[,$$

where $g : \mathbf{R} \to]0, 1[$ is a continuous, strictly increasing function, satisfying

$$\lim_{x \to -\infty} g(x) = 0, \qquad \lim_{x \to +\infty} g(x) = 1.$$

(Take for example $g(x) = (1 + e^{-x})^{-1}$.) The family $\mathcal{A}$ is a bounded path connecting $]a, b[$ and $]c, d[$. We have

$$a < g(x)(c - a) + a < c < b < g(x)(d - b) + b < d,$$

verifying Condition (1) in 2.20. Notice that we also have

$$A(x) \setminus]c, d[=]g(x)(c - a) + a, c],$$
$$A(x) \setminus]a, b[= [b, g(x)(d - b) + b[.$$

This yields for any $x \leq y$

$$A(y)\backslash\,]c,d[\,\subseteq\, A(x)\backslash\,]c,d[\quad \text{and} \quad A(y)\backslash\,]a,b[\,\supseteq\, A(x)\backslash\,]a,b[,$$

which verifies Condition (2) in 2.20. Finally, we observe that

$$A(x)\backslash\,]a,b[\,=\, \cup_{z<x}(A(z)\backslash\,]a,b[),$$
$$A(x)\backslash\,]c,d[\,=\, \cup_{x<z}(A(z)\backslash\,]c,d[),$$

establishing Condition (3).

As a partial extension of Theorem 2.10 in the case of possibly infinite knowledge spaces, we have:

2.22. Theorem. *For any knowledge space* $(Q, \mathcal{K})$, *the following two conditions are equivalent:*

(i) $(Q, \mathcal{K})$ *is well-graded;*

(ii) *all the learning paths in* $(Q, \mathcal{K})$ *are gradations.*

Moreover, Conditions (i) and (ii) are implied by

(iii) *for any two distinct states* K *and* L,

$$(K \,\triangle\, L) \cap K^{\mathcal{F}} \neq \varnothing.$$

PROOF. (i) $\Rightarrow$ (ii). Let $\mathcal{C}$ be any learning path, and take any K in $\mathcal{C}\backslash\{\varnothing\}$. Setting $U = \cup\{L \in \mathcal{C} \mid L \subset K\}$, let us assume $U \neq K$. Notice that $U \in \mathcal{C}$ follows from the closure of $\mathcal{K}$ under union and the maximality of $\mathcal{C}$. By (i), there is a bounded path $\mathcal{D}$ from U to K. Since $U \subset K$, we must have $U \subseteq D \subseteq K$ for all $D \in \mathcal{D}$. Suppose that there is some D in $\mathcal{D}$ such that $U \subset D \subset K$. Then $\mathcal{C} \cup \{D\}$ is a chain, which is impossible because $\mathcal{C}$ is maximal. Thus $\mathcal{D} = \{U, K\}$. Since $\mathcal{D}$ is a bounded path, we derive the existence of $q \in K$ such that $U \cup q^* = K$. This proves that $\mathcal{C}$ is a gradation.

(ii) $\Rightarrow$ (i). Let $K, L \in \mathcal{K}$, thus also $K \cup L \in \mathcal{K}$. Take a learning path $\mathcal{C}_1$ that contains K and $K \cup L$, and a learning path $\mathcal{C}_2$ that contains L and $K \cup L$. Then

$$\{D \in \mathcal{C}_1 \mid K \subseteq D \subseteq K \cup L\} \cup \{E \in \mathcal{C}_2 \mid L \subseteq E \subseteq K \cup L\}$$

is a bounded path from K to L.

(iii) $\Rightarrow$ (ii). Assume again that $\mathcal{C}$ is a learning path, $K \in \mathcal{C} \setminus \{\varnothing\}$, and $U = \cup\{L \in \mathcal{C} \mid L \subset K\} \neq K$. Thus $U \in \mathcal{K}$ and $U \in \mathcal{C}$. By (iii), there is an item q in $(K \setminus U) \cap U^{\mathcal{F}}$. Then $U \cup q^*$ is a state that must be equal to K. This shows that $\mathcal{C}$ is a gradation. $\qquad\square$

Example 2.19 shows that, even for discriminative spaces, Condition (iii) in Theorem 2.22 does not follow from Conditions (i) and (ii). There are also ordinal spaces that can be used as similar counter-examples; for instance, take $\mathbf{R}$ with its usual order.

2.23. Theorem. *Any quasi ordinal space $(Q, \mathcal{K})$ is well-graded.*

PROOF. We establish Condition (ii) in Theorem 2.22. Let K be a state of some learning path $\mathcal{C}$ in $(Q, \mathcal{K})$. Then $U = \cup\{L \in \mathcal{C} \mid L \subset K\}$ is a state of $\mathcal{C}$. It suffices to show that there cannot be two non equally informative items p and q in $K \setminus U$. If there were two such items p and q, we would find a state M with either $p \in M$ and $q \notin M$, or $q \in M$ and $p \notin M$. A contradiction follows by considering the state $U \cup (M \cap K)$. $\qquad\square$

Finite Learnability

Consider the discriminative knowledge structure

$$\mathcal{J} = \big\{ \varnothing, \{a\}, \{a, b, c\}, \{a, b, d\}, \{a, c, d\}, \{a, b, c, d, e\} \big\}.$$

Suppose that some individual in state $\{a\}$ wishes to acquire item d. Since there is no intermediate state between $\{a\}$ and $\{a, b, d\}$ or $\{a, c, d\}$, this can only be done by mastering simultaneously either b and d, or c and d. This situation does not arise in the knowledge structure

$$\mathcal{J} \cup \{\{a, b\}\},$$

in which the individual may progress from state $\{a\}$ to a state containing d by steps involving one new item at a time. This example illustrates a possible mechanism for some pedagogical difficulties that some students may experience. It suggests the following definition, which applies to any knowledge structure, infinite or not, discriminative or not. Notice that we count notions rather than items. (We recall that a notion has been defined as a class of equally informative items; cf. Definition 1.4).

2.24. Definition. A knowledge structure is *finitely learnable* if there is a positive integer ℓ such that, for any state K and any item $q \notin K$, there exists a positive integer h and a chain of states $K = K_0 \subset K_1 \subset \cdots \subset K_h$ satisfying

(1) $q \in K_h$;
(2) $e(K_i, K_{i+1}) \leq \ell$,   for $0 \leq i \leq h-1$.

A finitely learnable knowledge structure $(Q, \mathcal{K})$ necessarily has a smallest number ℓ satisfying these conditions, which is called the *learnstep number* of $(Q, \mathcal{K})$. We write then

$$\mathrm{lst}(\mathcal{K}) = \ell.$$

We have, for example

$$\mathrm{lst}(\mathcal{J} \cup \{\{a, b\}\}) = 2.$$

Indeed, from state $\{a, b, d\}$, item e can be mastered only by mastering simultaneously items c and e. Note that any finite well-graded structure has learning step number 1 (see Problem 11).

2.25. Remarks. a) Some infinite knowledge structures are finitely learnable. In fact, for every infinite set Q, we clearly have $\mathrm{lst}(2^Q) = 1$. (Obviously, any essentially finite knowledge structure is finitely learnable.)

b) A well-graded infinite knowledge structure may not be finitely learnable. Take, for example, the ordinal space derived on $\mathbf{R}$ from the usual order (in the sense of Definition 1.50).

c) A finite knowledge space $(Q, \mathcal{K})$ may satisfy $\mathrm{lst}(\mathcal{K}) = 1$ without being well-graded. An example is obtained by letting $\mathcal{K}$ consist of

$$\varnothing, \; \{a\}, \; \{a, b\}, \; \{a, b, c\}, \; \{a, c, d\}, \; \{a, b, c, d\}.$$

Original Sources and Related Works

The concepts of learning paths and well-graded knowledge structures were introduced by Falmagne and Doignon (1988b), at least in the finite case. Well-graded knowledge spaces are dual to the so-called 'antimatroids' or 'convex geometries' in the sense of Edelman and Jamison (1985). Specifically, a finite closure space is a convex geometry exactly when its dual $(Q, \mathcal{K})$ is a knowledge space in which all learning paths are gradations.

(Additional characterizations of finite, well-graded knowledge spaces can be derived from Edelman and Jamison, 1985). Definition 2.20 extends the concept of wellgradedness to the infinite case. This definition is a natural one in the context of knowledge structures in education, but would probably not be suitable for abstract convexity.

The application of the wellgradedness concept to families of relations, especially to biorders and semiorders, is taken from Doignon and Falmagne (1997; see also Falmagne and Doignon, 1997); Ovchinnikov (1983) was a forerunner for the special case of partial orders. Biorders appeared under other names in the literature: Guttman scales (Guttman, 1944), Ferrers relations (Riguet, 1951; Cogis, 1982), bi-quasi-series (Ducamp and Falmagne, 1969). Among the more recent papers, we mention Doignon, Ducamp and Falmagne (1984), which introduced the term, and Doignon, Monjardet, Roubens and Vincke (1986). Two important special cases of biorders are the semiorders introduced by Luce (1956; see also Scott and Suppes, 1959) and the interval orders due to Fishburn (1970). For an introduction with applications of these concepts in the social sciences, the reader is referred to Roberts (1979), Roubens and Vincke (1985), Suppes, Krantz, Luce and Tversky (1989), or Pirlot and Vincke (1997). Purely mathematical expositions can be found in Fishburn (1985) and Trotter (1992).

Problems

1. Check whether the equivalence between the three statements in Theorem 2.10 still holds, when the axiom of closure under union is replaced by Axiom [JS] of Problem 11 from Chapter 1. Prove your result.

2. Suppose that a knowledge structure is well-graded (resp. 1-connected, 1-learnable, i.e. has learnstep number equal to 1). Does that imply that the dual structure is also well-graded (resp. 1-connected, 1-learnable)?

3. Are wellgradedness and 1-connectedness hereditary properties in the sense of Definition 1.17?

4. Prove (ii) $\Rightarrow$ (iii) in Theorem 2.10.

5. The following definitions apply in Problems 5-8. A *quasi knowledge* *structure* is a pair $(Q, \mathcal{K})$ such that Q is a set, and $\mathcal{K}$ is a family of subsets of $\mathcal{K}$ satisfying the following two conditions: (1) $\varnothing \in \mathcal{K}$; (2) $\cup \mathcal{K} = Q$. The definitions of wellgradedness and (resp. inner, outer) fringes remain unchanged. Prove that the collection $\mathcal{P}$ of all partial orders (cf. 0.13) on a finite set X, regarded as sets of pairs, is a quasi knowledge structure (cf. Problem 5). Describe the inner and outer fringes of any partial order on X. Prove that $\mathcal{P}$ is well-graded.

6. A semiorder on a set X is an irreflexive biorder R between X and X satisfying the following additional condition:

$$[S] \quad R R \overline{R}^{-1} \subseteq R.$$

Prove that the collection $\mathcal{S}$ of all semiorders on a finite set X, regarded as sets of pairs, is a quasi knowledge structure. Compute the inner and outer fringes of any semiorder on X. (Doignon and Falmagne, 1997.)

7. (Difficult.) Prove that $\mathcal{S}$ is well-graded. (Doignon and Falmagne, 1997.)

8. Would the results regarding wellgradedness in Problems 5 and 7 still hold if we drop the requirement that X is finite? Provide proofs of your responses.

9. For the knowledge space $\mathcal{O}$ of Example 2.21, construct a bounded path connecting the two states $\{x \mid a < x < b \text{ or } c < x < d\}$ and $]e, f[$, with $a < e < b < f < c < d$.

10. Does a gradation $\mathcal{C}$ in a knowledge structure $(Q, \mathcal{K})$ necessarily satisfy the following property? For $K \in \mathcal{C} \setminus \{Q\}$, (i) or (ii) must hold, with:

(i) $K = K' \setminus q^*$, for some $K' \in \mathcal{C}$ and $q \in K'$;

(ii) $K = \cap \{L \in \mathcal{C} \mid K \subset L\}$.

11. Verify that a finite well-graded knowledge structure necessarily has its learnstep number equal to 1.

12. Prove the following equality for the fringe of a state K of a knowledge structure $(Q, \mathcal{K})$ (cf. Definition 2.7):

$$K^{\mathcal{F}} = (\cup N(K, 1)) \setminus (\cap N(K, 1)).$$

Chapter 3

Surmise Systems

When a knowledge structure is a quasi ordinal space, it can be faithfully represented by its surmise relation (cf. Theorem 1.49). In fact, as illustrated by Example 1.46, an ordinal space is completely recoverable from the Hasse diagram of the surmise relation. However, for knowledge structures in general, and even for knowledge spaces, the information provided by the surmise relation may be incomplete. In this chapter, we introduce the 'surmise system', a concept generalizing that of a surmise relation, and allowing more than one possible learning history for an item[1]. We then derive, in the style of Theorem 1.49, a one-to-one correspondence between knowledge spaces and surmise systems. The surmise systems are closely related to the AND/OR graphs encountered in artificial intelligence. A section of this chapter is devoted to clarifying the relationship between the two concepts. This chapter also contains a discussion of the particular surmise systems which arise from well-graded knowledge spaces. Other highlights are: a generalization of the concept of a Hasse diagram, and a study of intractable 'cyclic' histories which leads us to formulate conditions precluding such situations.

Basic Concepts

3.1. Example. Consider the knowledge structure

$$\mathcal{L} = \big\{ \varnothing, \{a\}, \{b,d\}, \{a,b,c\}, \{b,c,e\}, \{a,b,d\},$$
$$\{a,b,c,d\}, \{a,b,c,e\}, \{b,c,d,e\}, \{a,b,c,d,e\} \big\} \qquad (1)$$

on the domain $Q = \{a, b, c, d, e\}$. Actually, $\mathcal{L}$ is a discriminative knowledge space, with a surmise or precedence relation $\precsim$ represented in Figure 3.1 by its Hasse diagram.

Thus, by definition (cf. 1.43), we have for r, q in Q,

$$r \precsim q \Longleftrightarrow r \in \cap \mathcal{L}_q.$$

---

[1] The surmise relation only permits one history for any item q, which is formed by all the items preceding q in the surmise relation.

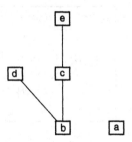

Fig. 3.1. Hasse diagram of the surmise relation of the knowledge structure $\mathcal{L}$ specified by Equation (1).

Note that $\{q \in Q \mid q \precsim b\} = \{b\}$: there are no questions in Q that must be mastered before b. This information, however, gives a distorted picture of the situation. Examining Equation (1) leads to the conclusion that b can be learned only if d was acquired in the same time, or both a and c, or both c and e. Indeed,

$$\{b, d\}, \quad \{a, b, c\}, \quad \{b, c, e\}$$

are the three atoms at b, i.e. the three minimal states of $\mathcal{L}$ containing b.

In the case of a quasi ordinal space $\mathcal{K}$ with surmise relation $\precsim$, the situation is simpler. For any item q, there is always exactly one atom at q which is specified by $\cap \mathcal{K}_q$ and contains all the precedents of q in $\precsim$ (cf. Theorem 1.40 and Definition 1.47). In fact, considering the equivalence $r \precsim q \Leftrightarrow r \in \cap \mathcal{K}_q$, the full information concerning the quasi ordinal space is obtained by listing the unique atom $\cap \mathcal{K}_q$ at each question q. In the general case, however, there may be several atoms at a question (see Table 3.1 below), or possibly no atoms (as in Examples 1.21 or 1.27). This is illustrated by the knowledge space $\mathcal{L}$ of Equation (1), whose atoms at each question are listed in Table 3.1.

This example suggests a generalization of the surmise relation into a 'surmise function' associating, to each item q of a domain Q, a collection of subsets of Q called the 'backgrounds of' q or the 'clauses for' q. Each of these backgrounds represents a possible history of the mastery of item q. We also formulate four axioms on this function which are consistent with this interpretation. First, there must be at least one clause for each item. Second, we ask that each clause for an item contains this item. Third, each

Questions	Atoms
a	$\{a\}$
b	$\{b,d\}$, $\{a,b,c\}$, $\{b,c,e\}$
c	$\{a,b,c\}$, $\{b,c,e\}$
d	$\{b,d\}$
e	$\{b,c,e\}$

Table 3.1. Items and their atoms in the knowledge space of Equation (1).

item in any clause C for an item has itself a clause included in C. (We give a pictorial illustration of this axiom in Figure 3.3). Finally, any two clauses for the same item must be incomparable with respect to set inclusion.

The concepts of a surmise system and its clauses are introduced in the next definition. Note that we do not assume the existence of a knowledge structure. However, it turns out that any surmise function uniquely defines a granular knowledge space (cf. Theorem 3.10). (We recall from Definition 1.35 that a knowledge space $\mathcal{K}$ is *granular* when for each state K in $\mathcal{K}$ and each item q in K there exists an atom A at q with $q \in A \subseteq K$.)

3.2. Definition. Let Q be a nonempty set of *items*, and let σ be a function mapping Q into 2^{2^Q}. Thus, every value of σ is a family of subsets of Q. We say that σ is an *attribution (function)* on the set Q if this family is always nonempty, that is

(1) if $q \in Q$, then $\sigma(q) \neq \varnothing$.

For each $q \in Q$, any $C \in \sigma(q)$ is said to be a *background of* q, or synonymously, a *clause for* q (in σ). We consider three additional conditions: for all $q, q' \in Q$, and $C, C' \subseteq Q$,

(2) if $C \in \sigma(q)$, then $q \in C$;
(3) if $q' \in C \in \sigma(q)$, then $C' \subseteq C$ for some $C' \in \sigma(q')$;
(4) if $C, C' \in \sigma(q)$ and $C' \subseteq C$, then $C = C'$.

When all four conditions are satisfied, the pair (Q, σ) is a *surmise system* and the function σ is called a *surmise function* on Q.

3.3. Remarks. Condition (1) is reasonable. (See Problem 1 in this connection. Note that we can have $\sigma(q) = \{\varnothing\}$.) Condition (2) is introduced for convenience and plays a minor role. Condition (3) is natural if a clause for an item q is interpreted as a possible minimal history of the mastery of q: if q' is in a clause C for q, there must be a path to the mastery of q' within C, and so there must be a background of q' included in C. Condition (4) ensures that the histories formalized by the clauses are not redundant: suppose that C is a background of q, and C' is also a clause for q, with C' included in C; then C' must be equal to C, since, otherwise, C would not be a minimal history of q.

The concept of a surmise function generalizes that of a quasi order. In particular, Conditions (2) and (3) correspond to reflexivity and transitivity, respectively. Actually, except for a trivial change in the encoding, any binary relation is a special case of an attribution.

3.4. Definition. Let $\mathcal{R}$ be any binary relation on a nonempty set Q. Define an attribution that has exactly one clause for each item q of Q by the equation:

$$\sigma(q) = \{\{r \in Q \mid r\mathcal{R}q\}\}.$$

We then say that $\mathcal{R}$ is *cast* as the attribution σ. Notice that σ and $\mathcal{R}$ contain exactly the same information. It is easily checked that

(i) $\mathcal{R}$ is reflexive iff σ satisfies Condition (2) of a surmise function;

(ii) $\mathcal{R}$ is transitive iff σ satisfies Condition (3) of a surmise function (cf. Problem 2).

Thus, the collection of all surmise functions on Q encompasses the collection of all quasi orders on Q. Also, the collection of all attributions on Q encompasses the collection of all binary relations on Q.

3.5. Example. To get a pictorial display of an attribution in the case of a small finite set Q, we extend the standard conventions used for the graph of a binary relation. For instance, Figure 3.2 displays the graph of the attribution σ on $Q = \{a, b, c, d, e\}$ with

$$\sigma(a) = \{\{a, b, c\}, \{c, d\}\}, \qquad \sigma(b) = \{\{e\}\},$$
$$\sigma(c) = \{\{c\}\}, \qquad \sigma(d) = \{\{d\}\}, \qquad \sigma(e) = \{\{a, d\}, \{b\}\}.$$

The rules governing such a representation are as follows. Consider a clause C for an element q of Q and suppose that C contains q and at least one item $q' \neq q$. Then C is represented by an ellipse surrounding all the

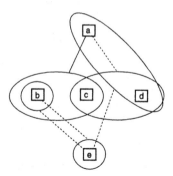

Fig. 3.2. Graph of the attribution in Example 3.5.

elements of C **but** q, and linked to q by a solid segment. This is illustrated
in Figure 3.2 by the ellipse surrounding the points b and c and linked to
the point a by a solid segment. Indeed, $\{a, b, c\}$ is a clause for a in the
attribution σ. If C is a clause for q which does not contain q, but contains
some $q' \neq q$, the representation of that clause is the same, but the segment
linking q to the ellipse is dashed. Four examples of such a linking are
drawn in Figure 3.2. Finally, if a clause for q contains only q, then no
ellipse is drawn, as in the case of the point c of the figure because we have
$\sigma(c) = \{c\}$. Note that there is no dashed lines in the representation of a
surmise function because of Condition 2 in Definition 3.2: for any item q,
any clause C for q contains q.

Such figures may become very intricate. (We invite the reader to graph
the surmise function of the knowledge structure in Example 3.1 or Table
3.1.) On the other hand, a given partial order can be encoded, in a minimal
efficient way, by its Hasse diagram. Surmise functions generalize quasi orders and thus partial orders. This evokes the potential concept of a 'Hasse
system' capable of faithfully summarizing the information in a surmise system. A section of this chapter will be devoted to a precise definition of such
a concept (see 3.23) and some of its consequences.

Figure 3.3, a display of the kind just introduced, illustrates Axiom (3) in
the definition of a surmise system (see 3.2).

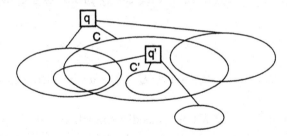

Fig. 3.3. Pictorial representation of Condition (3) in Definition 3.2: $q' \in C \in \sigma(q)$ and $C' \in \sigma(q')$ with $C' \subseteq C$.

Knowledge Spaces and Surmise Systems

The following four definitions and examples pave the way to a fundamental relationship between knowledge spaces and surmise systems that will be made precise in Theorem 3.10.

3.6. Definition. Let $(Q, \mathcal{K})$ be a granular knowledge structure (cf. 1.35). Accordingly, any item q has at least one atom. Let $\sigma : Q \to 2^{2^Q}$ be a function defined by the equivalence:

$$C \in \sigma(q) \quad \Longleftrightarrow \quad C \text{ is an atom at } q,$$

with $C \subseteq Q$ and $q \in Q$. It is easily seen that σ is a surmise function on Q. We shall say that σ is the surmise function on Q derived from $(Q, \mathcal{K})$.

Note that if a granular knowledge structure $(Q, \mathcal{K})$ is closed under intersection, its derived surmise function σ has only one clause for each item. There exists thus a well-defined relation $\mathcal{R}$ on Q that is cast as σ (cf. 3.4). If the knowledge structure $(Q, \mathcal{K})$ is a quasi ordinal space, this relation $\mathcal{R}$ is exactly the quasi order derived from $(Q, \mathcal{K})$ in the sense of Definition 1.50.

3.7. Example. Applying the construction of Definition 3.6 to the knowledge structure $\mathcal{L}$ of Equation (1), we obtain from Table 3.1 the surmise function σ specified by:

$$\sigma(a) = \{\, \{a\}\, \}, \qquad \sigma(b) = \{\, \{b,d\}, \{a,b,c\}, \{b,c,e\}\, \},$$
$$\sigma(c) = \{\, \{a,b,c\}, \{b,c,e\}\, \}, \quad \sigma(d) = \{\, \{b,d\}\, \}, \quad \sigma(e) = \{\, \{b,c,e\}\, \}.$$

3.8. Definition. Any attribution σ on a set Q produces a knowledge space $(Q, \mathcal{K})$ with

$$K \in \mathcal{K} \quad \Longleftrightarrow \quad \forall q \in K, \exists C \in \sigma(q) : C \subseteq K. \qquad (2)$$

We also say that the knowledge space $(Q, \mathcal{K})$ is *derived* from the attribution σ on Q.

The verification that $(Q, \mathcal{K})$ is indeed a knowledge space is left to the reader. When (Q, σ) is a surmise system, its derived knowledge structure is always a granular knowledge space. In particular, each clause for an item q is an atom at q in $\mathcal{K}$ (cf. Definition 3.2). Thus, the states in $\mathcal{K}$ are unions of clauses, and conversely, any union of clauses is a state.

This construction of $(Q, \mathcal{K})$ from (Q, σ) is a natural outcome of our interpretation of surmise systems: a set K of items forms a knowledge state when K includes, for each of its items q, a minimal history leading to q.

3.9. Example. With the surmise function obtained in Example 3.7, we verify that $\{a, b, c, e\}$ is a state in the sense of Definition 3.8. Indeed, remark that

$$
\begin{aligned}
\{a\} &\in \sigma(a) & \text{and} & & \{a\} &\subseteq \{a, b, c, e\}, \\
\{a, b, c\} &\in \sigma(b) & \text{and} & & \{a, b, c\} &\subseteq \{a, b, c, e\}, \\
\{a, b, c\} &\in \sigma(c) & \text{and} & & \{a, b, c\} &\subseteq \{a, b, c, e\}, \\
\{b, c, e\} &\in \sigma(e) & \text{and} & & \{b, c, e\} &\subseteq \{a, b, c, e\}.
\end{aligned}
$$

(Notice that we could have used the clause $\{b, c, e\}$ for b.) On the other hand, the subset $\{a, c, d, e\}$ is not a state because it does not include a clause for e. The family $\mathcal{K}$ of all states is easily constructed and coincides with the original family in Example 3.1.

In fact, the constructions of σ in Definition 3.6 and of $\mathcal{K}$ in Definition 3.8 are mutual inverses. We have the following general result:

3.10. Theorem. *A one-to-one correspondence exists between the collection of all granular knowledge spaces on a set Q and the collection of all surmise functions on Q. It is defined, for all granular knowledge space $\mathcal{K}$ and surmise function σ, by the following formula, where $S \subseteq Q$ and $q \in Q$:*

$$S \text{ is an atom at } q \text{ in } \mathcal{K} \quad \Longleftrightarrow \quad S \text{ is a clause for } q \text{ in } \sigma. \qquad (3)$$

PROOF. Let s be the function from the collection $\tilde{\mathbf{K}}^{\mathbf{G}}$ of all granular knowledge spaces on a set Q to the collection $\tilde{\mathbf{F}}^{\mathbf{S}}$ of all surmise functions on Q, defined as in Definition 3.6 by the equivalence (where $\mathcal{K} \in \tilde{\mathbf{K}}^{\mathbf{G}}$ and $\sigma \in \tilde{\mathbf{F}}^{\mathbf{S}}$)

$$s(\mathcal{K}) = \sigma \quad \Longleftrightarrow \quad \forall q \in Q: \ \sigma(q) = \{S \in 2^Q \mid S \text{ is an atom at } q\}. \quad (4)$$

Suppose now that $\mathcal{K}$ and $\mathcal{K}'$ are two distinct granular knowledge spaces on Q, with $\sigma = s(\mathcal{K})$ and $\sigma' = s(\mathcal{K}')$. Then, $\mathcal{K}$ and $\mathcal{K}'$ must have different bases. In particular, there must be an item q such that the set of all atoms at q in $\mathcal{K}$ differs from the set of all atoms at q in $\mathcal{K}'$ (Theorems 1.39 and 1.26). We thus have $\sigma(q) \neq \sigma'(q)$, and therefore $\sigma \neq \sigma'$. We conclude that s is an injective mapping of $\tilde{\mathbf{K}}^{\mathbf{G}}$ into $\tilde{\mathbf{F}}^{\mathbf{S}}$. The mapping s is actually surjective onto $\tilde{\mathbf{F}}^{\mathbf{S}}$. For take any σ in $\tilde{\mathbf{F}}^{\mathbf{S}}$. It follows easily from Definition 3.2 that any clause of σ cannot be a union of other clauses. Consequently, all the clauses of σ form the base of some granular knowledge space $\mathcal{K}$ with, automatically, $s(\mathcal{K}) = \sigma$. $\qquad\qquad\qquad\qquad\qquad\qquad\qquad\qquad\qquad\square$

Theorem 3.10 provides some additional motivation for the choice of the knowledge space as one of our core concepts. In the case of a finite set Q, the axiom of closure under union selects, among all knowledge structures, the families of knowledge states that can be derived from clauses for the items.

According to Definitions 3.6 and 3.8, a granular space $(Q, \mathcal{K})$ and a surmise system (Q, σ) related as in Theorem 3.10 are said to be derived one from the other. This terminology is coherent with Definition 1.50 in the sense that the correspondence in Theorem 3.10 extends the correspondence between quasi ordinal spaces and quasi orders obtained in Birkhoff's Theorem 1.49.

AND/OR Graphs

We show here how attributions can be regarded as a type of AND/OR graphs, and how these constructs generate knowledge spaces. The AND/OR graphs are used in the field of artificial intelligence, although often without a formal definition. They model the organization of a task into subtasks, for example in the resolution of a practical problem. Each of these subtasks might itself require the preliminary completion of one of some sets of other subtasks—or maybe no subtask at all. Here, the (sub)tasks are called 'OR-vertices.' A set of subtasks whose completion delivers the solution of another (sub)task is encoded as an 'AND-vertex.' An edge from an AND-vertex α to an OR-vertex a specifies that the combination α of subtasks gives a way to solve task a. An edge from an 'OR-vertex' b to an 'AND-vertex' α indicates that task b is involved in the combination α of tasks. To eliminate possible ambiguities, we will formulate as axioms some assumptions that are often left implicit elsewhere. The main difference between AND/OR graphs and surmise systems lies in the introduction of artificial AND-vertices representing combinations of subtasks; these 'AND-vertices' play the rôle of the clauses in the setting of attributions (cf. Definition 3.2).

Understanding the next definition will be facilitated by examining Example 3.12 and Figure 3.4.

3.11. Definition. An *AND/OR graph* is a directed graph $G = (V, E)$, where the nonempty set V of vertices is the disjoint union of two subsets V_{AND} of *AND-vertices* and V_{OR} of *OR-vertices*. An element of the set E is a (directed) *edge*, that is a pair of vertices. We also require that

(1) either the initial vertex of an edge belongs to V_{AND}, and the terminal vertex belongs to V_{OR}, or vice versa;

(2) each AND-vertex α belongs to exactly one edge (α, a), where $a \in V_{OR}$;

(3) each OR-vertex a belongs to at least one edge (α, a), where $\alpha \in V_{AND}$.

The interpretation of these three axioms relies on the meaning of the vertices and edges discussed before Definition 3.11. As a matter of fact, Axiom 1 essentially says that the edges admit exactly one of the two intended meanings. Axiom 2 requires that any combination of tasks relates to a definite task. Axiom 3 imposes that any task is accessible through some combination of (sub)tasks (which can be the empty combination).

3.12. Example. Suppose that $V_{AND} = \{\alpha_1, \alpha_2, \beta_1, \beta_2, \gamma_1, \gamma_2, \delta, \epsilon, \eta, \gamma, \theta\}$ and $V_{OR} = \{a, b, \ldots, h\}$. An AND/OR graph $V = V_{OR} \cup V_{AND}$ with 21 edges is displayed in Figure 3.4. We use $\vee$ and $\wedge$ to mark the OR-vertex and the AND-vertex, respectively.

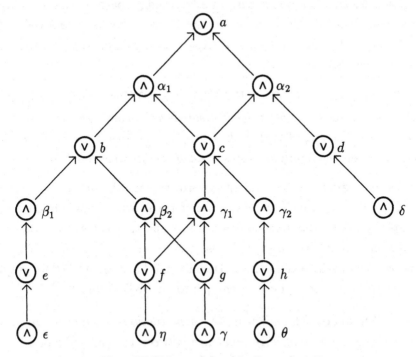

Fig. 3.4. The AND/OR graph used in Example 3.12.

As mentioned at the beginning of this section, the usual interpretation of such a graph is in terms of tasks represented by OR-vertices. For instance, task a requires the previous completion of (sub)tasks b and c, or of (sub)tasks c and d. Each AND-vertex specifies a set of tasks whose overall completion allows to start the unique task to which it is linked.

The 'exactly one' in Condition (2) from Definition 3.11 is replaced by some authors with 'at least one.' Our additional requirement is not a severe restriction, in the sense that it can be fulfilled after addition of AND-vertices without altering the intended meaning of the graph. Indeed, if some AND-vertex α belonged to several edges (α, b_1), (α, b_2), ..., (α, b_n), we could always replace α by clones, one for each b_i, with $i = 1$, 2, ..., n. Such a cloning process would become necessary in Example 3.11 if the AND-vertices β_2 and γ_1 were collapsed.

Notice that our Definition 3.11 does not rule out 'intractable' situations involving 'cycles' of subtasks. An analysis of these concepts will be given in this chapter (beginning with Theorem 3.30).

Each AND/OR graph generates a knowledge space on the set of its OR-vertices. This should be clear to the reader who has perceived the link between AND/OR graphs and attributions; this link will be made explicit in Theorem 3.14.

3.13. Theorem. *Let $G = (V, E)$ be an AND/OR graph. A subset K of V_{OR} is said to be a state of G if it satisfies the condition: for any a in K, there exists an edge (α, a) with $\alpha \in V_{AND}$ such that $b \in K$ for each edge (b, α). Then the family of states is a knowledge space on V_{OR}.*

As the arguments are straightforward, we skip the proof.

The next theorem shows how to transform an attribution into an AND/OR graph by taking the items as OR-vertices and the clauses as AND-vertices. Some care must be taken in a case where one set is a clause for several items. To obtain Axiom (2) in Definition 3.11 of an AND/OR graph, we transform a clause C for an item q into an AND-vertex (q, C).

3.14. Theorem. *The following two constructions are mutual inverses and establish a one-to-one correspondence between the collection of all attributions and the collection of all AND/OR graphs.*

If σ is an attribution on the set Q, construct an AND/OR graph by setting $V_{OR} = Q$, $V_{AND} = Q \times C$, and defining the edges as follows. Declare an edge $((q, C), a)$, whenever $q = a \in Q$ and C is a clause for item a in σ. Declare an edge $(b, (q, C))$ whenever C is a clause for q in σ, and $b \in C$.

Conversely, if $G = (V, E)$ is an AND/OR graph, let $Q = V_{OR}$, and define an attribution σ on Q as follows. A subset C of Q is a clause for item q if there exists an edge (α, q) with α some AND-vertex such that $C = \{b \in V_{OR} \,|\, (b, \alpha) \text{ is an edge}\}$.

The two constructions described in the theorem indicate that attributions on Q and AND/OR graphs with $V_{OR} = Q$ are obvious rephrasings of one another, the items corresponding to OR-vertices, and the clauses (or rather the pairs (C, q), with C a clause for the item q) corresponding to the AND-vertices. We leave the verifications to the reader.

Surmise Functions and Wellgradedness

As in Theorem 3.10, we consider a surmise system (Q, σ) and its derived granular knowledge space $(Q, \mathcal{K})$. Thus, for $K \subseteq Q$,

$$K \in \mathcal{K} \quad \Longleftrightarrow \quad \forall q \in K, \exists C \in \sigma(q) : C \subseteq K.$$

Remember from Theorem 2.23 that the space $(Q, \mathcal{K})$ is well-graded iff all its learning paths are gradations. We now investigate how wellgradedness is reflected in the surmise system (Q, σ). Remember also that any question q of the knowledge space $(Q, \mathcal{K})$ belongs to exactly one notion q^* (cf. Definition 1.4).

3.15. Theorem. *For a surmise system (Q, σ) and its derived knowledge space $(Q, \mathcal{K})$, the following three conditions are equivalent:*

(i) *the knowledge space $(Q, \mathcal{K})$ is well-graded;*

(ii) *two questions having at least one common clause are equally informative; formally:*

$$\forall q, q' \in Q : \quad \sigma(q) \cap \sigma(q') \neq \varnothing \Longrightarrow q^* = q'^*;$$

(iii) *any clause for a question, minus the notion containing this question, is a state, that is:*

$$\forall q \in Q, \forall C \in \sigma(q) : \quad C \setminus q^* \in \mathcal{K}.$$

The proof given below applies to finite or infinite sets Q. A reader having skipped the starred section on wellgradedness in Chapter 2, which deal with the infinite case (2.19 to 2.24) is invited to provide his own proof for the finite case.

PROOF. (i) $\Rightarrow$ (ii). Suppose $(Q, \mathcal{K})$ is well-graded, and that two questions q and q' have a common clause C, while $q^* \neq q'^*$. Consider any learning path $\mathcal{L}$ containing the knowledge state C (such a learning path exists by the maximality principle). As C is a knowledge state minimal for $q \in C$, we cannot have $C = K \cup x^*$ for any state K in $\mathcal{L} \setminus \{C\}$ and notion x^* with $x^* \neq q^*$. Similarly, as C is minimal also for $q' \in C$, we cannot have $C = K \cup q^*$ for any K in $\mathcal{L} \setminus \{C\}$. For the same reasons, we have

$$C \neq \bigcup \{L \in \mathcal{L} \mid L \subset C\}.$$

Hence the learning path $\mathcal{L}$ cannot be a gradation, in contradiction with the assumed wellgradedness.

(ii) $\Rightarrow$ (iii). Suppose that for some question q and some clause C for q, the subset $C \setminus q^*$ is not a state. Then there must be some q' in $C \setminus q^*$ such that no clause for q' is included in $C \setminus q^*$. On the other hand, there is some clause C' for q' included in the state C. Thus, $q \in C'$. As C is a state minimal for $q \in C$, we derive $C = C'$. We obtain in this way a common clause for q and q', the required contradiction.

(iii) $\Rightarrow$ (i). Assume Condition (iii), and also that the space $(Q, \mathcal{K})$ is not well-graded. Hence, by Theorem 2.23, there exists some learning path $\mathcal{L}$ which is not a gradation: in other terms, we can find some K in $\mathcal{L}$ satisfying both $K \neq \bigcup\{L \in \mathcal{L} \,|\, L \subset K\}$, and $K \setminus q^* \notin \mathcal{L}$ for each item $q \in K$. Set $K^\circ = \bigcup\{L \in \mathcal{L} \,|\, L \subset K\}$ and notice that $K^\circ \in \mathcal{L}$. There must exist some question r in $K \setminus K^\circ$ with a clause C for r included in K. By the definition of K° and because $r \notin K^\circ$, we get $C = K$. Condition (iii) implies that $K \setminus r^* \in \mathcal{K}$, and thus $K \setminus r^* \subseteq K^\circ$. But since r^* is a notion, we must have $K \setminus r^* = K^\circ \in \mathcal{L}$, contradicting our choice of K. $\square$

Hasse Systems

Let $\mathcal{R}$ be a relation on a set Q. If we cast $\mathcal{R}$ as an attribution function in the sense of Definition 3.4, then a subset K of Q is a state of $\mathcal{R}$ if it satisfies

$$\forall (p, q) \in \mathcal{R} : \; q \in K \Longrightarrow p \in K$$

(cf. Definition 3.8.) In the case where $\mathcal{R}$ is a quasi order, the states of $\mathcal{R}$ are exactly the states of the space derived from $\mathcal{R}$ in the sense of Definition 1.49. The Hasse diagram $\mathcal{P}^h$ of a partial order $\mathcal{P}$ on a finite set Q has the same states as $\mathcal{P}$. Moreover, $\mathcal{P}^h$ is the smallest relation having exactly those states (where 'smallest' means 'minimal for inclusion'). In that sense, $\mathcal{P}^h$ is a most economical summary of the partial order $\mathcal{P}$. The concern of this section is to develop a similar concept of 'most economical summary' for surmise systems. We shall define a 'Hasse system' as an 'economical' attribution, where the precise meaning of 'economical' relies on a comparison method for attributions which is introduced in the next definition.

3.16. Definition. A relation $\precsim$ is defined on the collection $\tilde{\mathbf{F}}$ of all attributions on a nonempty set Q by declaring, for σ, σ' in $\tilde{\mathbf{F}}$:

$$\sigma' \precsim \sigma \iff \forall q \in Q, \forall C \in \sigma(q), \exists C' \in \sigma'(q) : C' \subseteq C.$$

This relation $\precsim$ is always a quasi order, but not necessarily a partial order (however, see Problem 5). It will be referred to as the *attribution order* on $\tilde{\mathbf{F}}$. Note that the restriction of $\precsim$ to the set of all relations on Q (cast as attributions) is the usual inclusion comparison of relations.

In general, there may be several attributions producing the states of a particular granular knowledge space $(Q, \mathcal{K})$ (in the sense of Definition 3.8). A natural condition on any attribution σ providing an economical description of $\mathcal{K}$ is that σ be a minimal element in the subset of all attributions producing $\mathcal{K}$ (where minimal refers to the attribution order $\precsim$). In the case of an infinite domain Q, the existence of at least one such minimal element is not ensured. In this connection, remember that for infinite partial orders, Hasse diagrams do not necessarily exist (see e.g. the usual linear order on the set of real numbers).

3.17. Example. The knowledge space $\mathcal{L}$ of Equation (1),

$$\mathcal{L} = \{\varnothing, \{a\}, \{b, d\}, \{a, b, c\}, \{b, c, e\}, \{a, b, d\}, \{a, b, c, d\},$$
$$\{a, b, c, e\}, \{b, c, d, e\}, \{a, b, c, d, e\}\},$$

has a derived surmise function σ, which was described in Example 3.7 as

$$\sigma(a) = \{\{a\}\}, \qquad \sigma(b) = \{\{b, d\}, \{a, b, c\}, \{b, c, e\}\},$$
$$\sigma(c) = \{\{a, b, c\}, \{b, c, e\}\}, \quad \sigma(d) = \{\{b, d\}\}, \quad \sigma(e) = \{\{b, c, e\}\}.$$

This attribution σ is *not* minimal (for $\mathcal{L}$), since the following attribution ϵ satisfies $\epsilon \precsim \sigma$, but not $\sigma \precsim \epsilon$, while having the same knowledge states:

$$\epsilon(a) = \{\varnothing\}, \qquad \epsilon(b) = \{\{c\}, \{d\}\},$$
$$\epsilon(c) = \{\{a, b\}, \{b, e\}\}, \quad \epsilon(d) = \{\{b\}\}, \quad \epsilon(e) = \{\{c\}\}.$$

We leave to the reader to verify that ϵ is *not* a minimal attribution for $\mathcal{L}$. Any deletion of items from a clause would change the collection of states, but adding the clause $\{d, e\}$ to $\epsilon(c)$ gives a strictly 'smaller' attribution, in the sense of $\precsim$, which still produces $\mathcal{L}$. Notice that each state of $\mathcal{L}$ containing $\{c\} \cup \{d, e\}$ also contains another clause for c. Hence the addition of the clause $\{d, e\}$ for c would be superfluous in an 'economical' attribution that produces $\mathcal{L}$.

3.18. Example. The following knowledge space is ordinal

$$\mathcal{M} = \{ \varnothing, \{a\}, \{a, b\}, \{a, c\}, \{a, b, c\}, \{a, b, c, d\} \},$$

since it is derived from a partial order cast as the surmise function δ with

$$\delta(a) = \{ \{a\} \}, \qquad \delta(b) = \{ \{b, a\} \},$$
$$\delta(c) = \{ \{c, a\} \}, \qquad \delta(d) = \{ \{d, a, b, c\} \}.$$

The space $\mathcal{M}$ is also derived from the attribution γ, with

$$\gamma(a) = \{\varnothing\}, \qquad \gamma(b) = \{ \{a\}, \{c\} \},$$
$$\gamma(c) = \{ \{a\} \}, \qquad \gamma(d) = \{ \{b, c\} \}.$$

There is thus some minimal attribution μ for $\mathcal{M}$ with $\mu \precsim \gamma$ and $\{c\} \in \mu(b)$. (The construction of μ is not necessary for our argument). This is cumbersome in view of the extraneous clause $\{c\}$ in $\mu(b)$, which is not contained in the (unique) atom $\{a, b\}$ at b. Moreover, each state containing $\{b\} \cup \{c\}$ also contains the clause $\{a\}$ for b. The condition defined below rules out such extraneous clauses.

3.19. Definition. An attribution σ on the nonempty set Q is *tense* when for any item q and any clause C for q, there is some state K (of the knowledge space derived from σ in the sense of Definition 3.8) which contains q and includes C but no other clause for q.

Notice that any relation cast as an attribution is tense. Also, any surmise system is tense.

3.20. Theorem. *Any attribution σ on Q which is tense satisfies Condition (4) in the definition of a surmise system (see 3.2), namely*

$$\forall q \in Q, \; \forall C, C' \in \sigma(q): \; C \subseteq C' \implies C = C'.$$

PROOF. Assume $C, C' \in \sigma(q)$ with $C' \subset C$. Then any state containing q and C would also contains C', in contradiction with the tensity of σ. □

3.21. Theorem. *If an attribution is tense and produces a granular knowledge space $\mathcal{K}$, then any clause for any item q is contained in some atom of $\mathcal{K}$ at q.*

PROOF. Let σ be the attribution and suppose that $C \in \sigma(q)$. Select a state K containing $\{q\} \cup C$ but no clause for q distinct from C. By granularity, there exists some atom A at q with $A \subseteq K$. Since $A \in \mathcal{K}$, there is some clause D for q such that $D \subseteq A \subseteq K$. We must have $C = D$ because of the choice of K. □

3.22. Remarks. a) Granularity is automatically fulfilled when Q or $\mathcal{K}$ is finite. Our guess is that the conclusion of Theorem 3.21 does not hold when granularity is not assumed, but we have no counter-example. We shall return in Chapter 6 to attributions that produce granular knowledge spaces. (see Definition 6.24).

b) Notice also that the atom mentioned in Theorem 3.21 is not necessarily unique, as shown by the attribution ϵ in Example 3.17 with $q = b$.

Our preparation is now complete. Theorem 3.26 will show that the definition of a 'Hasse system' given in the next definition is a genuine generalization of the Hasse diagrams for partial orders.

3.23. Definition. A *Hasse system* for a granular knowledge space $(Q, \mathcal{K})$, or for its derived surmise system, is any attribution σ on Q which is minimal for $\precsim$ in the set of all attributions that

(1) are tense;
(2) produce $\mathcal{K}$.

In this situation, we also say that (Q, σ) is a *Hasse system* of $(Q, \mathcal{K})$.

3.24. Example. The attribution ϵ in Example 3.17 is a Hasse system.

3.25. Example. Consider the attribution γ that produces the ordinal knowledge space $\mathcal{M}$ in Example 3.18. There is no Hasse system α for $\mathcal{M}$ with $\alpha \precsim \gamma$. (There cannot be one, because γ is not a tense attribution in view of its clause $\{c\}$ for b.) On the other hand, the Hasse diagram of the partial order δ cast as an attribution β leads to the Hasse system

$$\beta(a) = \{\varnothing\}, \qquad \beta(b) = \{\,\{a\}\,\},$$
$$\beta(c) = \{\,\{a\}\,\}, \qquad \beta(d) = \{\,\{b,c\}\,\}.$$

It is clear that any finite knowledge space admits at least one Hasse system. Indeed, the number of tense attributions producing $(Q, \mathcal{K})$ is positive and finite. Thus one of them must be a minimal element with respect to the attribution order $\precsim$. Any such minimal element (Q, σ) is by definition a Hasse system.

3.26. Theorem. *Any finite, ordinal knowledge space $(Q, \mathcal{K})$ admits exactly one Hasse system. For each item q in Q, this system has a unique clause which contains all the items covered by q in the partial order on Q derived from $\mathcal{K}$.*

Thus the Hasse diagram of the partial order on Q derived from $\mathcal{K}$ is cast as the unique Hasse system mentioned in the Theorem.

PROOF. Let σ be a Hasse system for $(Q, \mathcal{K})$. (By the arguments given before the statement, we know there exists at least one Hasse system). We have to show that σ has exactly one clause for each item q, and that this clause consists of all items covered by q in the partial order P on Q derived from $\mathcal{K}$. Take any clause C in $\sigma(q)$ (by Definition 3.2, $\sigma(q) \neq \varnothing$). By Theorem 3.21, xPq holds for all items x in C. By the minimality of σ (together with Theorem 3.20), we have $q \notin C$. Now let K be the smallest state containing C; then it is easily checked that $K \cup \{q\}$ is also a state. Hence each question y covered by q belongs to K. Moreover, such a question y must be in C (otherwise, there would be an element z of C satisfying $yPzPq$, and q would not cover y). We have thus proved that each clause for q contains all the elements covered by q. From minimality again, no other element can belong to this clause. $\qquad\square$

3.27. Remarks. Had we not required tensity in the definition of a Hasse diagram, the last theorem would not hold (see Example 3.18). A quasi order has more than one Hasse system if it has at least three elements and a notion with more than one element. This is illustrated in the two examples below.

3.28. Example. Let $\{(b,a),(b,c),(c,b)\}$ be a relation on $Q = \{a,b,c\}$. This relation is cast as the attribution σ, with

$$\sigma(a) = \{\,\{b\}\,\}, \qquad \sigma(b) = \{\,\{c\}\,\}, \qquad \sigma(c) = \{\,\{b\}\,\}.$$

The derived knowledge space $\mathcal{K} = \{\varnothing, \{b,c\}, \{a,b,c\}\}$ is quasi ordinal. In fact, σ is a Hasse system of $\mathcal{K}$. It is not difficult to construct another one.

3.29. Example. Let $Q = \{a,b,c\}$ and let S be the relation consisting of the pairs (a,b), (b,c), and (c,a). The resulting quasi ordinal space $\mathcal{K}$ has only two states. It admits several Hasse systems, one of which is S cast as an attribution.

We do not know how to characterize efficiently the granular knowledge spaces that admit a unique Hasse system.

Resolubility and Acyclicity

In our discussion of AND/OR graphs, we mentioned that any such graph had an interpretation as an organizing device for subtasks of a main task. We also indicated that Definition 3.11 does not preclude intractable situations involving cycles of subtasks. Constraints ruling out such cases are considered in this section. In keeping with this motivation, we shall introduce the concepts of 'resoluble' attributions. In the context of knowledge assessment, the idea is that any item is accessible by way of one or more learning tracks (i.e., the prerequisites are not self-contradictory). This rules out intractable situations such as that of Example 3.28, in which each of items b and c is a prerequisite for the other.

A priori, two meanings can be given to 'resolubility': it can be either local (each individual question can be accessed), or global (a strategy for accessing the whole structure can be designed). We begin by showing the equivalence of the two conceptions in the finite case.

3.30. Theorem. *Consider the two following conditions for an attribution σ on a nonempty set Q:*

   (i) *for each item q in Q, there exists some natural number k and some sequence q_1, q_2, ..., q_k of items such that $q_k = q$, and moreover for each i in $\{1, 2, ..., k\}$,*

$$\exists C \in \sigma(q_i) : C \subseteq \{q_1, q_2, \ldots, q_i\};$$

(ii) *there exists a linear order T on Q satisfying, for each item q in Q:*
    (a) $\exists C \in \sigma(q) : C \subseteq T^{-1}(q)$;
    (b) $T^{-1}(q)$ *is finite.*

Then (ii) $\Rightarrow$ (i), and if Q is finite, (i) $\Rightarrow$ (ii).

In Condition (i), by allowing $i = 1$, we impose in particular that some clause for q_1 be included in $\{q_1\}$. The notation $T^{-1}(x)$ in Condition (ii) designates $\{y \in Q \mid yTx\}$.

PROOF. (ii) $\Rightarrow$ (i). Let T be the linear order of Condition (ii), and pick any item $q \in Q$. The sequence q_1, q_2, ..., q_k of Condition (i) is formed by the items that precede or equal q in the order T.

(i) $\Rightarrow$ (ii), when Q is finite. Consider all the subsets Y of Q that can be provided with a linear order T in such a way that (a) and (b) in Condition (ii) are satisfied for all $q \in Y$. There exists at least one such subset Y, namely a set $\{q_1\}$ as in Condition (i). Now take some maximal subset among all these subsets, and call it again Y with T the linear order as above. We prove $Y = Q$. Indeed, suppose that there is some $q \in Q \setminus Y$, and take a sequence q_1, q_2, ..., q_k as in Condition (i). There is a smallest index j such that $q_j \notin Y$. We may add q_j to Y and extend the linear order T to $Y \cup \{q_j\}$ by putting q_j after the elements of Y. The resulting linearly ordered set contradicts the maximality of Y. $\square$

That (i) $\Rightarrow$ (ii) is not true in general is shown by the following example. Take an uncountable set Q with the 'trivial' attribution σ defined by $\sigma(q) = \{\varnothing\}$ for all $q \in Q$. Then Condition (i) is satisfied (even with $k = 1$), but not Condition (ii). Notice that a linearly ordered set (Q, T), with T satisfying (b) in Theorem 3.30 (ii), is necessarily isomorphic to a subset of the natural numbers with the usual order.

3.31. Definition. An attribution σ on the nonempty set Q is *resoluble* when it satisfies Condition (ii) in Theorem 3.30. The order T is called a *resolution order*.

3.32. Theorem. *Let σ be an attribution on a nonempty set Q, and let $\mathcal{K}$ be the knowledge space produced by σ. Then σ is resoluble iff $\mathcal{K}$ contains some chain C of states such that*

(i) $\varnothing \in C$;

(ii) $\forall K \in \mathcal{C} \setminus \{Q\}, \exists q \in Q \setminus K : K \cup \{q\} \in \mathcal{C};$

(iii) $\forall K \in \mathcal{C} : K$ *is finite;*

(iv) $\bigcup \mathcal{C} = Q.$

PROOF. Let $\mathcal{T}$ be a resolution order for σ. The empty set plus all sets $\mathcal{T}^{-1}(q)$, for $q \in Q$, constitute a chain $\mathcal{C}$ satisfying Conditions (i) to (iv). Conversely, if we have a chain $\mathcal{C}$ satisfying (i) to (iv), then a resolution order $\mathcal{T}$ is obtained by setting

$$q \mathcal{T} r \iff (\forall K \in \mathcal{C} : r \in K \Rightarrow q \in K)$$

for all $q, r \in Q$.   □

3.33. Corollary. *If two attributions produce the same knowledge space, then both are resoluble, or neither is.*

3.34. Definition. A knowledge space is *resoluble* when it is produced by at least one resoluble attribution.

The definitions of a resoluble attribution and of a resoluble knowledge space are not very appealing, because they involve an existential quantifier on linear orders. We now give other sufficient and/or necessary conditions for resolubility. (The first proof is left as Problem 8).

3.35. Theorem. *An attribution on a finite, nonempty set is resoluble if the derived knowledge space is well-graded.*

The converse of Theorem 3.35 is not true (see Problem 8).

3.36. Definition. A relation $\mathcal{R}$ on a set Q is *acyclic* when there does not exist any finite sequence $x_1, x_2, \ldots, x_k$ of elements of Q such that $x_1 \mathcal{R} x_2$, $x_2 \mathcal{R} x_3, \ldots, x_{k-1} \mathcal{R} x_k, x_k \mathcal{R} x_1$ and $x_1 \neq x_k$. (Note that an acyclic relation may be reflexive.)

3.37. Theorem. *Any finite partially ordered set (cast as an attribution) is resoluble. More generally, a relation on a finite set is resoluble iff it is acyclic.*

PROOF. This derives from the existence of a linear order extending a given partial order (Szpilrajn's Theorem, 1930; see e.g. Trotter, 1992), and of a partial order extending a given acyclic relation.   □

3.38. Example. The space $\mathcal{L}$ of Example 3.1 does not include any gradation, and by Theorem 3.32 is thus not resoluble. We recall the surmise or
precedence relation $\precsim$ of $\mathcal{L}$, defined earlier by

$$r \precsim q \Longleftrightarrow r \in \cap \mathcal{L}_q \qquad \text{(in Definition 1.43)},$$
$$r \precsim q \Longleftrightarrow r \in \cap \sigma(q) \qquad \text{(in terms of the surmise function } \sigma).$$

This precedence relation $\precsim$ is represented in Figure 3.1 by its Hasse diagram. Notice that it is acyclic. The second characterization suggests
another relation $\mathcal{R}$, defined by

$$r\mathcal{R}q \Longleftrightarrow r \in \cup \sigma(q). \tag{5}$$

The relation $\mathcal{R}$ has many cycles. For instance, we have $b\mathcal{R}d\mathcal{R}b$ because
$\{b, d\} \in \sigma(b) \cap \sigma(d)$.

3.39. Notation. For an attribution σ on the set Q, we define the relation
$\mathcal{R}_\sigma$ on Q by setting, for $q, q' \in Q$:

$$q\mathcal{R}_\sigma q' \quad \Longleftrightarrow \quad \exists C \in \sigma(q') : q \in C.$$

3.40. Theorem. *Let σ be an attribution on a finite, nonempty set Q.
Consider the following three conditions:*

(i) the relation $\mathcal{R}_\sigma$ is acyclic;
(ii) the space $\mathcal{K}$ produced by σ is resoluble;
(iii) the precedence relation of $\mathcal{K}$ is acyclic.

Then (i) $\Rightarrow$ (ii) $\Rightarrow$ (iii).

PROOF. See Problem 10.   □

3.41. Definition. An attribution σ is *acyclic* when the relation $\mathcal{R}_\sigma$ is
acyclic.
   We shall run into acyclicity again in Chapter 6 (see Theorem 6.28).

Original Sources and Related Works

The link between knowledge spaces and surmise systems was established
in Doignon and Falmagne (1985) in the finite case. We have spelled out

in Theorem 3.14 the close relationship existing between attributions and AND/OR graphs. For the latter concept, the reader may consult textbooks on artificial intelligence, for instance Barr and Feigenbaum (1981), or Rich (1983).

The correspondence between a knowledge space and its derived surmise system (Theorem 3.10) is rephrased as follows for simple closure spaces (cf. Definition 1.11). For a nonempty set Q, consider a mapping $\gamma : Q \rightarrow 2^{2^Q}$ with $\gamma(x)$ being a family of subsets of Q called 'semi-spaces' or 'co-points at x.' In the case of convex subsets in real affine spaces, a semi-space would be a convex subset that is maximal for the property of not containing x. If C consists of all vector subspaces, a semi-space at x would be any hyperplane avoiding x. We formulate four axioms on semi-spaces, where $x, x' \in Q$:

(1) $\gamma(x) \neq \varnothing$;

(2) if $S \in \gamma(x)$, then $x \notin S$;

(3) if $x' \notin S \in \gamma(x)$, then $S' \supseteq S$ for some $S' \in \gamma(x')$;

(4) if $S, S' \in \gamma(x)$ and $S' \subseteq S$, then $S = S'$.

The attentive reader has surely noticed that each one of these axioms is the dual of one of the axioms defining a surmise function (see Definition 3.2). Functions γ on Q satisfying these four axioms are in a one-to-one correspondence with simple closure spaces on Q.

Theorem 3.10 (linking knowledge spaces and surmise systems) can also be inferred from Flament (1976). A related result, in a different context and stated in a very different language, can be found in Davey and Priestley (1990). Their Theorem 3.38 relates what, in our setting, would be on the one hand finitary spaces (cf. Definition 1.35), and on the other hand a variant of surmise systems.

The translation of wellgradedness into a property of surmise systems was obtained in the finite case by Koppen (1989, see also 1998). (More precisely, Koppen uses Condition (ii) in Theorem 3.15.) A closely related work can be found in the setting of 'convex geometries'; see e.g. Edelman and Jamison (1985) or Van de Vel (1993).

The definition of Hasse systems given in 3.23 relies on the tensity property, rather than on the Axiom [M] used by Doignon and Falmagne (1985). The new definition sharpens the focus on the minimality of an attribution. The concept of resolubility is new here, while acyclicity was already considered in Doignon and Falmagne (1985).

Problems

1. In Definition 3.2, the first condition on a surmise system (Q, σ) is that $\sigma(q)$ is never identical to the empty subfamily of 2^{2^Q}. Show that removing this condition would correspond to dropping the requirement that $Q \in \mathcal{K}$ for a knowledge space $(Q, \mathcal{K})$. In other words, state and prove a result analogous to Theorem 3.10 for the modified concepts of surmise system and knowledge space.

2. Let σ be an attribution on a set Q, and let $\mathcal{R}$ be a relation on Q defined by $r\mathcal{R}q \Leftrightarrow r \in \cup\sigma(q)$. Suppose that $|\sigma(q)| = 1$ for all q in Q. Show that the relation $\mathcal{R}$ is transitive iff σ satisfies Condition (2) of a surmise function (cf. 3.2).

3. Describe the surmise systems derived from the granular knowledge space $(Q, \mathcal{K})$, if

(1) $Q = \{1, 2, \ldots, 100\}$, and $\mathcal{K} = \{K \in 2^Q \,|\, |K| = 0 \text{ or } |K| \geq 50\}$;

(2) $Q = \{a, b, \ldots, z\}$, and $\mathcal{K} = \{K \in 2^Q \,|\, K = \emptyset \text{ or } a \in K\}$;

(3) $Q = \mathbf{R}^2$, and $\mathcal{K} = \{K \subseteq \mathbf{R}^2 \,|\, \mathbf{R}^2 \setminus K \text{ is an affine subspace}\}$ (an affine subspace is either the empty set, a subset formed with a single point, a (straight) line or the whole $\mathbf{R}^2$);

(4) $\mathcal{K} = \{\emptyset, Q\}$;

(5) Q is finite and $\mathcal{K}$ is a chain of subsets of Q.

4. Suppose $(Q, \mathcal{K})$ is the knowledge space derived from the surmise system (Q, σ). Give a necessary and sufficient condition, in terms of clauses of σ, ensuring that $(Q, \mathcal{K})$ is discriminative.

5. Show that the relation $\precsim$ on the collection $\tilde{\mathbf{F}}$ of all functions on a set Q (cf. Definition 3.16) is a quasi order, but that it is not necessarily a partial order. Show however that this relation, restricted to the collection $\tilde{\mathbf{F}}^S$ of all surmise functions on Q, is a partial order. Do you need all the axioms of a surmise function to prove this?

6. Let $Q = \{a, b, c, d\}$ and $\mathcal{K} = \{\emptyset, \{a\}, \{c\}, \{a, c\}, \{c, d\}, \{a, b, c\}, \{a, c, d\}, Q\}$. Is the knowledge structure $(Q, \mathcal{K})$ produced by some attribution? In the case of a positive response, find a surmise function producing $\mathcal{K}$; is this function unique? Solve the same problem for $(Q, \mathcal{K}')$, with

$\mathcal{K}' = \mathcal{K} \cup \{\{a,b\}, \{b,c\}, \{b,c,d\}\}$, and for $(Q, \mathcal{K}'')$, with $\mathcal{K}'' = \mathcal{K} \cup \{\{b\}, \{b,c\}, \{b,c,d\}\}$.

7. Describe the knowledge space produced by each of the attributions σ in the following cases:

    (1) $Q = \{1,2,3,\ldots,100\}$, and $\sigma(q) = \{\{q\}\}$;

    (2) $Q = \mathbf{N}$, $\sigma(0) = \{\emptyset\}$, and $\sigma(q) = \{\{q-1\}\}$ for $q \geq 1$;

    (3) Q is finite and $\sigma(q)$ consists of all infinite subsets of Q, plus the empty set;

    (4) $\sigma(q) = \{Q\}$.

8. Prove Theorem 3.35 and show that the converse does not hold.

9. Find all implications among the following conditions on a finite knowledge structure $(Q, \mathcal{K})$:

    (1) $(Q, \mathcal{K})$ has learnstep number equal to 1 (cf. Definition 2.25);

    (2) $(Q, \mathcal{K})$ is well-graded (cf. Definition 2.4);

    (3) $(Q, \mathcal{K})$ is produced by an acyclic attribution (cf. Definition 3.31).

Do the implications remain true for an infinite knowledge structure $(Q, \mathcal{K})$?

10. Prove Theorem 3.40. Are the converse implications also true? Does the answer change if we assume that the attribution σ is a surmise function?

Chapter 4

Skill Maps, Labels and Filters

So far, cognitive interpretations of our mathematical concepts have been limited to the use of mildly evocative words such as 'knowledge state', 'learning path' or 'gradation.' This makes sense since, as suggested by our Examples in 0.9, 0.10 and 0.11, many of our results are potentially applicable to widely different fields. It must be realized, however, that our basic concepts are consistent with traditional explanatory features of psychometric theory, such as 'skills' or 'latent trait' (cf. Lord and Novick, 1974; Weiss, 1983; Wainer and Messick, 1983). Some possible relationships between knowledge states and skills, and other features of the items, are explored in this chapter.

Skills

Following Marshall (1981) and others (e.g. Falmagne et al., 1990, Albert, Schrepp and Held, 1992, Lukas and Albert, 1993), we assume the existence of some basic set S of 'skills.' These skills may consist in methods, algorithms or tricks which could in principle be identified. The idea is to associate with each question q in the domain, the skills in S which are useful or instrumental to solve this problem, and to deduce from this association what the knowledge states are. Our discussion will be illustrated by an example of a question which could be included in a test of proficiency in the UNIX operating system.

4.1. Example. Question a: *How many lines of the file* **lilac** *contain the word 'purple'?* (*Only one command line is allowed.*)

This question can be solved by a variety of methods, three of which are listed below. For each method, we state the command line in typewriter style face, following the prompt '>.'

(1) > grep purple lilac | wc

The system responds by listing three numbers; the first one is the response to the question. (The command 'grep', followed by the two arguments 'purple' and 'lilac', extracts all the lines containing

the word 'purple' from the file **lilac**; the 'pipe' command '|' directs
this output to the '**wc**' (word count) command, which computes the
number of lines, words and characters in this output.)

(2) > cat lilac | grep purple | wc
A less efficient solution achieving the same result. (The 'cat' command requires a listing of the file **lilac**, which is unnecessary.)

(3) > more lilac | grep purple | wc
This is similar to the preceding solution.

Examining these three methods suggests several possible types of association between the skills and the questions, and corresponding ways of
constructing the knowledge states consistent with those skills. A simple
idea is to regard each one of the three methods as a skill. The complete set
S of skills would contain those three skills and some others. The linkage between the questions and skills is then formalized by a function $\tau : Q \to 2^S$
associating to each question q a subset $\tau(q)$ of skills. In particular, we
would have[1]:

$$\tau(a) = \{1, 2, 3\}.$$

Consider a subject endowed with a particular subset T of skills, containing some of the skills in $\tau(a)$ plus some other skills relevant to different
questions; for example,

$$T = \{1, 2, s, s'\}.$$

This subject is able to solve Question a because $T \cap \tau(a) = \{1, 2\} \neq \varnothing$.
In fact, the knowledge state K of this subject contains all those questions
that can be solved by at least one skill possessed by the subject; that is,

$$K = \{q \in Q \mid \tau(q) \cap T \neq \varnothing\}.$$

This linkage between skills and states is investigated in the next section,
under the name 'disjunctive model.' We shall see that the knowledge structure induced by the disjunctive model is necessarily a knowledge space (cf.
Theorem 4.4). We also briefly consider, for completeness, a model that we
call 'conjunctive' and that is the dual of the disjunctive model. In the disjunctive model, only one of the skills assigned to an item q suffices to master

---

[1] There are many ways of solving Question a in the UNIX system. We only list three of
them here to simplify our discussion.

that item. In the case of the conjunctive model, all the skills assigned to an item are required. Thus, K is a state if there is a set T of skills such that, for any item q, we have $q \in K$ exactly when $\tau(q) \subseteq T$ (rather than $\tau(q) \cap T \neq \varnothing$ as in the disjunctive model). The conjunctive model formalizes a situation in which, for any question q, there is a unique solution method represented by the set $\tau(q)$, which gathers all the skills required. The resulting knowledge structure is closed under intersection (cf. Theorem 4.14). We leave to the reader the analysis of a model producing a knowledge structure closed under both intersection and union (see Problem 1).

A different type of linkage between skills and states will also be discussed. The disjunctive and conjunctive models were obtained from a rather rudimentary analysis of Example 4.1, which regarded the three methods themselves as skills, even though several commands are required in each case. A more refined analysis would proceed by considering each command as a skill, including the 'pipe' command '|.'

The complete set S of skills would be of the form [2]

$$S = \{\texttt{grep}, \texttt{wc}, \texttt{cat}, |, \texttt{more}, s_1, \dots, s_k\}$$

where, as before, $s_1, \dots, s_k$ refer to skills relevant to the other questions in the domain under consideration. To solve Question a, a suitable subset of S may be used. For example, a subject equipped with the subset of skills

$$R = \{\texttt{grep}, \texttt{wc}, |, \texttt{more}, s_1, s_2\}$$

would be able to solve Question a using either Method 1 or Method 3. Indeed, the two relevant sets of commands are included in the subject's set of skills R; we have

$$\{\texttt{grep}, \texttt{wc}, |\} \subseteq R,$$
$$\{\texttt{more}, \texttt{grep}, \texttt{wc}, |\} \subseteq R.$$

This example suggests a more complicated association between questions and skills. We shall postulate the existence of a function $\mu : Q \to 2^{2^S}$

---

[2] It could be objected that this analysis omits the skill associated with the proper sequencing of the commands. However, it is reasonable to subsume this skill in the command |, whose sole purpose is to link two commands.

associating to each question q the collection of all the subsets of skills corresponding to the possible solutions. In the case of question a, we have

$$\mu(a) = \{\{\texttt{grep}, |, \texttt{wc}\}, \{\texttt{cat}, \texttt{grep}, |, \texttt{wc}\}, \{\texttt{more}, \texttt{grep}, |, \texttt{wc}\}\}.$$

In general, a subject having some set R of skills is capable of solving some question q if there exists at least one C in $\mu(q)$ such that $C \subseteq R$. Each of the subsets C in $\mu(q)$ will be referred to as a 'competency for' q. This particular linkage between skills and states will be discussed under the name 'competency model.' We shall see that this model is consistent with general knowledge structures (i.e. not necessarily closed under union or intersection; cf. Theorem 4.18).

Example 4.1 may suggest that the skills associated with a particular domain could always be identified easily. In fact, it is by no means obvious how such an identification might proceed in general. For most of this chapter, we shall leave the set of skills unspecified and regard S as an abstract set. Our focus will be the formal analysis of some possible linkages between questions, skills, and knowledge states, along the lines sketched above. Cognitive or educational interpretations of these skills will be postponed until the last section, where we discuss a possible systematic labeling of the items, which could lead to an identification of the skills, and more generally to a description of the content of the knowledge states themselves.

Skill Maps: The Disjunctive Model

4.2. Definition. A *skill map* is a triple (Q, S, τ), where Q is a nonempty set of *items*, S is a nonempty set of *skills*, and τ is a mapping from Q to $2^S \setminus \{\varnothing\}$. When the sets Q and S are specified by the context, we shall sometimes refer to the function τ itself as the *skill map*. For any q in Q, the subset $\tau(q)$ of S will be referred to as the set of skills *assigned* to q (by the skill map τ).

Let (Q, S, τ) be a skill map and T a subset of S. We say that $K \subseteq Q$ is the knowledge state *delineated* by T *(via the disjunctive model)* if

$$K = \{q \in Q \,|\, \tau(q) \cap T \neq \varnothing\}.$$

Notice that the empty subset of skills delineates the empty knowledge state (because $\tau(q) \neq \varnothing$ for each item q), and that S delineates Q. The family of all knowledge states delineated by subsets of S is the knowledge structure *delineated* by the skill map (Q, S, τ) *(via the disjunctive model)*. When the term 'delineate' is used in the framework of a skill map without reference to any particular model, it must always understood with respect to the disjunctive model. Occasionally, when all ambiguities are removed by the context, the family of all states delineated by subsets of S will be referred to as the *delineated knowledge structure*.

4.3. Example. With $Q = \{a, b, c, d, e\}$ and $S = \{s, t, u, v\}$, we define the function $\tau : Q \to 2^S$ by

$$\tau(a) = \{t, u\}, \qquad \tau(b) = \{s, u, v\}, \qquad \tau(c) = \{t\},$$
$$\tau(d) = \{t, u\}, \qquad \tau(e) = \{u\}.$$

Thus (Q, S, τ) is a skill map. The knowledge state delineated by $T = \{s, t\}$ is $\{a, b, c, d\}$. On the other hand, $\{a, b, c\}$ is *not* a knowledge state, since it cannot be delineated by any subset R of S. Indeed, such a subset R would necessarily contain t (because of item c); thus, the knowledge state delineated by R would also contain d. The knowledge structure delineated by the skill map τ is

$$\mathcal{K} = \{ \varnothing, \{b\}, \{a, c, d\}, \{a, b, c, d\}, \{a, b, d, e\}, Q \}.$$

Notice that $\mathcal{K}$ is a space. This is not an accident, for we have the following result:

4.4. Theorem. *Any knowledge structure delineated (via the disjunctive model) by a skill map is a knowledge space. Conversely, any knowledge space is delineated by at least one skill map.*

PROOF. Assume that (Q, S, τ) is a skill map, and let $(K_i)_{i \in I}$ be some arbitrary subcollection of delineated states. If, for any $i \in I$, the state K_i is delineated by a subset T_i of S, it is easily checked that $\bigcup_{i \in I} K_i$ is delineated by $\bigcup_{i \in I} T_i$; that is, $\bigcup_{i \in I} K_i$ is also a state. Thus, the knowledge structure delineated by a skill map is always a space.

Conversely, let $(Q, \mathcal{K})$ be a knowledge space. We build a skill map by taking $S = \mathcal{K}$, and letting $\tau(q) = \mathcal{K}_q$ for any $q \in Q$. (The knowledge states containing q are thus exactly the skills assigned to q; notice that $\tau(q) \neq \varnothing$

follows from $q \in Q \in \mathcal{K}$). For $T \subseteq S = \mathcal{K}$, we check that the state K delineated by T belongs to $\mathcal{K}$. Indeed,

$$
\begin{aligned}
K &= \{q \in Q \,|\, \tau(q) \cap T \neq \varnothing\} \\
&= \{q \in Q \,|\, \mathcal{K}_q \cap T \neq \varnothing\} \\
&= \{q \in Q \,|\, \exists K' \in \mathcal{K} : q \in K' \text{ and } K' \in T\} \\
&= \{q \in Q \,|\, \exists K' \in T : q \in K'\} \\
&= \cup T,
\end{aligned}
$$

yielding $K \in \mathcal{K}$ since $\mathcal{K}$ is a space. Finally, we show that any state K from $\mathcal{K}$ is delineated by some subset of S, namely by the subset $\{K\}$. Denoting by L the state delineated by the subset $\{K\}$, we get

$$
\begin{aligned}
L &= \{q \in Q \,|\, \tau(q) \cap \{K\} \neq \varnothing\} \\
&= \{q \in Q \,|\, \mathcal{K}_q \cap \{K\} \neq \varnothing\} \\
&= \{q \in Q \,|\, K \in \mathcal{K}_q\} \\
&= K.
\end{aligned}
$$

We conclude that the space $\mathcal{K}$ is delineated by $(Q, \mathcal{K}, \tau)$. □

Minimal Skill Maps

In the last proof, we constructed, for any knowledge space, a specific skill map that delineates this space. It is tempting to regard such a representation as a possible explanation of the organization of the collection of states, in terms of the skills used to master the items. In science, an explanation of a phenomena is typically not unique, and there is a tendency to favor 'economical' ones. The material in this section is inspired by such considerations. We begin by studying a situation in which two distinct skill maps only differ by a mere relabeling of the skills. In such a case, not surprisingly, we shall talk about 'isomorphic skill maps', and we shall sometimes say of such skill maps that they assign 'essentially the same skills' to any item q. This concept of isomorphism is introduced in the next definition.

4.5. Definition. Two skill maps (Q, S, τ) and (Q, S', τ') (with the same set Q of items) are said to be *isomorphic* if there exists a one-to-one mapping

f from S onto S' that satisfies, for any $q \in Q$:

$$\tau'(q) = f(\tau(q)) = \{f(s) \mid s \in \tau(q)\}.$$

The function f is called an *isomorphism* between (Q, S, τ) and (Q, S', τ').

Definition 4.5 defines 'isomorphism for skill maps' with the same set of items. A more general situation is considered in Problem 2.

4.6. Example. Let $Q = \{a, b, c, d, e\}$ and $S' = \{1, 2, 3, 4\}$. Define the skill map $\tau' : Q \to 2^{S'}$ by

$$\tau'(a) = \{1, 4\}, \qquad \tau'(b) = \{2, 3, 4\}, \qquad \tau'(c) = \{1\}$$
$$\tau'(d) = \{1, 4\}, \qquad \tau'(e) = \{4\}.$$

The skill map (Q, S', τ') is isomorphic to the one given in Example 4.3: an isomorphism $f : S' \to S$ obtains by setting

$$f(1) = t, \qquad f(2) = s, \qquad f(3) = v, \qquad f(4) = u.$$

The following result is clear:

4.7. Theorem. *Two isomorphic skill maps (Q, S, τ) and (Q, S', τ') delineate exactly the same knowledge space on Q.*

Two skill maps may delineate the same knowledge space without being isomorphic. As an illustration, notice that deleting skill v from the set S in Example 4.3 and redefining τ by setting $\tau(b) = \{s, u\}$ yields the same delineated space $\mathcal{K}$. Skill v is thus superfluous for the delineation of $\mathcal{K}$. As recalled in the introduction to this section, it is a standard practice in science to search for economical explanations of the phenomena under study. In our context, this translates into favoring skill maps with small, possibly minimal, sets of skills. Specifically, we shall call a skill map 'minimal' if the deletion of any of its skills modifies the delineated knowledge space. If this knowledge space is finite, a minimal skill map always exists, and has the smallest possible number of skills (this assertion follows from Theorem 4.11). In the infinite case, the situation is a bit more complicated because a minimal skill map does not necessarily exist. However, a skill map delineating the space and having a set of skills with minimum cardinality always exists (because the class of all cardinals is well-ordered, cf. Dugundji, 1966). It must be noted that such a skill map with minimum number of skills is not necessarily unique – even up to isomorphism (see Problem 10).

4.8. Example. Consider the family $\mathcal{O}$ of all open subsets of the set $\mathbf{R}$ of real numbers, and let $\mathcal{I}$ be any family of open intervals of $\mathbf{R}$, such that $\mathcal{I}$ spans $\mathcal{O}$. For $x \in \mathbf{R}$, set $\tau(x) = \mathcal{I}_x = \{I \in \mathcal{I} \mid x \in I\}$. Then the skill map $(\mathbf{R}, \mathcal{I}, \tau)$ delineates the space $(\mathbf{R}, \mathcal{O})$. Indeed, a subset T of $\mathcal{I}$ delineates $\{x \in \mathbf{R} \mid \mathcal{I}_x \cap T \neq \varnothing\} = \cup T$, and moreover an open subset O is delineated by $\{I \in \mathcal{I} \mid I \subseteq O\}$. It is well-known that there are countable families $\mathcal{I}$ satisfying the above conditions. Notice that such countable families will generate skill maps with minimum number of skills, that is, with a set of skills of minimum cardinality. Nevertheless, there is no minimal skill map. This can be proved directly, or inferred from Theorem 4.11.

With respect to uniqueness, minimal skill maps delineating a given knowledge space – if they exist – behave in a better way. In fact, any two of them are isomorphic. This will be shown in Theorem 4.11. This Theorem also provides a characterization of knowledge spaces having a base (in the sense of 1.19) as those that can be delineated by some minimal skill map.

4.9. Definition. The skill map (Q', S', τ') prolongs (resp. *strictly prolongs*) the skill map (Q, S, τ) if the following conditions hold:

(1) $Q' = Q$;
(2) $S' \supseteq S$ (resp. $S' \supset S$);
(3) $\tau(q) = \tau'(q) \cap S$, for all $q \in Q$.

A skill map (Q, S', τ') is *minimal* if there is no skill map delineating the same space while being strictly prolonged by (Q, S', τ').

4.10. Example. Deleting skill v in the skill map of Example 4.3, we now set $Q = \{a, b, c, d, e\}$, $S = \{s, t, u\}$, and

$$\tau(a) = \{t, u\}, \qquad \tau(b) = \{s, u\}, \qquad \tau(c) = \{t\},$$
$$\tau(d) = \{t, u\}, \qquad \tau(e) = \{u\}.$$

It can be checked that (Q, S, τ) is a minimal skill map.

4.11. Theorem. *A knowledge space is delineated by some minimal skill map iff it admits a base. In such a case, the cardinality of the base is* *equal to that of the set of skills. Moreover, any two minimal skill maps delineating the same knowledge space are isomorphic. Also, any skill map (Q, S, τ) delineating a space $(Q, \mathcal{K})$ having a base prolongs a minimal skill map delineating the same space.*

PROOF. Consider any (not necessarily minimal) skill map (Q, S, τ), and denote by $(Q, \mathcal{K})$ the delineated knowledge space. For any $s \in S$, we write $K(s)$ for the state from $\mathcal{K}$ delineated by $\{s\}$. We have thus

$$q \in K(s) \Longleftrightarrow s \in \tau(q). \tag{1}$$

Take any state $K \in \mathcal{K}$, and consider the subset T of skills that delineates it. For any item q, we have

$$
\begin{aligned}
q \in K \quad &\Longleftrightarrow \quad \tau(q) \cap T \neq \varnothing \\
&\Longleftrightarrow \quad \exists s \in T : s \in \tau(q) \\
&\Longleftrightarrow \quad \exists s \in T : q \in K(s) \qquad \text{[by (1)]} \\
&\Longleftrightarrow \quad q \in \bigcup_{s \in T} K(s)
\end{aligned}
$$

yielding $K = \bigcup_{s \in T} K(s)$. Consequently, $\mathcal{A} = \{K(s) \mid s \in S\}$ spans $\mathcal{K}$. If we now assume that the skill map (Q, S, τ) is minimal, then the spanning family $\mathcal{A}$ must be a base. Indeed, if $\mathcal{A}$ is not a base, then some $K(s) \in \mathcal{A}$ can be expressed as a union of other members of $\mathcal{A}$. Deleting s from S would result in a skill map strictly prolonged by (Q, S, τ) and still delineating $(Q, \mathcal{K})$, contradicting the supposition that (Q, S, τ) is minimal. We conclude that any knowledge space delineated by a minimal skill map has a base. Moreover the cardinality of the base is equal to that of the set of skills. (When (Q, S, τ) is minimal, we have $|\mathcal{A}| = |S|$.)

Suppose now that the space $(Q, \mathcal{K})$ has a base $\mathcal{B}$. From Theorem 4.4, we know that $(Q, \mathcal{K})$ has at least one skill map, say (Q, S, τ). By Theorem 1.20, the base $\mathcal{B}$ of $(Q, \mathcal{K})$ must be included in any spanning subset of $\mathcal{K}$. We have thus, in particular, $\mathcal{B} \subseteq \mathcal{A} = \{K(s) \mid s \in S\}$, where again $K(s)$ is delineated by $\{s\}$. Defining $S' = \{s \in S \mid \exists B \in \mathcal{B} : K(s) = B\}$ and $\tau'(q) = \tau(q) \cap S'$, it is clear that (Q, S', τ') is a minimal skill map.

Notice that a minimal skill map (Q, S, τ) for a knowledge space with base $\mathcal{B}$ is isomorphic to the minimal skill map $(Q, \mathcal{B}, \psi)$ with $\psi(q) = \mathcal{B}_q$. The isomorphism is $s \mapsto K(s) \in \mathcal{B}$, with as above, $K(s)$ delineated by $\{s\}$. Two minimal skill maps are thus always isomorphic to each other.

Finally, let (Q, S, τ) be any skill map delineating a knowledge space $\mathcal{K}$ having a base $\mathcal{B}$. Defining $K(s)$, S' and τ' as before, we obtain a minimal skill map prolonged by (Q, S, τ).   □

Skill Maps: The Conjunctive Model

In the conjunctive model, the knowledge structures that are delineated by skill maps are the simple closure spaces in the sense of Definition 1.11 (see Theorem 4.14 below). Since these structures are dual to the knowledge spaces delineated via the disjunctive model, we will not go into much detail.

4.12. Definition. Let (Q, S, τ) be a skill map and let T be a subset of S. The knowledge state K delineated by T via *the conjunctive model* is specified by
$$K = \{q \in Q \mid \tau(q) \subseteq T\}.$$

The resulting family of all knowledge states is the knowledge structure *delineated via the conjunctive model* by the skill map (Q, S, τ).

4.13. Example. As in 4.3, let $Q = \{a, b, c, d, e\}$, and $S = \{s, t, u, v\}$, with $\tau : Q \to S$ defined by

$$\tau(a) = \{t, u\}, \qquad \tau(b) = \{s, u, v\}, \qquad \tau(c) = \{t\},$$
$$\tau(d) = \{t, u\}, \qquad \tau(e) = \{u\}.$$

Then $T = \{t, u, v\}$ delineates via the conjunctive model the knowledge state $\{a, c, d, e\}$. On the other hand, $\{a, b, c\}$ is not a knowledge state. Indeed if $\{a, b, c\}$ were a state delineated by some subset T of S, then T would include $\tau(a) = \{t, u\}$ and $\tau(b) = \{s, u, v\}$; thus, d and e would also belong to the delineated knowledge state. The knowledge structure delineated by the given skill map is

$$\mathcal{L} = \{\varnothing, \{c\}, \{e\}, \{b, e\}, \{a, c, d, e\}, Q\}.$$

Notice that $\mathcal{L}$ is a simple closure space. The dual knowledge structure $\overline{\mathcal{L}}$ coincides with the knowledge space $\mathcal{K}$ delineated by the same skill map via the disjunctive model; this space $\mathcal{K}$ was obtained in Example 4.3.

4.14. Theorem. *The knowledge structures delineated via the disjunctive and conjunctive model by the same skill map are dual one to the other. As a consequence, the knowledge structures delineated via the conjunctive model are exactly the simple closure spaces.*

The verification of these simple facts is left to the reader.

4.15. Remark. In the finite case, Theorems 4.4 and 4.14 are mere rephrasing of a known result on 'Galois lattices' of relations; for 'Galois lattice', see Chapter 6, especially Definition 6.16. We can reformulate a skill map (Q, S, τ), with Q and S finite, as a relation R between the sets Q and S: for $q \in Q$ and $s \in S$, we define

$$q \, R \, s \quad \Longleftrightarrow \quad s \notin \tau(q).$$

Then the knowledge state delineated via the conjunctive model by a subset T of S is the set $K = \{q \in Q \,|\, \forall s \in S \setminus T : q \, R \, s\}$. These sets K can be regarded as the elements of the 'Galois lattice' of the relation R. It is also well-known that any finite family of finite sets that is closed under intersection can be obtained as the elements of the 'Galois lattice' of some relation. Theorems 4.4 and 4.14 restate this result and extend it to infinite sets[3]. There is of course a direct analogue of Theorem 4.11 for families of sets closed under intersection.

Skill Multimaps: The Competency Model

The last two sections dealt with the delineation of knowledge structures closed under union or under intersection. We still need to discuss the general case. The delineation of arbitrary knowledge structures will be achieved by generalizing the concept of skill map. The intuition behind this generalization is natural enough. To each item q, we associate a collection $\mu(q)$ of subsets of skills. Any subset C of skills in $\mu(q)$ can be viewed as a method—called 'competency' in the next definition—for solving question q. Thus, possessing just one of these competencies is sufficient to solve question q.

4.16. Definition. By a *skill multimap*, we mean a triple $(Q, S; \mu)$, where Q is a nonempty set of *items*, S is a nonempty set of *skills*, and μ is a mapping

---

[3] This extension is straightforward.

that associates to any item q a nonempty family $\mu(q)$ of nonempty subsets of S. Thus, the mapping μ is from the set Q to the set $(2^{2^S \setminus \{\emptyset\}}) \setminus \{\emptyset\}$. We call *competency* for the item q any set belonging to $\mu(q)$.

A subset K of Q is said to be *delineated* by some subset T of skills if K contains all the items having at least one competency included in T; formally

$$q \in K \iff \exists C \in \mu(q) : C \subseteq T.$$

Taking $T = \emptyset$ and $T = S$, we see that $\emptyset$ is delineated by the empty set of skills, and that Q is delineated by S. The set $\mathcal{K}$ of all delineated subsets of Q forms thus a knowledge structure. We say then that the knowledge structure $(Q, \mathcal{K})$ is *delineated* by the skill multimap $(Q, S; \mu)$. This model is referred to as the *competency model*.

4.17. Example. With $Q = \{a, b, c, d\}$ and $S = \{s, t, u\}$, define the mapping $\mu : Q \to 2^S$ by listing the competencies for each item in Q:

$$\mu(a) = \{ \{s, t\}, \{s, u\} \}, \qquad \mu(b) = \{ \{u\}, \{s, u\} \},$$
$$\mu(c) = \{ \{s\}, \{t\}, \{s, u\} \}, \quad \mu(d) = \{ \{t\} \}.$$

Applying Definition 4.16, we see that this skill multimap delineates the knowledge structure:

$$\mathcal{K} = \{ \emptyset, \{b\}, \{c\}, \{c, d\}, \{a, b, c\}, \{a, c, d\}, \{b, c, d\}, Q \}.$$

Notice that $\mathcal{K}$ is closed neither under union nor under intersection.

4.18. Theorem. *Each knowledge structure is delineated by at least one skill multimap.*

PROOF. Let $(Q, \mathcal{K})$ be a knowledge structure. A skill multimap is defined by setting $S = \mathcal{K}$, and for $q \in Q$,

$$\mu(q) = \{ \mathcal{K} \setminus \{M\} \mid M \in \mathcal{K}_q \}.$$

Thus, for each knowledge state M containing question q, we create the competency $\mathcal{K} \setminus \{M\}$ for q. Notice that $\mathcal{K} \setminus \{M\}$ is nonempty, because it has the empty subset of Q as a member. To show that $(Q, S; \mu)$ delineates

$\mathcal{K}$, we apply Definition 4.16. For any $K \in \mathcal{K}$, we consider the subset $\mathcal{K} \backslash \{K\}$ of $\mathcal{K}$ and compute the state L that it delineates:

$$
\begin{aligned}
L &= \{q \in Q \mid \exists M \in \mathcal{K}_q : \mathcal{K} \setminus \{M\} \subseteq \mathcal{K} \setminus \{K\}\} \\
  &= \{q \in Q \mid \exists M \in \mathcal{K}_q : M = K\} \\
  &= \{q \in Q \mid K \in \mathcal{K}_q\} \\
  &= K.
\end{aligned}
$$

Thus, each state in $\mathcal{K}$ is delineated by some subset of S. Conversely, if $T \subseteq S = \mathcal{K}$, the state L delineated by T is defined by

$$
\begin{aligned}
L &= \{q \in Q \mid \exists M \in \mathcal{K}_q : \mathcal{K} \setminus \{M\} \subseteq T\} \\
  &= \begin{cases}
     Q, & \text{when } T = \mathcal{K}, \\
     K, & \text{when } T = \mathcal{K} \setminus \{K\} \text{ for some } K \in \mathcal{K}, \\
     \varnothing, & \text{when } |\mathcal{K} \setminus T| \geq 2,
     \end{cases}
\end{aligned}
$$

and we see that L belongs to $\mathcal{K}$. Thus $\mathcal{K}$ is indeed delineated by the skill multimap $(Q, S; \mu)$. $\square$

We shall not pursue any further the study of the skill multimaps $(Q, S; \mu)$. As in the case of the simple skill map, we could investigate the existence and uniqueness of a minimum skill multimap for a given knowledge structure. Other variants of delineation are conceivable. For example, we could define a knowledge state as a subset K of Q consisting of all items q whose competencies all meet a particular subset of S (depending on K). These developments will be left to the interested reader.

Labels and Filters

On any question in a genuine domain of knowledge (such as arithmetic or grammar) there typically is a wealth of information which could have a bearing on the relevant skills and on the associated knowledge structure. This background information could also be used to paraphrase the knowledge state of a student in a description intended for a parent or for a teacher. Indeed, the complete list of items contained in the student knowledge state may have hundreds of items and may be hard to assimilate, even for an expert. An intelligent summary must be provided, which could rely on the information available on the items forming the knowledge state of the student. Note that such a summary might cover much more than the skills possessed (or lacked) by a student, and may include such features as a prediction of success in a future test, a recommendation of a course of study, or an assignment of some remedial work.

This section outlines a program of description (labeling) of the items and integration (filtering) of the corresponding background information contained in the knowledge states. We begin with some examples taken from the Aleks system outlined in Chapter 0.

4.19. Examples of labels. Suppose that a large pool of questions has been selected, covering all the main concepts of the high school mathematics curriculum in some country. Detailed information concerning each of these items can be gathered under 'labels' such as

(1) Descriptive name of the item.
(2) Grade where the item is to be mastered.
(3) Topic (section of a standard book) to which the item belongs.
(4) Chapter (of a standard book) where the item is presented.
(5) Division of the curriculum to which the item belongs.
(6) Concepts and skills involved in the mastery of the item.
(7) Type of the item (word problem, computation, reasoning, etc.).
(8) Type of response required (word, sentence, formula, number).

Needless to say, the above list is only meant as an illustration. The actual list could be much longer, and would evolve from an extensive collaboration with experts in the field (in this case, experienced teachers). Two examples of items with their associated labels are given in Table 4.1.

Each of the items in the pool would be labeled in the same manner. The task is to develop a collection of computer routines permitting the analysis of knowledge states in terms of the labels. In other words, suppose that a particular knowledge state K has been diagnosed by some assessment routine like those described in Chapters 10 and 11. The labels associated with the items specifying that knowledge state will be passed through a collection of 'filters', resulting in a number of statements expressed in everyday language in terms of educational concepts. Some blueprint examples of filters are given below.

Specification list	Item

(1) Measure of missing angle in a triangle
 (2) 7
 (3) Sum of angles in a plane triangle
 (4) Geometry of the triangle
 (5) Elementary euclidean geometry
 (6) Measure of an angle,
    sum of the angles in a triangle,
    addition, subtraction, deduction
 (7) Computation
 (8) Numerical
	In a triangle ABC, the measure of angle A is $X°$ and the measure of angle B is $Y°$. What is the number of degrees in the measure of angle C?

(1) Addition and subtraction of
    2-place decimal numbers with carry
 (2) 5
 (3) Addition and subtraction of
    decimal numbers
 (4) Decimal numbers
 (5) Arithmetic
 (6) Addition, subtraction, decimals,
    carry, currency
 (7) Word problem and computation
 (8) Numerical
	Mary bought two books costing $\$X$ and $\$Y$. She gave the clerk $\$Z$. What amount of change should she receive?

Table 4.1. Two examples of items and their associated lists of labels.

4.20. The grade reflected by the assessment. Suppose that, at the beginning of a school year, a teacher wishes to know which grade (in mathematics, say) is best suited for a student newly arrived from a foreign country. A knowledge assessment routine has been used, which has determined that the state of a student is K. A suitable collection of filters could be designed along the following lines. As before, we write Q for the domain. For each grade n, $1 \leq n \leq 12$, a filter computes the subset G_n of Q containing all the items to be mastered at that grade or earlier (label (2) in the list above). If the educational system is sensible, we should have

$$G_1 \subset G_2 \subset \cdots \subset G_{12}.$$

We may find

$$G_{n-1} \subset K \subset G_n \tag{2}$$

for some n, in which case the student could be assigned to grade $n - 1$. However, this would not be the best solution when $G_n \setminus K$ is very small. We need more information. Moreover, we must provide for situations in which (2) does not hold for any n. Next, the filter calculates the standard distance $|K \bigtriangleup G_n|$ for all grades n, and retains the set

$$S(K) = \{n_j \mid |K \bigtriangleup G_{n_j}| \leq |K \bigtriangleup G_n|, 1 \leq n \leq 12\}. \qquad (3)$$

Thus, $S(K)$ contains all the grades which minimize the distance to K. Suppose that $S(K)$ contains a single element n_j, and that we also have $G_{n_j} \subset K$. It would seem reasonable then to place the student in grade $n_j + 1$. But $S(K)$ may very well contain more than one member. We still need more information. In particular, the content of K, with its strengths and weaknesses relative to its closest sets G_{n_j} must be summarized in some useful way. Without going into the technical details of such summary, we outline an example of a report that the system might produce at this juncture:

> *The closest match for Student X is grade 5. However, X would be an unusual student in that grade. Her knowledge of elementary geometry far exceeds that of a representative student in grade 5. For example, X is aware of the Pythagorean Theorem and capable of using it in applications. On the other hand, X has surprising weaknesses in arithmetic. For example, etc...*

Descriptions of this type would require the development of a varied collection of new filters, beyond those involved in the computation of $S(K)$ in Equation (3). Moreover, the system must have the capability of transforming, via a natural language generator, the output of such filters into grammatically correct statements in everyday language. We shall not pursue this discussion here. The point of this paragraph was to illustrate how the labeling of the items, vastly extending the concept of skills, could lead to refined descriptions of the knowledge states that could be useful for diverse purposes.

Original Sources and Related Works

Skill maps were not introduced from the start in the theory of knowledge structures. As indicated in the introductory paragraph of this chapter, we originally eschewed cognitive interpretations of our concepts since we believed that the overall machinery of knowledge spaces had potential use in a variety of empirical contexts quite different from mental testing. Nevertheless, traditional interpretations of tests results could not be permanently ignored. In fact, inquiries were often raised about the possibility of 'explaining' these states from a small number of 'basic' aptitudes (see e.g. Albert, Schrepp and Held, 1992, or Lukas and Albert, 1993). A first pass at establishing such a linkage was made in Falmagne, Koppen, Villano, Doignon and Johannesen (1990). Many of the details about skill maps were provided by Doignon (1994b). Further recent results on skill maps can also be found in Düntsch and Gediga (1995). To this date, nothing is published concerning the role of labels and filters in knowledge space theory.

Problems

1. For which type of relation $\mathcal{Q}$ on a set Q is it true that there exist some set S and some mapping $\tau : Q \to S$ such that $q\mathcal{Q}r \iff \tau(q) \subseteq \tau(r)$?

2. Definition 4.5 of isomorphism between skill maps was formulated for two maps defined on the same set of items. Drop this assumption and propose a more general type of isomorphic skill assignment. Show then that the knowledge spaces delineated according to the disjunctive model by two isomorphic skill assignments (in this new sense) are isomorphic.

3. Following up on Example 4.8, prove that no minimal skill map exists without making reference to Theorem 4.11.

4. Verify that the skill map of Example 4.10 is minimal.

5. Give a proof of Theorem 4.14.

6. Under which condition on a skill multimap (Definition 4.16) is the delineated structure a knowledge space? Construct an example.

7. Solve a similar problem for the case of a knowledge structure closed under intersection.

8. Design an appropriate set of filters capable of listing all the items that a student in state K would not know, but would be ready to master.

9. Find a necessary and sufficient condition on a disjunctive model ensuring that the delineated knowledge space is discriminative.

10. Prove the assertion in the last sentence before Example 4.8 (hint: use in Example 4.8 two countable families $\mathcal{I}$ having different properties).

Chapter 5

Entailments and the Maximal Mesh

In practice, how can we build a knowledge structure for a specific body of information? The first step is to select the items forming a domain Q. For real-life applications, we will typically assume this domain to be finite. The second step is then to construct a list of all the subsets of Q that are knowledge states. To secure such a list, we could in principle rely on one or more experts in the particular body of information. However, if no assumption is made on the structure to be uncovered, the only exact method consists in the presentation of all subsets of Q to the expert, so that he can point out the states. As the number of subsets of Q grows exponentially with the size $|Q|$ of Q, this method becomes impractical even for relatively small sets Q (e.g., for only 20 items we would have to present $2^{20} - 2 = 1,048,574$ subsets to the expert).

Three complementary solutions were investigated for building knowledge structures in practical situations. The first one relies on supposing that the knowledge structure under consideration satisfies some conditions, the closure under union or intersection being prime examples. Such assumptions may result in a considerable reduction of the number of questions to be asked from an expert. An empirical example is discussed in Chapter 12 where it is shown that, at least for some empirical domains, a practical technique is feasible with 50 items[1]. The first part of this chapter is devoted to some theoretical results obtained in that line.

The second solution is also described in this chapter. The idea is to build a large knowledge structure by combining a number of small ones. Suppose that we have obtained—using experts and the method of the first solution, for example—all the structures on subdomains of at most seven items, say. These structures on subdomains can be regarded as substructures (in the sense of Definition 1.17) of some unknown structure on the full domain. They can then be combined into a global one on the whole domain Q.

---

[1] An even more convincing case is provided by the ALEKS system (see Chapter 0) which uses, in particular, a knowledge structure for 100 items in the arithmetic curriculum. This structure has been built in part by a technique elaborating on the methods described in this Chapter. (See also Chapter 12.)

Here, combining means selecting in a sensible way, if possible, one structure on Q that is the parent structure of all the substructures on the subsets of seven items. Theoretical results will isolate the situations allowing for such a construction. Moreover, we study how properties of structures are preserved by the construction.

The third solution is based on collecting the responses of a large number of subjects to the items of the domain. By an appropriate statistical analysis of such data, the knowledge states can in principle be uncovered. This has been demonstrated by Villano (1991) for small domains, using real data. The technique is also applicable to large domains provided that the statistical analysis of the data has been preceded by a 'pruning down' of the collection of potential states by the methods of the first and/or second solutions (see Cosyn and Thiéry, 1996). This particular technique relies on a heavy use of some stochastic procedure for knowledge assessment (such as those presented in Chapters 10 an 11) and is described in Chapter 12.

Entailments

We begin by examining the case of a quasi ordinal knowledge space $(Q, \mathcal{K})$. From Birkhoff Theorem 1.49, we know that this space is completely specified by its derived quasi order $\mathcal{Q}$, with

$$p\mathcal{Q}q \quad \Longleftrightarrow \quad (\forall K \in \mathcal{K} : q \in K \Rightarrow p \in K),$$

where $p, q \in Q$ (cf. Definition 1.50). As a practical application, we may uncover a quasi ordinal space on a given domain by asking an (ideal) expert all queries of the form

[Q0] *Suppose a student has failed to solve item p. Do you believe this student will also fail item q? Answer this query under the assumption that chance factors, such as lucky guesses and careless errors, play no role in the student's performance.*

Assuming that the expert's answers are consistent with the unknown, quasi ordinal space $(Q, \mathcal{K})$, we form a relation $\mathcal{Q}$ on Q by collecting all pairs (p, q) for which the expert gave a positive answer to query [Q0]. The family $\mathcal{K}$ is then obtained by applying Theorem 1.49, since

$$K \in \mathcal{K} \quad \Longleftrightarrow \quad (\forall (p, q) \in \mathcal{Q} : p \notin K \Rightarrow q \notin K).$$

If we drop the assumption that the unknown knowledge space $(Q, \mathcal{K})$ is quasi ordinal, the answers to all queries of the form [Q0] do not suffice to construct the space. As announced in 0.6, we consider in that case the more general type of query:

[Q1] *Suppose that a student has failed to solve items p_1, p_2, ..., p_n. Do you believe this student would also fail to solve item q? You may assume that chance factors, such as lucky guesses and careless errors, do not interfere in the student's performance.*

Such a query is summarized by the nonempty set $\{p_1, p_2, \ldots, p_n\}$ of items, plus the single item q. Thus, all positive answers to the queries form a relation $\mathcal{P}$ from 2^Q to Q. The expert is consistent with the (unknown) knowledge space $(Q, \mathcal{K})$ exactly when the following implication is satisfied, for $A \in 2^Q \setminus \{\varnothing\}$ and $q \in Q$:

$$A \mathcal{P} q \iff (\forall K \in \mathcal{K} : A \cap K = \varnothing \Rightarrow q \notin K). \tag{1}$$

5.1. Example. For the knowledge space $(Q, \mathcal{K})$ defined by $Q = \{a, b, c\}$ and $\mathcal{K} = \{\varnothing, \{a, b\}, \{a, c\}, Q\}$, the queries (A, q) with $q \notin A$ which call for a positive answer are listed below:

$$(\{a\}, b), \quad (\{a\}, c), \quad (\{a, b\}, c), \quad (\{a, c\}, b), \quad (\{b, c\}, a).$$

5.2. Example. Let $k, m \in \mathbb{N}$ with $0 < k < n$, and consider the space $(Q, \mathcal{K})$, where Q has n elements, and $\mathcal{K}$ is the family of all subsets of Q having either 0 or at least k elements. We have for the corresponding relation $\mathcal{P}$: for all $A \in 2^Q \setminus \{\varnothing\}$ and $p \in Q$,

$$A \mathcal{P} q \iff (q \in A \text{ or } |A| > n - k).$$

We return to the general situation. To design an efficient procedure for questioning the expert, we need to examine the relations $\mathcal{P}$ obtained through Equation (1) from all knowledge spaces $\mathcal{K}$ on Q.

5.3. Theorem. *Let $(Q, \mathcal{K})$ be a knowledge structure, and suppose that $\mathcal{P}$ is the relation from 2^Q to Q defined by Equation (1). Then, necessarily:*

  (i) *$\mathcal{P}$ extends the reverse membership relation, that is: if $p \in A \subseteq Q$, then $A \mathcal{P} p$;*

  (ii) *if A, $B \in 2^Q \setminus \{\varnothing\}$ and $p \in Q$, then $A \mathcal{P} b$ for all $b \in B$ and $B \mathcal{P} p$ imply $A \mathcal{P} p$.*

PROOF. The first condition is immediate. Suppose that A, B and p are as in Condition (ii) with $A\mathcal{P}b$ for all $b \in B$ and $B\mathcal{P}p$. We have to show that for all $K \in \mathcal{K}$, $A \cap K = \varnothing$ implies $p \notin K$. Take any $K \in \mathcal{K}$ with $A \cap K = \varnothing$. Thus, by Equation (1), we have $b \notin K$, for all $b \in B$. This means that $B \cap K = \varnothing$. Using (1) again and the fact that $B\mathcal{P}p$, we get $p \notin K$, which yields $A\mathcal{P}p$.   □

The next Theorem shows that all relations from $2^Q \setminus \{\varnothing\}$ to Q satisfying Conditions (i) and (ii) are obtained as in Theorem 5.3 from some knowledge space. Since these relations will play a fundamental rôle in the sequel, we give them a name.

5.4. Definition. An *entailment* for the nonempty domain Q (which may be infinite) is a relation $\mathcal{P}$ from $2^Q \setminus \{\varnothing\}$ to Q that satisfies Conditions (i) and (ii) in Theorem 5.3.

5.5. Theorem. *There is a one-to-one correspondence between the family of all knowledge spaces $\mathcal{K}$ on the same domain Q, and the family of all entailments $\mathcal{P}$ for Q. It is expressed by the two equivalences*

$$A\mathcal{P}q \iff (\forall K \in \mathcal{K} : A \cap K = \varnothing \Rightarrow q \notin K), \qquad (2)$$
$$K \in \mathcal{K} \iff (\forall (A,p) \in \mathcal{P} : A \cap K = \varnothing \Rightarrow p \notin K). \qquad (3)$$

PROOF. To each knowledge space $(Q, \mathcal{K})$, we associate through Equation (2) the relation $f(\mathcal{K}) = \mathcal{P}$. By Theorem 5.3 and Definition 5.4, $\mathcal{P}$ is an entailment. Conversely, let $\mathcal{P}$ be any entailment for Q. We define then a family $g(\mathcal{P}) = \mathcal{K}$ of subsets of Q by Equation (3). We show that the family $\mathcal{K}$ is a space on Q. It is clear that $\varnothing, Q \in \mathcal{K}$. Suppose that $K_i \in \mathcal{K}$ for i in some index set I. We have to show that $\cup_{i \in I} K_i \in \mathcal{K}$. Assume $A\mathcal{P}p$ and $A \cap (\cup_{i \in I} K_i) = \varnothing$. Then $A \cap K_i = \varnothing$ for all $i \in I$, thus $p \notin K_i$. It follows that $p \notin \cup_{i \in I} K_i$. Applying the equivalence (2), we obtain $\cup_{i \in I} K_i \in \mathcal{K}$.

We now show that f and g are inverse maps of one another. We proceed in two steps.

(1) We prove that $(g \circ f)(\mathcal{K}) = \mathcal{K}$. Let $\mathcal{K}$ be a space on Q and let $\mathcal{P} = f(\mathcal{K})$. Setting $\mathcal{L} = g(\mathcal{P})$, we show $\mathcal{L} = \mathcal{K}$. By definition:

$$L \in \mathcal{L} \iff (\forall A \in 2^Q \setminus \{\varnothing\}, p \in Q : (A\mathcal{P}p \text{ and } A \cap L = \varnothing) \Rightarrow p \notin L).$$

Writing $A\mathcal{P}b$ in the right member explicitly in terms of $\mathcal{K}$ and omitting the quantifiers for A and p, we obtain

$$
\begin{aligned}
L \in \mathcal{L} \quad &\Longleftrightarrow \quad (((\forall K \in \mathcal{K} : A \cap K = \varnothing \Rightarrow p \notin K) \\
&\qquad\qquad \text{and } A \cap L = \varnothing) \Rightarrow p \notin L) \\
&\Longleftarrow \quad L \in \mathcal{K}.
\end{aligned} \tag{5}
$$

To prove the converse of the last implication, assume $L \in \mathcal{L}$ together with $L \notin \mathcal{K}$. Denote by L° the largest state contained in L. (Because $\mathcal{K}$ is a space, L° is well defined: it is equal to the union of all the states contained in L.) Since $L \notin \mathcal{K}$, there must exist some item p with $p \in L \setminus L^\circ$. Setting $A = Q \setminus L$, we have for any $K \in \mathcal{K}$:

$$
\begin{aligned}
A \cap K = \varnothing \quad &\Longrightarrow \quad K \subseteq L, \\
&\Longrightarrow \quad K \subseteq L^\circ, \\
&\Longrightarrow \quad p \notin K.
\end{aligned}
$$

As also $A \cap L = \varnothing$, the right member of (5) gives $p \notin L$, a contradiction. This completes the proof that $\mathcal{K} = \mathcal{L}$. We conclude that $(g \circ f)(\mathcal{K}) = \mathcal{K}$ for each space $\mathcal{K}$ on Q.

(2) We prove that $(f \circ g)(\mathcal{P}) = \mathcal{P}$. Take any entailment $\mathcal{P}$ for Q. With $\mathcal{K} = g(\mathcal{P})$ and $\mathcal{Q} = f(\mathcal{K})$, we show that $\mathcal{Q} = \mathcal{P}$. For $A \in 2^Q \setminus \{\varnothing\}$ and $p \in Q$, it is easily checked that

$$
\begin{aligned}
A \mathcal{Q} p \quad &\Longleftrightarrow \quad (\forall K \in \mathcal{K} : A \cap K = \varnothing \Rightarrow p \notin K) \\
&\Longleftrightarrow \quad \Big(\forall K \in 2^Q : ((\forall B \in 2^Q \setminus \{\varnothing\}, \forall q \in Q : (B\mathcal{P}q \text{ and } \\
&\qquad\qquad B \cap K = \varnothing) \Rightarrow q \notin K) \text{ and } A \cap K = \varnothing) \Rightarrow p \notin K \Big).
\end{aligned}
$$

Let us write $\mathbf{X}$ for the right member in the last equivalence. It is clear that $A\mathcal{P}p \Rightarrow \mathbf{X}$. To prove that we also have $\mathbf{X} \Rightarrow A\mathcal{P}p$, we proceed by contradiction. Suppose that $\mathbf{X}$ holds and that $A\mathcal{P}p$ does not hold. Set $K = \{q \in Q \,|\, \text{not } A\mathcal{P}q\}$. For any $B \in 2^Q \setminus \{\varnothing\}$ and $q \in Q$, we see that $B\mathcal{P}q$ together with $B \cap K = \varnothing$ implies $q \notin K$. (Indeed, $B \cap K = \varnothing$ implies $A\mathcal{P}b$, $\forall b \in B$; as also $B\mathcal{P}q$, Condition (ii) of Definition 5.4 implies $A\mathcal{P}q$, thus $q \notin K$.) Moreover, by Condition (i) of Definition 5.4, we have $A \cap K = \varnothing$. As $p \in K$ by the definition of K, we have reached a contradiction with $\mathbf{X}$. We have thus proved $\mathcal{Q} = \mathcal{P}$, that is $(f \circ g)(\mathcal{P}) = \mathcal{P}$. $\qquad\square$

5.6. Definition. When an entailment $\mathcal{P}$ for Q and a knowledge space $\mathcal{K}$ on Q correspond to each other as in Equations (2) and (3) in Theorem 5.5, we say that they are *derived* from one another.

The correspondence obtained in Theorem 5.5 can be reformulated in an intuitive way. Starting from the space $(Q, \mathcal{K})$, it can be checked that $A \mathcal{P} q$ holds exactly when q does not belong to the largest state L_A disjoint from A. That is, for $A \in 2^Q \setminus \{\varnothing\}$ and $q \in Q$, Equation (2) is equivalent to

$$A \mathcal{P} q \quad \Longleftrightarrow \quad q \notin L_A \tag{6}$$

(the proof of the equivalence is left as Problem 1). In terms of the closure space dual to $\mathcal{K}$ (cf. Definition 1.11), $A \mathcal{P} p$ holds exactly when p belongs to the closure of A. On the other hand, for $K \in 2^Q$, Equation (3) is equivalent to

$$K \in \mathcal{K} \quad \Longleftrightarrow \quad K = \{p \in Q \,|\, \text{not } (Q \setminus K)\mathcal{P}p\}, \tag{7}$$

(see Problem 2). This equivalence is rephrased as follows in terms of closed sets (i.e., complements of states): a subset F of Q is closed iff it contains all items p satisfying $F \mathcal{P} p$.

Entail Relations

Condition (ii) in Theorem 5.3 is the key requirement for entailments. It must be recognized as a disguised form of a transitivity condition for a relation. To see this, we associate to any relation $\mathcal{P}$ from $2^Q \setminus \{\varnothing\}$ to Q a relation $\mathcal{Q}$ on $2^Q \setminus \{\varnothing\}$ by setting

$$A \mathcal{Q} B \quad \Longleftrightarrow \quad (\forall b \in B : A \mathcal{P} b). \tag{8}$$

Condition (ii) in Theorem 5.3 for $\mathcal{P}$ can be restated in terms of $\mathcal{Q}$, yielding

$$(A \mathcal{Q} B \text{ and } B \mathcal{Q} \{p\}) \quad \Longrightarrow \quad A \mathcal{Q} \{p\}, \tag{9}$$

for $A, B \in 2^Q \setminus \{\varnothing\}$ and $p \in Q$. In the last formula, we can replace the singleton set $\{p\}$ with any subset C of Q. Thus, Equation (9) essentially states that $\mathcal{Q}$ is transitive. The following Theorem characterizes such relations $\mathcal{Q}$. (Note that Equation (8) implies $A \mathcal{P} b \Leftrightarrow A \mathcal{Q} \{b\}$.)

5.7. Theorem. *Equation (8) establishes a one-to-one correspondence between the family of all entailments $\mathcal{P}$ for Q and the family of all relations $\mathcal{Q}$ on $2^Q \setminus \{\varnothing\}$ that satisfy*

(i) *$\mathcal{Q}$ extends reverse inclusion, that is: for $A, B \in 2^Q \setminus \{\varnothing\}$, we have $A\mathcal{Q}B$ when $B \subseteq A$;*

(ii) *$\mathcal{Q}$ is a transitive relation;*

(iii) *if $A, B_i \in 2^Q \setminus \{\varnothing\}$ for i in some index set I, then $A\mathcal{P}B_i$ for all $i \in I$ implies $A\mathcal{P}(\cup_{i \in I}B_i)$.*

The proof is left to the reader, to whom we also leave to establish the following additional assertions (see Problems 3 and 4). For a relation $\mathcal{Q}$ on $2^Q \setminus \{\varnothing\}$ which is a transitive extension of reverse inclusion, it can be checked that Condition (iii) in the last Theorem is equivalent to any of the following two conditions:

(iv) for each $A \in 2^Q \setminus \{\varnothing\}$, there is a maximum subset B of Q such that $A\mathcal{Q}B$ (here, maximum means maximum for the inclusion);

(v) for all $A, B \in 2^Q \setminus \{\varnothing\}$, we have $A\mathcal{Q}B$ iff $A\mathcal{Q}\{b\}$ holds for each $b \in B$.

If the domain Q is essentially finite Condition (iii) in Theorem 5.7 is also equivalent to

(vi) $A\mathcal{Q}B$ implies $A\mathcal{Q}(A \cup B)$ for all $A, B \in 2^Q \setminus \{\varnothing\}$.

5.8. Definition. A relation $\mathcal{Q}$ on $2^Q \setminus \{\varnothing\}$ satisfying Conditions (i), (ii) and (iii) in Theorem 5.5 is called an *entail relation* for Q.

Meshability of Knowledge Structures

We now turn to a second approach for building knowledge structures. In many situations, the straightforward procedure of securing an entailment from a particular expert and building the associated space is not practical. For one thing, the knowledge structure does not need to be a space. For another, no matter how competent the expert may be, his reliability in the course of many hours of questioning may not be perfect, and the resulting space may be partly erroneous[2]. Finally, the domain may simply be so

---

[2] Some of the issues related to a possible unreliability of experts have been analyzed in detail by Cosyn and Thiéry (1996). Their results are reviewed in Chapter 12.

large that the number of questions required to obtain an entailment would
be unacceptable. These objections call for other strategies.

One possibility discussed here is based on combining a number of small
structures into a big one. These small structures may for example be obtained from several different experts, each of them being questioned for a
short time on a small subset of the domain; or they may result from a statistical analysis of the responses from a large number of subjects, as in the
work of Villano (1991). The origin of the small structures is not relevant
here. We simply suppose that a number of substructures of some unknown
knowledge structure are available, and we consider ways of assembling these
pieces into a coherent whole. Before getting into the theoretical background
of such a construction, we need a further look at the notion of substructure
or trace already encountered in Chapter 1. No finiteness assumption will
be made, except when otherwise mentioned.

From Definition 1.17 and Theorem 1.16, we recall that the trace of a
knowledge structure $(Q, \mathcal{K})$ on a nonempty subset A of Q is the knowledge
structure $(A, \mathcal{H})$ characterized by:

$$\mathcal{H} = \{H \in 2^A \mid H = A \cap K \text{ for some } K \in \mathcal{K}\}. \tag{10}$$

We also say that the state $A \cap K$ of $\mathcal{H}$ is the trace of the knowledge state
K on the subset A. The knowledge structure $\mathcal{H}$ is called a substructure of
the knowledge structure $(Q, \mathcal{K})$.

A property of knowledge structures is hereditary when its validity for
a knowledge structure $(Q, \mathcal{K})$ implies its validity for all the substructures
of $(Q, \mathcal{K})$ (Definition 1.17). It is not difficult to check that many of the
properties we encountered are hereditary, as for instance the property for a
structure of being a space (cf. Theorem 1.16) or being discriminative, quasi
ordinal, ordinal, well-graded, or 1-connected (cf. Problem 14 in Chapter
1 and Problem 3 in Chapter 2). By contrast, none of these properties
necessarily holds for the whole structure when it is valid for some of its
traces. Positive results for some of these properties obtain when validity is
assumed for all proper traces (cf. Problem 7 in Chapter 1).

We now turn to the combination of two structures on possibly overlapping
sets into a structure on the union of these two sets.

5.9. Definition. The knowledge structure $(X, \mathcal{K})$ is called a *mesh* of the knowledge structures $(Y, \mathcal{F})$ and $(Z, \mathcal{G})$ if

(1) $X = Y \cup Z$;

(2) $\mathcal{F}$ and $\mathcal{G}$ are the traces of $\mathcal{K}$ on X and Y, respectively.

As shown by the following examples, two knowledge structures may have more than one mesh, or no mesh at all. Two knowledge structures having a mesh are *meshable*; if this mesh is unique, they are *uniquely meshable*.

5.10. Example. The two knowledge structures (which are ordinal spaces)

$$\mathcal{F} = \big\{ \varnothing, \{a\}, \{a, b\} \big\}, \qquad \mathcal{G} = \big\{ \varnothing, \{c\}, \{c, d\} \big\}$$

admit the two meshes (which are also ordinal spaces)

$$\mathcal{K}_1 = \big\{ \varnothing, \{a\}, \{a, b\}, \{a, b, c\}, \{a, b, c, d\} \big\},$$
$$\mathcal{K}_2 = \big\{ \varnothing, \{a\}, \{a, b\}, \{a, c\}, \{a, b, c\}, \{a, b, c, d\} \big\}.$$

5.11. Example. Suppose that $(\{a, b, c, d\}, \mathcal{K})$ is a mesh of the two ordinal knowledge spaces

$$\mathcal{F} = \big\{ \varnothing, \{a\}, \{a, b\}, \{a, b, c\} \big\}, \qquad \mathcal{G} = \big\{ \varnothing, \{c\}, \{b, c\}, \{b, c, d\} \big\}.$$

Then $\mathcal{K}$ has a state K such that $K \cap \{b, c, d\} = \{c\} \in \mathcal{G}$. Hence either $K = \{a, c\}$ or $K = \{c\}$, and since $K \subseteq \{a, b, c\}$, either $\{a, c\}$ or $\{c\}$ must be a state of $\mathcal{F}$, a contradiction. Thus, $\mathcal{F}$ and $\mathcal{G}$ have no mesh.

5.12. Example. The two knowledge structures

$$\mathcal{F} = \big\{ \varnothing, \{a\}, \{a, b\} \big\}, \qquad \mathcal{G} = \big\{ \varnothing, \{b\}, \{b, c\} \big\}$$

are uniquely meshable. Indeed, they have the unique mesh

$$\mathcal{K} = \big\{ \varnothing, \{a\}, \{a, b\}, \{a, b, c\} \big\}.$$

(If a state contains c, it has to contain both of a and b; if it contains b, it has to contain a).

Notice that, in this example, the mesh $\mathcal{K}$ does not include the union of the two component knowledge structures $\mathcal{F}$ and $\mathcal{G}$, since $\{b, c\} \notin \mathcal{K}$.

We shall first investigate conditions under which a mesh exists.

5.13. Definition. A knowledge structure $(Y, \mathcal{F})$ is *compatible with a knowl-edge structure* $(Z, \mathcal{G})$ if, for any $F \in \mathcal{F}$, the intersection $F \cap Z$ is the trace on Y of some state of $\mathcal{G}$. When two knowledge structures are compatible with each other, we shall simply say that they are *compatible*.

In other words, two knowledge structures $(Y, \mathcal{F})$ and $(Z, \mathcal{G})$ are compatible if and only if they have the same trace on $Y \cap Z$.

5.14. Theorem. *Two knowledge structures are meshable if and only if they are compatible.*

PROOF. Let $(Y \cup Z, \mathcal{K})$ be a mesh of the two knowledge structures $(Y, \mathcal{F})$ and $(Z, \mathcal{G})$, and suppose that $F \in \mathcal{F}$. By the definition of a mesh, there is $K \in \mathcal{K}$ such that $K \cap Y = F$. Thus, $K \cap Z \in \mathcal{G}$ and $(K \cap Z) \cap Y = F \cap Z$. Hence $(Y, \mathcal{F})$ is compatible with $(Z, \mathcal{G})$. The other case follows by symmetry.

Conversely, suppose that $(Y, \mathcal{F})$ and $(Z, \mathcal{G})$ are compatible. Define

$$\mathcal{K} = \{K \in 2^{Y \cup Z} \mid K \cap Y \in \mathcal{F}, \ K \cap Z \in \mathcal{G}\}. \tag{11}$$

Clearly, $(Y \cup Z, \mathcal{K})$ is a knowledge structure. For any $F \in \mathcal{F}$, we have $F \cap Z = G \cap Y$ for some $G \in \mathcal{G}$. Setting $K = F \cup G$, we obtain $K \cap Y = F$ and $K \cap Z = G$, yielding $K \in \mathcal{K}$. Thus $\mathcal{F}$ is included in the trace of $\mathcal{K}$ on Y. By the definition of $\mathcal{K}$, the reverse inclusion is trivial, so $\mathcal{F}$ is this trace. Again, the other case results from symmetry. We conclude that $\mathcal{K}$ is a mesh of $\mathcal{F}$ and $\mathcal{G}$. □

The construction of the mesh used in the above proof is of interest and deserves a separate investigation.

The Maximal Mesh

5.15. Definition. Let $(Y, \mathcal{F})$ and $(Z, \mathcal{G})$ be two compatible knowledge structures. The knowledge structure $(Y \cup Z, \mathcal{F} \star \mathcal{G})$ defined by the equation

$$\mathcal{F} \star \mathcal{G} = \{K \in 2^{Y \cup Z} \mid K \cap Y \in \mathcal{F}, \ K \cap Z \in \mathcal{G}\}$$

is the *maximal mesh* of $\mathcal{F}$ and $\mathcal{G}$. Indeed, we have $\mathcal{K} \subseteq \mathcal{F} \star \mathcal{G}$ for any mesh $\mathcal{K}$ of $\mathcal{F}$ and $\mathcal{G}$. The operator $\star$ will be referred to as the *maximal meshing operator.* An equivalent definition of the maximal mesh is as follows:

$$\mathcal{F} \star \mathcal{G} = \{F \cup G \mid F \in \mathcal{F}, G \in \mathcal{G} \text{ and } F \cap Z = G \cap Y\}.$$

Obviously, we always have $\mathcal{F} \star \mathcal{G} = \mathcal{G} \star \mathcal{F}$. Notice in passing that, if $F \in \mathcal{F}$ and $F \subseteq Y \setminus Z$, then $F \in \mathcal{F} \star \mathcal{G}$. A corresponding property holds of course for the knowledge structure $\mathcal{G}$.

5.16. Example. The maximal mesh of the two ordinal, knowledge spaces from Example 5.10 is the ordinal space

$$\mathcal{F} \star \mathcal{G} = \big\{ \varnothing, \{a\}, \{c\}, \{a,b\}, \{a,c\}, \{c,d\}, \{a,b,c\}, \{a,c,d\}, \{a,b,c,d\} \big\}.$$

5.17. Theorem. *If $\mathcal{F}$ and $\mathcal{G}$ are compatible knowledge structures, then $\mathcal{F} \star \mathcal{G}$ is a space (respectively discriminative space) if and only if both $\mathcal{F}$ and $\mathcal{G}$ are spaces (respectively discriminative spaces).*

The proof is left to the reader (see Problem 7).

If $\mathcal{F}$ and $\mathcal{G}$ are compatible knowledge structures, and $\mathcal{F} \star \mathcal{G}$ is well-graded, then $\mathcal{F}$ and $\mathcal{G}$ are both well-graded. The maximal mesh of well-graded knowledge structures, or even spaces, is not necessarily well-graded, however. The counterexample below establishes this fact.

5.18. Example. Consider the two well-graded knowledge spaces

$$\mathcal{F} = \big\{ \varnothing, \{a\}, \{b\}, \{a,b\}, \{a,c\}, \{b,c\}, \{a,b,c\} \big\}$$

and

$$\mathcal{G} = \big\{ \varnothing, \{c\}, \{d\}, \{b,c\}, \{b,d\}, \{c,d\}, \{b,c,d\} \big\},$$

which are compatible. Their maximal mesh (necessarily a space)

$$\mathcal{F} \star \mathcal{G} = \big\{ \varnothing, \{a\}, \{d\}, \{a,c\}, \{a,d\}, \{b,c\}, \{b,d\}, \{a,b,c\},$$
$$\{b,c,d\}, \{a,b,d\}, \{a,c,d\}, \{a,b,c,d\} \big\}$$

is not well-graded since it contains $\{b,c\}$, but neither $\{b\}$ nor $\{c\}$.

A sufficient condition for wellgradedness is contained in the next definition.

5.19. Definition. A mesh $\mathcal{K}$ of two knowledge structures $\mathcal{F}$ and $\mathcal{G}$ is called *inclusive* if $F \cup G \in \mathcal{K}$ for any $F \in \mathcal{F}$ and $G \in \mathcal{G}$.

5.20. Theorem. *Consider the following three conditions on two knowledge structures* $(Y, \mathcal{F})$ *and* $(Z, \mathcal{G})$:

  (i) $\mathcal{F}$ *and* $\mathcal{G}$ *admit some inclusive mesh;*
  (ii) $\mathcal{F} \star \mathcal{G}$ *is inclusive;*
  (iii) $(\forall F \in \mathcal{F} : F \cap Z \in \mathcal{G})$ *and* $(\forall G \in \mathcal{G} : G \cap Y \in \mathcal{F})$.

Then (i)$\Leftrightarrow$(ii)$\Rightarrow$(iii). *Moreover, if* $\mathcal{F}$ *and* $\mathcal{G}$ *are spaces, then* (ii)$\Leftrightarrow$(iii).

We leave the proof to the reader (see Problem 8). The following example shows that in general, Condition (iii) does not imply Condition (ii).

5.21. Examples. a) Consider $\mathbf{R}^3$ and the two families $\mathcal{F}$ and $\mathcal{G}$, where $\mathcal{F}$ contains all the convex subsets of the plane $y = 0$, and $\mathcal{G}$ contains all the convex subsets of the plane $z = 0$. Let thus Y and Z denote the planes $y = 0$ and $z = 0$, respectively. Obviously, we do not in general have $F \cup G$ in $\mathcal{F} \star \mathcal{G}$ for any F in $\mathcal{F}$ and G in $\mathcal{G}$.

b) An example with a finite domain is easily constructed. Still in $\mathbf{R}^3$, take $Y = \{(0,0,1), (0,0,0), (1,0,0), (2,0,0)\}$ and $Z = \{(0,1,0), (0,0,0), (1,0,0), (2,0,0)\}$, with the states being the traces of the convex sets on Y and Z respectively. The maximal mesh $\mathcal{F} \star \mathcal{G}$ is not inclusive since $\{(0,0,1),(2,0,0)\} \in \mathcal{F}$ and $\{(0,0,0)\} \in \mathcal{G}$ but the union of these two states is not in $\mathcal{F} \star \mathcal{G}$.

5.22. Theorem. *If the maximal mesh* $\mathcal{F} \star \mathcal{G}$ *of two knowledge structures* $\mathcal{F}$ *and* $\mathcal{G}$ *is inclusive, then* $\mathcal{F} \cup \mathcal{G} \subseteq \mathcal{F} \star \mathcal{G}$. *When* $\mathcal{F}$ *and* $\mathcal{G}$ *are spaces,* $\mathcal{F} \cup \mathcal{G} \subseteq \mathcal{F} \star \mathcal{G}$ *implies that* $\mathcal{F} \star \mathcal{G}$ *is inclusive.*

We omit the proof (see Problem 9). The examples in 5.21 shows that we cannot replace "spaces" by "structures" in Theorem 5.22.

5.23. Theorem. *If the maximal mesh of two finite, compatible, well-graded knowledge structures is inclusive, then it is necessarily well-graded.*

Example 5.12 shows that the inclusiveness condition is not necessary for two well-graded knowledge structures (or even spaces) to have a maximal mesh that is also well-graded.

PROOF. Let $(Y, \mathcal{F})$ and $(Z, \mathcal{G})$ be two well-graded knowledge structures, and suppose that $\mathcal{F} \star \mathcal{G}$ is inclusive. To prove that $\mathcal{F} \star \mathcal{G}$ is well-graded, we use Theorem 2.9(iii). Take any $K, K' \in \mathcal{F} \star \mathcal{G}$. As $K \cap Y$ and $K' \cap Y$ are two states of the well-graded knowledge structure $\mathcal{F}$, there exist a positive integer m and some sequence of states in $\mathcal{F}$

$$K \cap Y = Y_0, \quad Y_1, \quad \ldots, \quad Y_m = K' \cap Y$$

such that for $i = 0, 1, \ldots, m-1$:

$$|Y_i \triangle Y_{i+1}| = 1 \quad \text{and} \quad Y_i \cap K' \subseteq Y_{i+1} \subseteq Y_i \cup K'.$$

Similarly, there exist a positive integer n and some sequence of states in $\mathcal{G}$

$$K \cap Z = Z_0, \quad Z_1, \quad \ldots, \quad Z_n = K' \cap Z$$

such that for $j = 0, 1, \ldots, n-1$:

$$|Z_j \triangle Z_{j+1}| = 1 \quad \text{and} \quad Z_j \cap K' \subseteq Z_{j+1} \subseteq Z_j \cup K'.$$

We then form the sequence

$$X_0 = Y_0 \cup (K \cap Z), \quad X_1 = Y_1 \cup (K \cap Z), \quad \ldots,$$
$$X_m = Y_m \cup (K \cap Z) = (K' \cap Y) \cup Z_o,$$
$$X_{m+1} = (K' \cap Y) \cup Z_1, \quad X_{m+2} = (K' \cap Y) \cup Z_2, \quad \ldots,$$
$$X_{m+n} = (K' \cap Y) \cup Z_n.$$

Clearly, $X_0 = K$ and $X_{m+n} = K'$. Since $\mathcal{F} \star \mathcal{G}$ is inclusive, we also have $X_k \in \mathcal{F} \star \mathcal{G}$, for $k = 0, 1, \ldots, m+n$. On the other hand, for $i = 0, 1, \ldots, m-1$:

$$
\begin{aligned}
X_i \cap K' &= (Y_i \cup (K \cap Z)) \cap K' \\
&\subseteq (Y_i \cap K') \cup (K \cap Z) \\
&\subseteq Y_{i+1} \cup (K \cap Z) \\
&= X_{i+1},
\end{aligned}
$$

and

$$
\begin{aligned}
X_{i+1} &= Y_{i+1} \cup (K \cap Z) \\
&\subseteq Y_i \cup K' \cup (K \cap Z) \\
&= X_i \cup K'.
\end{aligned}
$$

In a similar way, one proves for $j = m, m+1, \ldots, m+n-1$:

$$X_j \cap K' \subseteq X_{j+1} \subseteq X_j \cup K'.$$

Finally, it is easy to show that $|X_i \triangle X_{i+1}|$ equals 0 or 1. Thus, after deletion of repeated subsets in the sequence X_i, we obtain a sequence as in Theorem 2.9(iii).   □

We also indicate a simple result, which is very useful from the point of view of practical applications.

5.24. Theorem. *Suppose that* $(\mathcal{F},\mathcal{G})$, $(\mathcal{F}\star\mathcal{G},\mathcal{K})$, $(\mathcal{G},\mathcal{K})$ *and* $(\mathcal{F},\mathcal{G}\star\mathcal{K})$ *are four pairs of compatible knowledge structures. Then, necessarily*

$$(\mathcal{F}\star\mathcal{G})\star\mathcal{K} = \mathcal{F}\star(\mathcal{G}\star\mathcal{K}).$$

PROOF: Let X, Y and Z be the domains of $\mathcal{K}$, $\mathcal{F}$ and $\mathcal{G}$, respectively. The result follows immediately from the following string of equivalences:

$K \in (\mathcal{F}\star\mathcal{G})\star\mathcal{K}$

$\Longleftrightarrow \quad K \cap (Y \cup Z) \in \mathcal{F}\star\mathcal{G}$ and $K \cap X \in \mathcal{K}$

$\Longleftrightarrow \quad K \cap (Y \cup Z) \cap Y \in \mathcal{F}$ and $K \cap (Y \cup Z) \cap Z \in \mathcal{G}$ and $K \cap X \in \mathcal{K}$

$\Longleftrightarrow \quad K \cap Y \in \mathcal{F}$ and $K \cap Z \in \mathcal{G}$ and $K \cap X \in \mathcal{K}$.

□

Original Sources and Related Works

Entail relations were independently investigated in Koppen and Doignon (1988), and in Müller (1989, under the name of "implication relations"). Both sets of authors acknowledge an initial suggestion from Falmagne (see also Falmagne, Koppen, Villano, Doignon and Johanessen, 1990). Müller-Dowling obtains a version of Theorem 5.5 formulated in terms of implication relations, while our presentation follows Koppen and Doignon (1988). We recently learned from Bernard Monjardet that much similar results were obtained by Armstrong (1974); see also Wild (1994). Related algorithmic implementations will be discussed in Chapter 12.

Another interesting question concerns the description of a knowledge space by a 'minimal' part of its entailment. It is nicely treated by Guigues and Duquenne (1986) in terms of closure spaces and 'maximal informative implications' (see also Ganter, 1984).

The results on meshing are due to Falmagne and Doignon (1998). For additional results on entail relations, see Dowling (1994) or Düntsch and Gediga (1996).

Problems

1. Show the equivalence of Equation (2) with Equation (6).

2. Show the equivalence of Equation (3) with Equation (7).

3. Prove Theorem 5.7.

4. For a relation $\mathcal{Q}$ on $2^Q \setminus \{\varnothing\}$ which is a transitive extension of reverse inclusion, show that Condition (iii) in Theorem 5.7 is equivalent to any of the following two conditions:

(iv) for each $A \in 2^Q \setminus \{\varnothing\}$, there is a maximum subset B of Q such that $A\mathcal{Q}B$ (here, maximum means maximum for the inclusion);

(v) for all $A, B \in 2^Q \setminus \{\varnothing\}$, we have $A\mathcal{Q}B$ iff $A\mathcal{Q}\{b\}$ for all $b \in B$,

and in case the domain Q is essentially finite, also to

(vi) $A\mathcal{Q}B$ implies $A\mathcal{Q}(A \cup B)$ for all $A, B \in 2^Q \setminus \{\varnothing\}$.

In general, does Condition (vi) implies Condition (v)?

5. Let $\mathcal{P}$ be an entailment for the domain Q, and let $\mathcal{K}$ be the derived knowledge space on Q. State and prove a necessary and sufficient condition on $\mathcal{P}$ for the

(i) quasi ordinality of $\mathcal{K}$;
(ii) wellgradedness of $\mathcal{K}$;
(iii) granularity of $\mathcal{K}$.

6. Any knowledge space $\mathcal{K}$ on the finite domain Q is derived from exactly one surmise system σ on Q, and is also derived from exactly one entailment for Q. Make explicit the resulting one-to-one correspondence between surmise systems on Q and entailments for Q. Try to extend the result to the infinite case by considering granular knowledge spaces.

7. Show that the maximal mesh of two compatible knowledge spaces is again a space (cf. Theorem 5.17). If the two given spaces are (quasi) ordinal, is the maximal mesh also (quasi) ordinal?

8. Prove Theorem 5.20.

9. Prove Theorem 5.22.

10. Let $\mathcal{B}$ be the base of the maximal mesh $(X \cup Y, \mathcal{F} \star \mathcal{G})$ of two finite, compatible knowledge spaces $(X, \mathcal{F})$ and $(Y, \mathcal{G})$ with bases $\mathcal{C}$ and $\mathcal{D}$. Is there a simple construction of $\mathcal{B}$ from $\mathcal{C}$ and $\mathcal{D}$ (taking into account the intersection $Y \cap Z$)?

11. From Theorems 5.5 and 5.7, there exists a one-to-one correspondence between the family of knowledge spaces $\mathcal{K}$ on Q and the family of entail relations $\mathcal{P}$ for Q. State explicitly when $\mathcal{K}$ and $\mathcal{P}$ correspond to each other. In particular, spell out the interpretation of $A \mathcal{P} B$, for $A, B \subseteq Q$, in terms of closed sets (i.e. complements of states in $\mathcal{K}$).

Chapter 6

Galois Connections⋆

In various preceding chapters, several one-to-one correspondences were established between particular collections of mathematical structures. For instance, Birkhoff's Theorem 1.49 asserts the existence of a one-to-one correspondence between the collection of all quasi ordinal spaces on a domain Q, and the collection of all quasi orders on Q. All these correspondences will be shown to derive from natural constructions. Each derivation will be obtained from the application of a general result about 'Galois connections.' A compendium of the notation for the various collections and the three 'Galois connections' of main interest to us is given at the end of the chapter, before the Sources section. We star the whole chapter because its content is more abstract, and not essential to the rest of this book.

Three Exemplary Correspondences

Table 6.1 summarizes three correspondences, gives references to relevant theorems and recalls some notations.

1	2	3	4	5	6	7
1.49	quasi ordinal space	$\mathcal{K}$	$\tilde{\mathbf{K}}^{\mathbf{O}}$	$\tilde{\mathbf{R}}^{\mathbf{O}}$	$\mathcal{Q}$	quasi order
3.10	granular space	$\mathcal{K}$	$\tilde{\mathbf{K}}^{\mathbf{G}}$	$\tilde{\mathbf{F}}^{\mathbf{S}}$	σ	surmise function
5.5	space	$\mathcal{K}$	$\tilde{\mathbf{K}}$	$\tilde{\mathbf{E}}^{\mathbf{e}}$	$\mathcal{P}$	entailment

Table 6.1. References, terminology and notations for three one-to-one correspondences encountered earlier. Columns heading in the table are as follows:

1:	Theorem number,
2 and 7:	name of mathematical structure,
3 and 6:	typical symbol for this structure,
4 and 5:	notation for the collection.

We assumed throughout the chapter that the domain Q is a fixed nonempty set, which may be infinite. Line 1 of Table 6.1 refers to Birkhoff's Theorem.

The existence of the one-to-one correspondences in this table proves that the corresponding collections have the same cardinality. A close examination of these correspondences reveals a more interesting situation. First, each one of the correspondences can be canonically derived from a construction relating two respectively larger collections of structures. Second, these larger collections and thus the original ones can be naturally (quasi) ordered, and the correspondence, as well as the constructions, are 'order reversing' between the (quasi) ordered sets.

6.1. Definition. Given two quasi ordered sets $(Y, \mathcal{U})$ and $(Z, \mathcal{V})$, a mapping $f : Y \to Z$ is *order reversing* when for all $x, y \in Y$,

$$ x\mathcal{U}y \quad \Longrightarrow \quad f(x)\mathcal{V}^{-1}f(y). $$

The mapping f is an *anti-isomorphism* if it is bijective and satisfies the stronger condition

$$ x\mathcal{U}y \quad \Longleftrightarrow \quad f(x)\mathcal{V}^{-1}f(y), $$

again for all $x, y \in Y$.

   Taking the correspondence in the upper line of Table 6.1 as an example, the two larger collections are: on the one hand, the family of all knowledge structures on the fixed set Q, and on the other hand, the family of all relations on Q. Definition 1.43 associates to any knowledge structure $\mathcal{K}$ a relation, the surmise relation $\precsim$ (which happens to be a quasi order, cf. Theorem 1.44): we have, for $r, q \in Q$,

$$ r \precsim q \quad \Longleftrightarrow \quad r \in \cap \mathcal{K}_q. $$

Conversely, Theorem 1.51 shows how to construct, for any given relation $\mathcal{Q}$ on Q, a derived knowledge structure $\mathcal{K}$ on Q (cf. also Definition 1.52): a subset K of Q is a state of this structure when

$$ \forall q \in K, \forall r \in Q : rRq \quad \Longrightarrow \quad r \in K. $$

It can be checked that both of the resulting mappings are inclusion reversing; moreover, they form a so-called 'Galois connection' in the sense defined below. As shown by Monjardet (1970), the one-to-one correspondence in the upper line of Table 6.1 consists of appropriate restrictions of these mappings.

Closure Operators and Galois Connections

A closure space is defined in 1.11 as a collection of subsets of a domain Q that is closed under intersection and contains Q; typical examples were given in 1.12, such as the Euclidean space $\mathbf{R}^3$ provided with the family of all its affine subspaces, or provided with the family of all its convex subsets. Another example consists of the power set 2^E of a given set E, together will all knowledge spaces on E; it is indeed quickly verified that the intersection $\cap_{i \in I} \mathcal{K}_i$ of any family $(\mathcal{K}_i)_{i \in I}$ of spaces on E is again a space on E, and also that 2^E is a knowledge space (see Problem 1).

For any closure space $(Q, \mathcal{L})$, we built in Theorem 1.14 a mapping $2^Q \to 2^Q : A \mapsto A'$, with A' the closure of A (cf. Definition 1.15). In the three examples just mentioned, we obtain the affine closure, the convex closure and the 'spatial closure.' To be more explicit about the third example, any knowledge structure $\mathcal{K}$ on a domain E admits a 'spatial closure', which is the smallest knowledge space on E containing $\mathcal{K}$, or in the terms of Definition 1.19, the space spanned by $\mathcal{K}$ (see Example 6.3(a) below).

These situations have in common that the domain of the 'closure operator' (the power set of $\mathbf{R}^3$, or the family of all knowledge structures on E) can be ordered by inclusion, and the resulting partial order is tightly intertwined with that operator. Given a closure space $(Q, \mathcal{L})$, we denote by $h(A) = A'$ the closure of the subset A of Q, and recall the fundamental properties of the 'closure operator' h (cf. Theorem 1.14): for all A, B in 2^Q,

(1) $A \subseteq B$ implies $h(A) \subseteq h(B)$;
(2) $A \subseteq h(A)$;
(3) $h^2(A) = h(A)$;
(4) $A \in \mathcal{L}$ iff $A = h(A)$.

We now consider a fairly abstract setting, taking any quasi ordered set $(X, \precsim)$ as the domain of the 'closure operator.'

6.2. Definition. Let $(X, \precsim)$ be a quasi ordered set, and let h be a mapping of X into itself. Then h is a *closure operator* on $(X, \precsim)$ if it satisfies the following three conditions: for all x, y in X,

(1) $x \precsim y$ implies $h(x) \precsim h(y)$;
(2) $x \precsim h(x)$;
(3) $h^2(x) = h(x)$.

Moreover, any x in X is *closed* when $h(x) = x$.

6.3. Examples. a) Let $\tilde{\mathbf{K}}$ be the set of all knowledge structures on a set Q, with $\tilde{\mathbf{K}}$ ordered by the inclusion relation. For any $\mathcal{K} \in \tilde{\mathbf{K}}$, let $s(\mathcal{K})$ be the smallest space including $\mathcal{K}$. Then, s is a closure operator on $(\tilde{\mathbf{K}}, \subseteq)$, and the closed elements are the spaces (Problem 1).

b) More generally, let $(Q, \mathcal{L})$ be a closure space in the sense of Definition 1.11. For any $A \in 2^Q$, let $h(A) = A'$ be the smallest member of $\mathcal{L}$ including A. It is easily checked that the mapping $h : 2^Q \to 2^Q$ is a closure operator on $(2^Q, \subseteq)$, the closed elements being precisely the members of $\mathcal{L}$ (see Problem 3). This example covers many of fundamental settings in mathematics; a few of them are indicated here by the name given to the elements of $\mathcal{L}$ (together with the resulting closure operator): all affine subsets of an affine space (affine closure); all convex subsets of an affine space over an ordered skew field (convex closure); all sublattices of a given lattice (generated sublattice); all subgroups of a group (generated subgroup); all closed sets in a topological space (topological closure); all ideals of a ring (generated ideal). Two further examples are contained in Definitions 6.16 and 6.26.

Our next definition extends to quasi orders a standard concept of ordered set theory (see e.g. Birkhoff, 1967). To help the reader grasp this abstract concept, we recall the example mentioned in the upper line of Table 6.1: take as a first quasi ordered set the family of all knowledge structures on a fixed domain Q, and as a second quasi ordered set the family of all relations on Q, both families being ordered by inclusion. Then, consider the mappings mentioned in the previous section.

6.4. Definition. Let $(Y, \mathcal{U})$ and $(Z, \mathcal{V})$ be two quasi ordered sets, and let $f : Y \to Z$ and $g : Z \to Y$ be two mappings.

The pair (f, g) is a *Galois connection* between $(Y, \mathcal{U})$ and $(Z, \mathcal{V})$ if the following six conditions hold: for all $y, y' \in Y$ and all $z, z' \in Z$,

    (1) $y\mathcal{U}y'$ and $y'\mathcal{U}y$ imply $f(y) = f(y')$;

    (2) $z\mathcal{V}z'$ and $z'\mathcal{V}z$ imply $g(z) = g(z')$;

    (3) $y\mathcal{U}y'$ implies $f(y)\mathcal{V}^{-1}f(y')$;

    (4) $z\mathcal{V}z'$ implies $g(z)\mathcal{U}^{-1}g(z')$;

    (5) $y\mathcal{U}(g \circ f)(y)$;

    (6) $z\mathcal{V}(f \circ g)(z)$.

The following facts will be useful, and are easily verified. (We shall leave parts of the proof to the reader; see Problem 4.)

6.5. Theorem. *Let $(Y, \mathcal{U})$, $(Z, \mathcal{V})$, f and g be as in Definition 6.4. Then:*

(i) *$g \circ f$ and $f \circ g$ are closure operators, respectively on $(Y, \mathcal{U})$ and $(Z, \mathcal{V})$;*

(ii) *there is at most one closed element in every equivalence class of the quasi ordered set $(Y, \mathcal{U})$ (resp. $(Z, \mathcal{V})$);*

(iii) *the set Y_0 of all the closed elements of Y (resp. Z_0, Z) is partially ordered by $\mathcal{U}_0 = \mathcal{U} \cap (Y_0 \times Y_0)$ (resp. $\mathcal{V}_0 = \mathcal{V} \cap (Z_0 \times Z_0)$);*

(iv) *if $z \in f(Y)$, there exists z_0 in Z_0 such that $z \mathcal{V} z_0$ and $z_0 \mathcal{V} z$. Similarly, if $y \in g(Z)$, there exists y_0 in Y_0 with $y \mathcal{U} y_0$ and $y_0 \mathcal{U} y$;*

(v) *the restriction f_0 of f to Y_0 is an anti-isomorphism between $(Y_0, \mathcal{U}_0)$ and $(Z_0, \mathcal{V}_0)$. Moreover $f_0^{-1} = g_0$, where g_0 is the restriction of g to Z_0.*

PROOF. We only prove parts (i) to (iii). In view of the symmetry of the statements, we only need to establish the facts concerning the quasi order $(Y, \mathcal{U})$ and the mapping $g \circ f$.

(i) We have to verify that, with $\mathcal{U} = \precsim$ and $g \circ f = h$, Conditions (1) to (3) in Definition 6.2 are satisfied. Suppose that $x \mathcal{U} y$. Applying 6.4(3) and (4) yields, successively, $f(x) \mathcal{V}^{-1} f(y)$ and $(g \circ f)(x) \mathcal{U} (g \circ f)(y)$, establishing 6.2(1). Up to a change of notation, Conditions 6.4(5) and 6.2(2) are identical here. Finally, we have to show that, for all $x \in Y$, we have $h^2(x) = h(x)$, or more explicitly

$$g\big((f \circ g \circ f)(x)\big) = g\big(f(x)\big). \tag{1}$$

In view of 6.4(2), Equation (1) holds if we have

$$(f \circ g \circ f)(x) \mathcal{V} f(x) \tag{2}$$

and

$$f(x) \mathcal{V} (f \circ g \circ f)(x). \tag{3}$$

Both of these formulas are true. By 6.4(5), we have $x \mathcal{U} (g \circ f)(x)$. Applying 6.4(3) gives Equation (2). From 6.4(6), we derive Equation (3). We conclude that $g \circ f$ is a closure operator on $(Y, \mathcal{U})$.

(ii) and (iii). Suppose that x and y are in the same class of $(Y, \mathcal{U})$. Thus, $x \mathcal{U} y$ and $y \mathcal{U} x$. By 6.4(1), this implies $f(x) = f(y)$. If both x and y are closed elements (for the closure operator $g \circ f$), we obtain

$$x = (g \circ f)(x) = (g \circ f)(y) = y.$$

This argument also shows that $\mathcal{U}_0 = \mathcal{U} \cap (Y_0 \times Y_0)$ is antisymmetric, and thus establishes (iii).

Parts (iv) and (v) are left to the reader (see Problem 4).   □

When the quasi order $\mathcal{V}$ on Z happens to be a partial order, the first part of Condition (iv) in Theorem 6.5 can be reformulated as $f(Y) = Z_0$. The following Example shows that this simplification does not hold in general.

6.6. Example. We build two weakly ordered sets $(Y, \mathcal{U})$ and $(Z, \mathcal{V})$, each having two classes, by setting

$$Y = \{a, b\}, \qquad x\mathcal{U}y \text{ iff } (x = a \text{ or } y = b),$$
$$Z = \{u, v, w\}, \qquad z\mathcal{V}t \text{ iff } (z = u \text{ or } t = v \text{ or } t = w).$$

Define two mappings $f : Y \to Z$ and $g : Z \to Y$ by

$$f(a) = w, \qquad f(b) = v,$$
$$g(u) = g(v) = g(w) = b.$$

The pair (f, g) is a Galois connection between $(Y, \mathcal{U})$ and $(Z, \mathcal{V})$. The only closed elements are b in Y, and v in Z. However $f(\{a, b\}) = \{v, w\} \neq \{v\}$.

Lattices and Galois Connections

When the quasi ordered sets between which a Galois connection is defined are 'lattices', the collections of closed elements are themselves 'lattices.' Before stating the relevant definition and results, we briefly expose an important application of Theorem 6.5 in the field of ordinal data analysis. Let $\mathcal{R}$ be a relation from a set X to a set Y; a Galois connection (f, g) between $(2^X, \subseteq)$ and $(2^Y, \subseteq)$ will be built starting from $\mathcal{R}$. For $A \in 2^X$, we set

$$f(A) = \{y \in Y \,|\, \forall a \in A : a\mathcal{R}y\}, \tag{4}$$

and similarly for $B \in 2^Y$, we set

$$g(B) = \{x \in X \,|\, \forall b \in B : x\mathcal{R}b\}. \tag{5}$$

The following, natural concept will be used in the next theorem.

6.7. Definition. A *maximal rectangle* of a relation $\mathcal{R}$ from X to Y is a pair of subsets A of X and B of Y that satisfies:

(1) for all $a \in A$, $b \in B$, we have $a\mathcal{R}b$;
(2) for each x in $X \setminus A$ there is some b in B for which not $x\mathcal{R}b$;
(3) for each y in $Y \setminus B$, there is some a in A for which not $a\mathcal{R}y$.

The term "maximal rectangle" becomes natural for a relation $\mathcal{R}$ between two finite sets when $\mathcal{R}$ is encoded in a 0-1 array; see next example.

6.8. Examples. Let $X = \{a, b, c, d, e\}$ and $Y = \{p, q, r, s\}$; a relation $\mathcal{R}$ from X to Y is specified by its 0-1 array in Table 6.2.

	p	q	r	s
a	1	1	0	1
b	1	0	1	1
c	1	0	0	0
d	0	1	0	1
e	0	0	1	1

Table 6.2. The 0-1 array for the relation $\mathcal{R}$ in Example 6.8.

For this particular relation $\mathcal{R}$, here are some maximal rectangles (A, B), with

$$A = \{a, b, c\}, \qquad B = \{p\},$$
$$\text{or} \qquad A = \{b, e\}, \qquad B = \{r, s\},$$
$$\text{or} \qquad A = \{a, b, d, e\}, \qquad B = \{s\},$$
$$\text{or} \qquad A = \varnothing, \qquad B = \{p, q, r, s\}.$$

All the maximal rectangles of $\mathcal{R}$ will be listed in Example 6.17.

6.9. Theorem. *Given a relation $\mathcal{R}$ from X to Y, the pair (f, g) of mappings defined in Equations (4) and (5) forms a Galois connection between the ordered sets $(2^X, \subseteq)$ and $(2^Y, \subseteq)$. The pairs (A, B) such that A is a closed set in 2^X and B is a closed set in 2^Y with $B = f(A)$ (and thus also $A = g(B)$) are exactly the maximal rectangles of $\mathcal{R}$.*

PROOF. The first two requirements in Definition 6.4 for a Galois connection are automatically satisfied, since the domains of f and of g are ordered. If A_1, $A_2 \in 2^X$, then $A_1 \subseteq A_2$ implies $f(A_1) \supseteq f(A_2)$ because of the quantification in Equation (4); this establishes Condition (3) in Definition 6.4, and Condition (4) is similarly obtained from Equation (5). Condition (5) states here $A \subseteq g(f(A))$ for all $A \in 2^X$. It is a consequence of the definition of f and g, and so is also Condition (6). Finally, proving the assertion that pairs of related closed sets coincide with the maximal rectangles is easy and left to the reader. $\qquad\qquad\qquad\qquad\qquad\qquad\qquad\qquad\qquad\qquad\qquad\quad$ $\square$

6.10. Definition. The Galois connection built in Theorem 6.9 is called the *Galois connection of $\mathcal{R}$.*

As for any Galois connection, the two ordered collections of closed sets are anti-isomorphic (cf. Theorem 6.5(v)). In the situation of Theorem 6.9, they are moreover 'lattices.' After recalling some terminology, we will derive this assertion from a general result on Galois connections between 'lattices.'

6.11. Definition. An ordered set $(X, \mathcal{P})$ is a *lattice* if any two of its elements x, y admit a 'greatest lower bound' and a 'smallest upper bound'; the *greatest lower bound* of x and y is the element $x \wedge y$ in X satisfying $(x \wedge y)\mathcal{P}x$, $(x \wedge y)\mathcal{P}y$ and for all $l \in X$, $(l\mathcal{P}x$ and $l\mathcal{P}y)$ implies $l\mathcal{P}(x \wedge y)$. Similarly, the *smallest upper bound* of x and y is the element $x \vee y$ in X such that $x\mathcal{P}(x \vee y)$, $y\mathcal{P}(x \vee y)$, and for all $u \in X$, $(x\mathcal{P}u$ and $y\mathcal{P}u)$ implies $(x \vee y)\mathcal{P}u$.

Many examples of lattices appear as particular cases of the fairly general situation in the next example.

6.12. Examples. Let $(Q, \mathcal{L})$ be a closure space; then $(\mathcal{L}, \subseteq)$ is a lattice in which $x \wedge y = x \cap y$ and $x \vee y = h(x \cup y)$, for x, $y \in \mathcal{L}$ (and as usually $h(z)$ is the closure of z). This example will be generalized in the next theorem to any closure operator.

6.13. Theorem. *For a closure operator h on a lattice $(X, \preceq)$, the collection X_0 of all closed elements is itself a lattice for the induced order $\preceq_0 \; = \; \preceq \cap (X_0 \times X_0)$. For x, $y \in X_0$, the smallest upper bound $x \vee_0 y$ in $(X_0, \preceq_0)$ is equal to $h(x \vee y)$, where $x \vee y$ denotes the smallest upper bound in $(X, \preceq)$; on the other hand, the greatest lower bound of x and y in $(X_0, \preceq_0)$ and in $(X, \preceq)$ coincide.*

PROOF. We will use the axioms for a closure operator h without mentioning them explicitly. Let x, y be two elements closed for h, that is $h(x) = x$, and $h(y) = y$. As $x \precsim x \vee y$, we get $x = h(x) \precsim h(x \vee y)$, and similarly $y = h(y) \precsim h(x \vee y)$. Now if $z \in X_0$ satisfies $x \precsim_0 z$ and $y \precsim_0 z$, we infer $x \vee y \precsim z$, so also $h(x \vee y) \precsim_0 h(z) = z$. In all, this shows that $h(x \vee y)$ is the greatest lower bound in $(X_0, \precsim_0)$ of x and y.

Now, $x \wedge y \precsim x$ implies $h(x \wedge y) \precsim_0 h(x) = x$; as we have similarly $h(x \wedge y) \precsim_0 y$, we infer $h(x \wedge y) \precsim_0 x \wedge y$. Since we always have $x \wedge y \precsim h(x \wedge y)$, we conclude that $x \wedge y = h(x \wedge y)$, and hence $x \wedge y$ is a closed element, thus also the greatest lower bound of x and y in $(X_0, \precsim_0)$.   □

6.14. Corollary. *Let (f,g) be a Galois connection between the lattice $(Y, \mathcal{U})$ and the quasi ordered set $(Z, \mathcal{V})$. Then the two collections Y_0 and Z_0 of closed elements in respectively Y and Z are anti-isomorphic lattices for the induced orders $\mathcal{U}_0 = \mathcal{U} \cap (Y_0 \times Y_0)$ and $\mathcal{V}_0 = \mathcal{V} \cap (Z_0 \times Z_0)$. For x, $y \in Y_0$, the smallest upper bound $x \vee_0 y$ equals $(g \circ f)(x \vee y)$, and the greatest lower bound is $x \wedge y$, where $\vee$ and $\wedge$ indicate that bounds are taken in Y.*

PROOF. As $g \circ f$ is a closure operator on the set $g(Y)$ of closed elements in X, the result is indeed a corollary to the previous theorem.   □

There is of course a similar statement for the case in which $(Z, \mathcal{V})$ is a lattice. We now turn back to the Galois connection of a relation.

6.15. Theorem. *Given a relation $\mathcal{R}$ from X to Y, the anti-isomorphic ordered sets of closed elements of the Galois connection of $\mathcal{R}$ are lattices.*

PROOF. This is a direct application of Corollary 6.14, since both $(2^X, \subseteq)$ and $(2^Y, \subseteq)$ are lattices.   □

The last result has important applications in a class of situations where the data can be represented by a relation between two sets. Without going into much detail, let us remark that the lattice of closed elements in 2^X obtained in Theorem 6.15 admits a description in other, maybe more appealing, terms. Its elements can be identified with the maximal rectangles of the relation $\mathcal{R}$, one of these rectangles, say (A, B), being smaller in the lattice than another, say (C, D), iff $A \subseteq C$ iff $B \supseteq D$. For a particular case, notice that $\mathcal{R}$ is a biorder (in the sense of Definition 2.13) iff its lattice is a chain.

6.16. Definition. Let $\mathcal{R}$ be a relation from X to Y. The lattice of closed elements in 2^X of the Galois connection of $\mathcal{R}$ is the *Galois lattice* or *concept lattice* of $\mathcal{R}$.

The first term appears in Birkhoff (1967), Matalon (1965), and Barbut and Monjardet (1970), while the second was popularized by the Darmstadt school, see Ganter and Wille (1996).

6.17. Examples. Turning back to the relation $\mathcal{R}$ in Example 6.8, we display in Figure 6.1 the Hasse diagram of the Galois lattice of $\mathcal{R}$. Each node designates a maximal rectangle (A, B), specified by listing the elements in the subset A of $X = \{a, b, c, d, e\}$ and the elements in the subset B of $Y = \{p, q, r, s\}$.

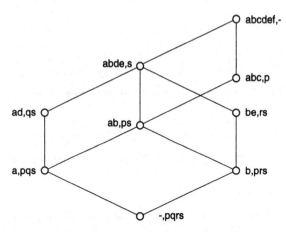

Fig. 6.1. The Galois lattice of relation $\mathcal{R}$ in Example 6.17.

Knowledge Structures and Binary Relations

As will be shown in Corollary 6.8, Birkhoff's Theorem 1.49 arises as a case of Theorem 6.5 in which the sets Y and Z are, respectively, the set $\tilde{K}$ of all knowledge structures on a domain Q, and the set $\tilde{R}$ of all binary relations on Q. This formulation is due to Monjardet (1970). It clarifies the special rôle played by the quasi orders, and the exact correspondence between the concepts involved.

6.18. Definition. Consider the set $\tilde{\mathbf{R}}$ of all binary relations on a set Q, ordered by inclusion. Let $\mathcal{R} \mapsto t(\mathcal{R})$ be the mapping of $\tilde{\mathbf{R}}$ to itself, defined by

$$t(\mathcal{R}) = \bigcup_{k=0}^{\infty} \mathcal{R}^k.$$

According to 0.16, $t(\mathcal{R})$ is the (reflexo-)transitive closure of $\mathcal{R}$. It can be shown directly that t is a closure operator on $(\tilde{\mathbf{R}}, \subseteq)$ (Problem 8). The closed elements are the quasi orders. The mapping t will be called the *transitive closure operator on* $(\tilde{\mathbf{R}}, \subseteq)$.

As before, let $\tilde{\mathbf{K}}$ be the set of all knowledge structures on a set Q. We associate to any $\mathcal{K} \in \tilde{\mathbf{K}}$ the smallest quasi ordinal space $u(\mathcal{K})$ including $\mathcal{K}$. The closure operator u will be referred to as the *quasi ordinal closure operator on* $(\tilde{\mathbf{K}}, \subseteq)$. The closed elements are the quasi ordinal spaces. (See Problem 8).

6.19. Theorem. *Consider the set $\tilde{\mathbf{K}}$ of all knowledge structures on a nonempty set Q, and the set $\tilde{\mathbf{R}}$ of all binary relations on Q, both being ordered by inclusion. Let $\mathcal{K} \mapsto r(\mathcal{K})$ be a mapping from $\tilde{\mathbf{K}}$ to $\tilde{\mathbf{R}}$ defined by*

$$p\,r(\mathcal{K})\,q \quad \Longleftrightarrow \quad \mathcal{K}_p \supseteq \mathcal{K}_q \tag{6}$$

where $p, q \in Q$. For $\mathcal{R} \in \tilde{\mathbf{R}}$, let $k(\mathcal{R})$ be the knowledge structure on Q defined from $\mathcal{R}$ by

$$K \in k(\mathcal{R}) \quad \Longleftrightarrow \quad (\forall (p, q) \in \mathcal{R} : q \in K \Rightarrow p \in K). \tag{7}$$

Thus, $\mathcal{R} \mapsto k(\mathcal{R})$ is a mapping from $\tilde{\mathbf{R}}$ to $\tilde{\mathbf{K}}$. Then, the pair (r, k) is a Galois connection between $(\tilde{\mathbf{K}}, \subseteq)$ and $(\tilde{\mathbf{R}}, \subseteq)$. Moreover, $k \circ r$ is the quasi ordinal closure operator on $(\tilde{\mathbf{K}}, \subseteq)$ and $r \circ k$ is the transitive closure operator on $(\tilde{\mathbf{R}}, \subseteq)$. The closed elements are respectively the quasi ordinal spaces in $\tilde{\mathbf{K}}$, and the quasi orders in $\tilde{\mathbf{R}}$.

The proof of this Theorem is contained in 6.21. In this framework, the corollary below is a slight improvement on Birkhoff's Theorem 1.49. It shows that the one-to-one correspondence is actually an anti-isomorphism.

6.20. Corollary. *Let (r, k) be the Galois connection of Theorem 6.19. Then, the restriction r_0 of r to the set $\tilde{K}^O$ of all quasi ordinal spaces on Q is anti-isomorphism from the lattice $(\tilde{K}^O, \subseteq)$ onto the lattice $(\tilde{R}^O, \subseteq)$ of all quasi orders on Q. The inverse mapping r_0^{-1} is the restriction of k to $\tilde{R}^O$. Moreover, the image $r_0(\mathcal{K})$ of any ordinal knowledge space $\mathcal{K}$ is a partial order.*

This immediately results from Theorems 6.19 and 6.5(v), together with Corollary 6.14 (notice that $(\tilde{K}, \subseteq)$ and $(\tilde{R}, \subseteq)$ are lattices).

6.21. Proof of 6.19. We first establish that (r, k) is a Galois connection, which amounts to prove Conditions (a)-(d) below, because $(\tilde{K}^O, \subseteq)$ and $(\tilde{R}^O, \subseteq)$ are ordered sets, and so Conditions (1) and (2) in Definition 6.4 are trivially true:

(a) $\mathcal{K} \subseteq \mathcal{K}' \implies (\forall p, q \in Q : \mathcal{K}'_p \supseteq \mathcal{K}'_q \implies \mathcal{K}_p \supseteq \mathcal{K}_q)$;

(b) $\mathcal{R} \subseteq \mathcal{R}' \Rightarrow (\forall S \subseteq Q : S \in k(\mathcal{R}') \Rightarrow S \in k(\mathcal{R}))$;

(c) $\mathcal{K} \subseteq (k \circ r)(\mathcal{K})$;

(d) $\mathcal{R} \subseteq (r \circ k)(\mathcal{R})$.

(a) Take any $p, q \in Q$ and suppose that $K \in \mathcal{K}_q$, with $\mathcal{K} \subseteq \mathcal{K}'$ and $\mathcal{K}'_p \supseteq \mathcal{K}'_q$. Successively, $K \in \mathcal{K}'$, $K \in \mathcal{K}'_q$ (since $q \in K$), $K \in \mathcal{K}'_p$, $p \in K$, yielding $K \in \mathcal{K}_p$ (since $K \in \mathcal{K}$).

(b) Suppose that $S \in k(\mathcal{R}')$, with $\mathcal{R} \subseteq \mathcal{R}'$. By Equation (7), S is a state of $k(\mathcal{R}')$ if and only if whenever $p\mathcal{R}'q$ then $q \in S \Rightarrow p \in S$. We must show that S is also a state of $k(\mathcal{R})$. Take any $p, q \in Q$ and suppose that $p\mathcal{R}q$; thus $p\mathcal{R}'q$, which implies $q \in S \Rightarrow p \in S$ (since S is a state of $k(\mathcal{R}')$). Applying Equation (7), we obtain $S \in k(\mathcal{R})$.

(c) Successively,

$$
\begin{aligned}
K \in \mathcal{K} \implies & \forall p, q \in Q : (\mathcal{K}_p \supseteq \mathcal{K}_q, q \in K) \Rightarrow p \in K & \\
\Longleftrightarrow & \forall p, q \in Q : (pr(\mathcal{K})q, q \in K) \Rightarrow p \in K & \text{[by (6)]} \\
\Longleftrightarrow & K \in (k \circ r)(\mathcal{K}). & \text{[by (7)]}
\end{aligned}
$$

(d) For all $p, q \in Q$,

$$
\begin{aligned}
p\mathcal{R}q \quad &\Longrightarrow \quad \forall K \in 2^Q : (K \in k(\mathcal{R}), \, q \in K) \Rightarrow p \in K \qquad \text{[by (7)]} \\
&\Longleftrightarrow \quad \forall K \in 2^Q : K \in (k(\mathcal{R}))_q \Rightarrow K \in (k(\mathcal{R}))_p \\
&\Longleftrightarrow \quad (k(\mathcal{R}))_p \supseteq (k(\mathcal{R}))_q \\
&\Longleftrightarrow \quad p(r \circ k)(\mathcal{R})q. \qquad\qquad\qquad\qquad\qquad \text{[by (6)]}
\end{aligned}
$$

Since (r, k) is a Galois connection, by Theorem 6.5(i), $k \circ r$ and $r \circ k$ are closure operators on $(\tilde{\mathbf{K}}, \subseteq)$ and $(\tilde{\mathbf{R}}, \subseteq)$, respectively. The following two conditions derive from Definition 6.2(1):

(e) $\mathcal{K} \subseteq \mathcal{K}' \Rightarrow (k \circ r)(\mathcal{K}) \subseteq (k \circ r)(\mathcal{K}')$;

(f) $\mathcal{R} \subseteq \mathcal{R}' \Rightarrow (r \circ k)(\mathcal{R}) \subseteq (r \circ k)(\mathcal{R}')$.

Now, from Equation (6), it is clear that, for any knowledge structure $\mathcal{K}$, $r(\mathcal{K})$ is a quasi order on Q. By Equation (7), $k(\mathcal{R})$ is a quasi ordinal space for any relation $\mathcal{R}$ on Q. In particular, $(k \circ r)(\mathcal{K})$ is a quasi ordinal space on Q. Moreover, it is the smallest quasi ordinal space including $\mathcal{K}$. Indeed, for any quasi ordinal space $\mathcal{K}'$ on Q, it is easily seen that $(k \circ r)(\mathcal{K}') = \mathcal{K}'$. Hence, if $\mathcal{K}'$ includes $\mathcal{K}$, we derive from Condition (e)

$$
(k \circ r)(\mathcal{K}) \subseteq (k \circ r)(\mathcal{K}') = \mathcal{K}'.
$$

Thus, $k \circ r$ is the quasi ordinal closure on $(\tilde{\mathbf{K}}, \subseteq)$.

We turn to the closure operator $r \circ k$. By Condition (d) and Equation (6), $(r \circ k)(\mathcal{R})$ is a transitive relation including $\mathcal{R}$. To prove that $r \circ k$ is the transitive closure operator on $(\tilde{\mathbf{R}}, \subseteq)$, we have to show that, for any $\mathcal{R} \in \tilde{\mathbf{R}}$ and any quasi order $\mathcal{R}'$ including $\mathcal{R}$, we have

$$
(r \circ k)(\mathcal{R}) \subseteq \mathcal{R}'.
$$

If $\mathcal{R}'$ is a quasi order, then $(r \circ k)(\mathcal{R}') = \mathcal{R}'$ (as can be checked easily). Thus, $\mathcal{R} \subseteq \mathcal{R}'$ implies

$$
\begin{aligned}
(r \circ k)(\mathcal{R}) &\subseteq (r \circ k)(\mathcal{R}') \qquad\qquad \text{[by Condition (f)]} \\
&= \mathcal{R}'.
\end{aligned}
$$

Finally, the fact that $\tilde{\mathbf{R}}^{\mathbf{O}}$ and $\tilde{\mathbf{K}}^{\mathbf{O}}$ are the closed elements of $\tilde{\mathbf{R}}$ and $\tilde{\mathbf{K}}$, respectively, results from Theorem 6.5(iv). $\qquad\qquad\qquad\qquad\qquad\qquad \square$

We now rephrase Definitions 1.52 and 1.43.

6.22. Definition. Referring to the mappings described in Theorem 6.19, we say that the quasi ordinal space $k(\mathcal{R})$ is *derived* from the relation $\mathcal{R}$, and similarly that the quasi order $r(\mathcal{K})$ is *derived* from the knowledge structure $\mathcal{K}$. Notice that $r(\mathcal{K})$ is the surmise relation (or precedence relation) of $\mathcal{K}$.

Granular Knowledge Structures and Granular Attributions

In Theorem 1.49, quasi ordinal knowledge spaces were put in a one-to-one correspondence with quasi orders. We just showed that this correspondence could be derived from a Galois connection (cf. Corollary 6.20). Another Galois connection will be exhibited here, from which we will derive a different proof of the one-to-one correspondence already established in Chapter 3 between the collection $\tilde{\mathbf{K}}^{\mathbf{G}}$ of all granular knowledge spaces on a set Q and the collection $\tilde{\mathbf{F}}^{\mathbf{S}}$ of all surmise functions on Q (cf. Theorem 3.10).

The starting point is the construction of Definition 3.6. For any granular knowledge structure $(Q, \mathcal{K})$, we defined there a derived surmise function σ on the set Q by setting, for any q in Q,

$$C \in \sigma(q) \quad \Longleftrightarrow \quad C \text{ is an atom at } q.$$

On the other hand, according to Definition 3.8, each attribution σ on the nonempty set Q produces a derived collection $\mathcal{K}$ of knowledge states on Q. This collection $\mathcal{K}$ consists of all subsets K of Q satisfying

$$\forall q \in K, \exists C \in \sigma(q) : C \subseteq K.$$

However, the resulting knowledge space is not necessarily granular.

6.23. Example. Let Q be an infinite set of items, and let σ be the attribution mapping each item to the collection of all infinite subsets of Q. (Thus, the attribution σ is constant.) The knowledge space derived from σ consists of all infinite subsets of Q, plus the empty set. This space has no atom whatsoever.

6.24. Definition. An attribution is *granular* when the knowledge space it produces is granular. We denote by $\tilde{\mathbf{F}}^{\mathbf{g}}$ the set of all granular attributions on the nonempty set Q.

We do not have a direct characterization of granular attributions. Such a characterization would be useful for the next Galois connection, which is between the collection $\tilde{\mathbf{K}}^{\mathbf{g}}$ of all granular knowledge structures on Q and the collection $\tilde{\mathbf{F}}^{\mathbf{g}}$ of all granular attributions on Q. The definition of a Galois connection (cf. 6.4) requires that these sets be first provided with a quasi order. Here, $\tilde{\mathbf{K}}^{\mathbf{g}}$ will be taken with the inclusion relation. On $\tilde{\mathbf{F}}^{\mathbf{g}}$, we take a relation $\precsim$ completely similar to the one introduced in Definition 3.16, with for σ, σ' in $\tilde{\mathbf{F}}^{\mathbf{g}}$:

$$\sigma' \precsim \sigma \iff \forall q \in Q, \forall C \in \sigma(q), \exists C' \in \sigma'(q) : C' \subseteq C.$$

6.25. Theorem. *Let $\tilde{\mathbf{K}}^{\mathbf{g}}$ be the collection of all granular knowledge structures on some nonempty domain Q, and let $\tilde{\mathbf{F}}^{\mathbf{g}}$ be the collection of all granular attributions on Q. Consider two mappings $a : \tilde{\mathbf{K}}^{\mathbf{g}} \to \tilde{\mathbf{F}}^{\mathbf{g}}$ and $g : \tilde{\mathbf{F}}^{\mathbf{g}} \to \tilde{\mathbf{K}}^{\mathbf{g}}$ defined as follows. The image $a(\mathcal{K})$ of a granular knowledge structure $\mathcal{K}$ is the attribution that associates to any question q the set of all atoms at q in $\mathcal{K}$. For any granular attribution σ, the image $g(\sigma)$ is the knowledge structure consisting of all subsets K of Q such that*

$$\forall q \in K, \exists C \in \sigma(q) : C \subseteq K.$$

Then the pair (a, g) of mappings forms a Galois connection between the quasi ordered sets $(\tilde{\mathbf{K}}^{\mathbf{g}}, \subseteq)$ and $(\tilde{\mathbf{F}}^{\mathbf{g}}, \precsim)$. The closed elements of this Galois connection are respectively in $\tilde{\mathbf{K}}^{\mathbf{g}}$ the granular knowledge spaces, and in $\tilde{\mathbf{F}}^{\mathbf{g}}$ the surmise functions. Moreover, the Galois connection induces between the two sets of closed elements the one-to-one correspondence that was obtained in Theorem 3.10.

PROOF. First notice that for $\mathcal{K} \in \tilde{\mathbf{K}}^{\mathbf{g}}$, we have $a(\mathcal{K}) \in \tilde{\mathbf{F}}^{\mathbf{g}}$ (see Problem 9). All six conditions in Definition 6.4 for a Galois connection are easily established (Problem 10). It is also straightforward to check that all closed elements in $\tilde{\mathbf{K}}^{\mathbf{g}}$ are spaces, and that all closed elements in $\tilde{\mathbf{F}}^{\mathbf{g}}$ are surmise functions. To show that any granular knowledge space $\mathcal{K}$ is a closed element in $\tilde{\mathbf{K}}^{\mathbf{g}}$, we have to show that $(g \circ a)(\mathcal{K}) = \mathcal{K}$. The inclusion $(g \circ a)(\mathcal{K}) \supseteq \mathcal{K}$ is true because (a, g) is a Galois connection. For the reverse inclusion, notice that if $K \in (g \circ a)(\mathcal{K})$, then K is a union of clauses of $a(\mathcal{K})$, thus a union of elements (in fact, atoms) of $\mathcal{K}$. As $\mathcal{K}$ is a space, it must contain K.

Now let us prove that $(a \circ g)(\sigma) = \sigma$ for any surmise function σ. If $q \in Q$, any σ-clause C for q is a state in $g(\sigma)$; moreover, there can be no element K' of $g(\sigma)$ such that $q \in K' \subset C$ (since no σ-clause for q could be included in K'). Thus $C \in ((a \circ g)(\sigma))(q)$. Conversely, if $C \in ((a \circ g)(\sigma))(q)$, then C is an element of $g(\sigma)$ which is minimal for the property $q \in C$. We leave to the reader to verify that $C \in \sigma(q)$.

The last sentence of the statement is also left to the reader.   □

6.26. Definition. In the notations of Theorem 6.25, $(g \circ a)(\mathcal{K})$ is the *spatial closure* of the granular knowledge structure $\mathcal{K}$ (it coincides with the space spanned by $\mathcal{K}$), while $(a \circ g)(\sigma)$ is the *surmise closure* of the granular attribution σ.

For relations cast as (necessarily granular) attributions (cf. Definition 3.4), it is easy to check that the surmise closure of a relation is exactly the transitive closure of this relation. We now state two properties of the spatial closure with respect to resolubility or acyclicity (in the sense of Definitions 3.31 and 3.41).

6.27. Theorem. *A granular attribution is resoluble iff its surmise closure is resoluble.*

The proof of Theorem 6.27 is left as Problem 11.

6.28. Theorem. *If σ is a granular, acyclic attribution on a nonempty, finite set Q, then its surmise closure $(a \circ g)(\sigma)$ is also acyclic.*

PROOF. Setting $\tau = (a \circ g)(\sigma)$, we assume that the relation $\mathcal{R}_\sigma$ (cf. Definition 3.39) is acyclic and we prove by contradiction that the relation $\mathcal{R}_\tau$ is also acyclic. If $x_1, x_2, \ldots, x_k$ is a cycle for $\mathcal{R}_\tau$, there is (by definition of $\mathcal{R}_\tau$) a clause C_i in $\tau(x_{i+1})$ that contains x_i (for a cyclic index i with $i = 1, 2, \ldots, k$). The thesis will result from the existence, for each value of i, of items $y_1^i, y_2^i, \ldots, y_{\ell_i}^i$ such that

$$
\begin{array}{llll}
x_1 \mathcal{R}_\sigma y_1^1, & y_1^1 \mathcal{R}_\sigma y_2^1, & \ldots, & y_{\ell_1}^1 \mathcal{R}_\sigma x_2, \\
x_2 \mathcal{R}_\sigma y_1^2, & y_1^2 \mathcal{R}_\sigma y_2^2, & \ldots, & y_{\ell_2}^2 \mathcal{R}_\sigma x_3, \\
& \ldots, & & \\
x_k \mathcal{R}_\sigma y_1^k, & y_1^k \mathcal{R}_\sigma y_2^k, & \ldots, & y_{\ell_n}^n \mathcal{R}_\sigma x_1
\end{array}
$$

(because we have here a cycle of $\mathcal{R}_\sigma$ in contradiction with our assumption). To construct the finite sequence $y_1^i, y_2^i, \ldots, y_{\ell_i}^i$, we first define a mapping η on a certain subset D of C_i. By the definition of $\tau = (a \circ g)(\sigma)$, the clause

C_i in $\tau(x_{i+1})$ is a minimal element in $g(\sigma)$ among the elements containing x_{i+1}. In particular, $C_i \setminus \{x_i\}$ is not a state of σ. There must be some item y in $C_i \setminus \{x_i\}$ such that no clause in $\sigma(y)$ is included in $C_i \setminus \{x_i\}$. On the other hand, C_i being an element of $g(\sigma)$, there is some clause C_1^i in $\sigma(y)$ included in C_i. Thus $x_i \in C_1^i$. We set $\eta(y) = x_i$. If $y = x_{i+1}$, then the construction of η is completed, with $D = \{x_i, x_{i+1}\}$. If $y \neq x_{i+1}$, we initialize D to $\{x_i, y\}$. Then $C_i \setminus D$ is not a state of σ, but contains x_{i+1}. Again since $C_i \in g(\sigma)$ but $C_i \setminus D \notin g(\sigma)$, there is some item y' such that y or x_i belongs to a clause in $\sigma(y')$. We add y' to D, and set $\eta(y')$ equal to y or x_i accordingly. The same construction for an increasing set D is repeated until D contains x_{i+1} (which must happen at some time since C_i is finite). Clearly, there exists then a sequence $z_1 = \eta(x_{i+1})$, $z_2 = \eta(z_1)$, $\ldots$, $z_\ell = \eta(z_{\ell-1})$ with moreover $\eta(z_\ell) = x_i$. We set $y_j^i = z_{\ell-j+1}$ for $j = 1$, $2, \ldots, \ell$.

The construction of the finite sequence y_1^i, y_2^i, $\ldots$, $y_{\ell_i}^i$ will be carried out for each value of the cyclic index i. $\square$

The example below shows that the converse of Theorem 6.28 does not hold.

6.29. Example. Define an attribution σ on $Q = \{a, b, c, d\}$ by

$$\sigma(a) = \{\,\{a\}\,\}, \qquad \sigma(b) = \{\,\{a, b\}, \{b, d\}\,\},$$
$$\sigma(c) = \{\,\{a, c\}\,\}, \qquad \sigma(d) = \{Q\}.$$

The relation $\mathcal{R}_\sigma$ is *not* acyclic (since b, d is a cycle), and so σ is not an acyclic attribution. The surmise closure τ of σ is acyclic, however. It is given by

$$\tau(a) = \{\,\{a\}\,\}, \qquad \tau(b) = \{\,\{a, b\}\,\},$$
$$\tau(c) = \{\,\{a, c\}\,\}, \qquad \tau(d) = \{Q\}.$$

The same example shows that the relation $\mathcal{R}_\tau$ attached to the surmise closure τ of an attribution σ can differ from the transitive closure of the relation $\mathcal{R}_\sigma$.

Knowledge Structures and Associations

We have exhibited a one-to-one correspondence α between the collection of all knowledge spaces on Q and the collection of all entailments for Q (Theorem 5.5); another one-to-one correspondence links the latter collection with that of entail relations for Q (Theorem 5.7 and Definition 5.8), so that we also have a one-to-one correspondence β between the collection of all knowledge spaces on Q and the collection of all entail relations for Q. Both correspondences α and β can be produced from a specific Galois connection. We will treat here the case of α, leaving the case of β to the reader (see Problem 12).

6.30. Definition. Let Q be a nonempty set. Define as follows a mapping v from the collection $\tilde{\mathbf{K}}$ of all knowledge structures on Q to the collection $\tilde{\mathbf{E}}$ of all relations from 2^Q to Q, where for $\mathcal{K} \in \tilde{\mathbf{K}}$:

$$v(\mathcal{K}) = \mathcal{P} \iff$$
$$\left(\forall A \in 2^Q, \forall q \in Q : A\mathcal{P}q \Leftrightarrow (\forall K \in \mathcal{K} : q \in K \Rightarrow K \cap A \neq \varnothing)\right). \quad (8)$$

Relations from 2^Q to Q will be called *association relations*, or in short *associations*. Notice that if $\mathcal{K}$ is a knowledge space, $v(\mathcal{K})$ is nothing else than its derived entailment (cf. Definition 5.6). When $\mathcal{K}$ is a general knowledge structure, we also say that $v(\mathcal{K})$ is *derived from* $\mathcal{K}$.

Define then a mapping $\ell : \tilde{\mathbf{E}} \to \tilde{\mathbf{K}}$ by setting for $\mathcal{P} \in \tilde{\mathbf{E}}$:

$$\ell(\mathcal{P}) = \{K \in 2^Q \,|\, \forall r \in Q, \forall B \subseteq Q : (B\mathcal{P}r \text{ and } r \in K) \Rightarrow B \cap K \neq \varnothing\}. \quad (9)$$

The knowledge structure $\ell(\mathcal{P})$ is said to be *derived from* $\mathcal{P}$. A similar construction was encountered in Equation (3) of Theorem 5.5.

6.31. Theorem. *Let Q be a nonempty set. The mappings $v : \tilde{\mathbf{K}} \to \tilde{\mathbf{E}}$ and $\ell : \tilde{\mathbf{E}} \to \tilde{\mathbf{K}}$ form a Galois connection if both $\tilde{\mathbf{K}}$ and $\tilde{\mathbf{E}}$ are ordered by inclusion. The closed elements in $\tilde{\mathbf{K}}$ form the lattice $(\tilde{\mathbf{K}}, \subseteq)$ of all knowledge spaces on Q; the closed elements in $\tilde{\mathbf{E}}$ form the lattice $(\tilde{\mathbf{E}}^e, \subseteq)$ of all entailments. The anti-isomorphisms induced by v and ℓ between these two lattices provides the one-to-one correspondence in Theorem 5.5.*

PROOF. Conditions (1) and (2) in the Definition 6.5 of a Galois connection are automatically satisfied because the inclusion relation, either on $\tilde{\mathbf{K}}$ or on $\tilde{\mathbf{E}}$, is a partial order. The other conditions are easily derived from the definitions of v and ℓ in Equations (8) and (9).

To show that the closed elements in $\tilde{\mathbf{K}}$ constitute the collection $\tilde{\mathbf{K}}$ of all knowledge spaces on Q, it suffices to establish $\ell(\tilde{\mathbf{E}}) = \tilde{\mathbf{K}}$. The inclusion $\ell(\tilde{\mathbf{E}}) \subseteq \tilde{\mathbf{K}}$ is easily obtained. The opposite inclusion follows from $\mathcal{K} = (\ell \circ v)(\mathcal{K})$ for any space $\mathcal{K}$ on Q; notice that $\mathcal{K} \subseteq (\ell \circ v)(\mathcal{K})$ holds because (v, ℓ) is a Galois connection, while $(\ell \circ v)(\mathcal{K}) \subseteq \mathcal{K}$ is proved as follows. If $L \in (\ell \circ v)(\mathcal{K}) \setminus \mathcal{K}$, let K be the largest state of the space $\mathcal{K}$ included in L. Picking r in $L \setminus K$ and setting $B = Q \setminus L$, we get both $B\,v(\mathcal{K})\,r$ and $r \in L$, contradicting $L \in (\ell \circ v)(\mathcal{K})$.

Similarly, to prove that the closed elements in $\tilde{\mathbf{E}}$ constitute the collection $\tilde{\mathbf{E}}^{\mathbf{e}}$ of all entailments, it suffices to show that $v(\tilde{\mathbf{K}}) = \tilde{\mathbf{E}}^{\mathbf{e}}$. We leave this to the reader.

Finally, as $(\tilde{\mathbf{K}}, \subseteq)$ clearly is a lattice, we may apply Corollary 6.14. $\quad\square$

All the Galois connections introduced in the chapter for the various structures at the focus of this monograph are recorded in Table 6.2.

1	2	3	4	5
6.19	knowledge structure / quasi ordinal space	$\tilde{\mathbf{K}}$ / $\tilde{\mathbf{K}}^{\mathbf{O}}$	$\tilde{\mathbf{R}}$ / $\tilde{\mathbf{R}}^{\mathbf{O}}$	relations / quasi order
6.25	granular knowledge structure / granular knowledge space	$\tilde{\mathbf{K}}^{\mathbf{g}}$ / $\tilde{\mathbf{K}}^{\mathbf{G}}$	$\tilde{\mathbf{F}}^{\mathbf{g}}$ / $\tilde{\mathbf{F}}^{\mathbf{S}}$	granular attribution / surmise function
6.31	knowledge structures / knowledge space	$\tilde{\mathbf{K}}$ / $\tilde{\mathbf{K}}$	$\tilde{\mathbf{F}}^{\mathbf{g}}$ / $\tilde{\mathbf{E}}^{\mathbf{e}}$	association / entailment

Table 6.3. Galois connections for various structures of this monograph. Columns headings are as follows:

1:          Theorem number,
2 and 5:   name of mathematical structure,
3 and 4:   notation for the collection.

Original Sources and Related Works

For background on Galois connections, we refer the reader to Birkhoff (1967) or Barbut and Monjardet (1970), for instance. The notion is closely related with that of a 'residuated mapping', see e.g. Blyth and Janowitz (1972). Definition 6.4 slightly extends the concept of a Galois connection, by allowing quasi ordered sets instead of ordered sets. It comes from Doignon and Falmagne (1985; a mistake in Proposition 2.7(iv) in this paper was pointed out by J. Heller).

The idea of deriving Birkhoff's Theorem from a Galois connection is due to Monjardet (1970), see our Corollary 6.20. The other applications given here of the theory of Galois connections, namely Theorems 6.25 and 6.31, come respectively from Doignon and Falmagne (1985) (in the finite case), and Koppen and Doignon (1988).

We have approached in passing the field of concept lattices, or Galois lattices. The reader is referred to Matalon (1965) or Barbut and Monjardet (1970) for references in French, and to Ganter and Wille (1996) for a wide coverage in German (an English translation of the latter was undertaken in 1997). Rusch and Wille (1996) have pointed out several relationships between the study of concept lattices and our investigation of knowledge spaces. Another link was mentioned in the Sources section of Chapter 1: Dowling (1993b) and Ganter (1984, 1987; see also Ganter and Reuter, 1991) algorithms can be restated in order to perform each other task (see Problem 6).

Problems

1. Let $\tilde{\mathbf{K}}$ be the set of all knowledge structures on a set Q. Prove that any intersection of knowledge spaces in $\tilde{\mathbf{K}}$ is a knowledge space, and that the mapping s of Example 6.3(a) is a closure operator on $(\tilde{\mathbf{K}}, \subseteq)$, the knowledge spaces being the closed elements.

2. If a knowledge structure is closed under intersection, is the same true for its spatial closure?

3. Check that the mapping $h : 2^Q \to 2^Q : A \mapsto h(A) = A'$ of Example 6.3(b) is a closure operator, with the members of $\mathcal{L}$ being the closed elements.

4. Complete the proof of Theorem 6.5.

5. Construct the Galois lattice of the following relations:

   (1) the identity relation from a set to the same set;

   (2) the membership relation from a set Q to its power set 2^Q;

   (3) the relation given in 6.4.

	p	q	r	s
a	1	1	0	1
b	1	0	1	1
c	1	0	0	0
d	1	1	0	1

Table 6.4. The 0-1 array for the relation in Problem 5(3).

6. Design an algorithm to compute the Galois lattice of a relation between two finite sets. Show that such an algorithm can be directly derived from Algorithm 1.31 (which constructs a knowledge space from any of its spanning families).

7. Does the Galois lattice of a relation determine this relation? Consider the case in which the elements of the lattice are marked as in Figure 6.1.

8. Prove that the function t of Definition 6.18 is a closure operator, with the quasi orders on Q forming the closed elements. Prove that u is also a closure operator, with the quasi ordinal spaces on Q being the closed elements.

9. Let a be the mapping defined in Theorem 6.25. Show that for any granular knowledge structure $\mathcal{K}$, the image $a(\mathcal{K})$ is a granular attribution.

10. Complete the proof of Theorem 6.25.

11. Prove Theorem 6.27.

12. Establish a Galois connection between knowledge structures and entail relations from which follows the one-to-one correspondence β mentioned before Definition 6.30.

13. Show that mappings $f : Y \to Z$ and $g : Z \to Y$ between ordered sets $(Y, \mathcal{U})$ and $(Z, \mathcal{V})$ form a Galois connection iff for all y in Y and z in Z:

$$y \mathcal{U} g(z) \quad \Longleftrightarrow \quad z \mathcal{V} f(y)$$

(Schmidt, see Birkhoff, 1967, p. 124). Give a similar result for the general case in which $(Y, \mathcal{U})$ and $(Z, \mathcal{V})$ are quasi ordered sets. Use this other characterization of Galois connections to work out different proofs of results of this chapter.

14. A lattice $(L, \precsim)$ is *complete* if any family (possibly infinite) of elements of L admit a greatest lower bound and a smallest upper bound (define these terms). Does Theorem 6.13 extend to complete lattices? Find examples of complete lattices among the collections of objects studied in this monograph.

Chapter 7

Probabilistic Knowledge Structures

The concept of a knowledge structure is a deterministic one. As such, it does not provide realistic predictions of subjects' responses to the problems of a test. There are two ways in which probabilities must enter in a realistic model. For one, the knowledge states will certainly occur with different frequencies in the population of reference. It is thus reasonable to postulate the existence of a probability distribution on the collection of states. For another, a subject's knowledge state does not necessarily specify the observed responses. A subject having mastered an item may be careless in responding, and make an error. Also, in some situations, a subject may be able to guess the correct response to a question not yet mastered. In general, it makes sense to introduce conditional probabilities of responses, given the states. A number of simple probabilistic models will be described in this chapter. They will be used to illustrate how probabilistic concepts can be introduced within knowledge space theory. These models will also provide a precise context for the discussion of some technical issues related to parameter estimation and statistical testing. In general, this material must be regarded as a preparation for the stochastic theories discussed in Chapters 8, 10 and 11.

Basic Concepts and Examples

7.1. Example. As an illustration, we consider the knowledge structure

$$\mathcal{H} = \big\{ \varnothing, \{a\}, \{b\}, \{a,b\}, \{a,b,c\}, \{a,b,d\},$$
$$\{a,b,c,d\}, \{a,b,c,e\}, \{a,b,c,d,e\} \big\} \qquad (1)$$

with domain $Q = \{a,b,c,d,e\}$ (see Figure 7.1). This example will be referred to many times in this chapter, under the name 'standard example.' The knowledge structure $\mathcal{H}$ is actually an ordinal knowledge space (in the sense of Definitions 1.7 and 1.47), with nine states. We suppose that any subject sampled from a given population of reference will necessarily be in one of these nine states. More specifically, we assume that to each knowledge state $K \in \mathcal{H}$ is attached a probability $p(K)$ measuring the likelihood that a sampled subject is in that state. We enlarge thus our theoretical framework by a probability distribution p on the family of all knowledge states. In practice, the parameters $p(K)$ must be estimated from the data.

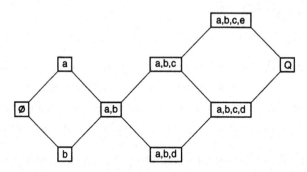

Fig. 7.1. The knowledge space $\mathcal{H}$ of Equation (1).

Notice that the states may not be directly observable. If careless errors or lucky guesses are committed, all kinds of 'response patterns' may be generated by the states in $\mathcal{H}$. A convenient coding will be adopted for these response patterns. Suppose that a subject has correctly solved questions c and d, and failed to solve a, b and e. We shall denote such a result by the subset $\{c, d\}$ of Q. In general, we shall represent a response pattern by the subset R of Q containing all the questions correctly solved by the subject. There are thus $2^{|Q|}$ possible response patterns.

For any $R \subseteq Q$ and $K \in \mathcal{H}$, we write $r(R, K)$ to denote the conditional probability of response pattern R, given state K. For example, $r(\{c, d\}, \{a, b, c\})$ denotes the probability that a subject in state $\{a, b, c\}$ responds correctly to questions c and d, and fails to solve a, b, and e. For simplicity, let us suppose in this example that a subject never responds correctly to a question not in his state. Writing $\rho(R)$ for the probability of response pattern R, we obtain for instance

$$\rho(\{c, d\}) = r(\{c, d\}, \{a, b, c, d\}) p(\{a, b, c, d\}) + r(\{c, d\}, Q) p(Q). \quad (2)$$

Indeed, the only two states capable of generating the response pattern $\{c, d\}$ are the states including it, namely, $\{a, b, c, d\}$ and Q. We shall also assume that, given the state, the response to the questions are independent. Writing β_q for the probability of an incorrect response to a question q in the subject's state, and using this independence assumption,

$$r(\{c,d\}, \{a,b,c,d\}) = \beta_a \beta_b (1 - \beta_c)(1 - \beta_d) \qquad (3)$$

$$r(\{c,d\}, Q) = \beta_a \beta_b \beta_e (1 - \beta_c)(1 - \beta_d) \qquad (4)$$

and thus, from Equation (2)

$$\rho(\{c,d\}) = \beta_a \beta_b (1 - \beta_c)(1 - \beta_d) p(\{a,b,c,d\}) + \beta_a \beta_b \beta_e (1 - \beta_c)(1 - \beta_d) p(Q).$$

In psychometrics, the type of assumption exemplified by Equations (3)–(4) is often labeled 'local independence.' We follow this usage in the definition below which generalizes the example and includes the case in which a correct response to some question q may be elicited by a state not containing q.

7.2. Definition. A *(finite) probabilistic knowledge structure* is a triple $(Q, \mathcal{K}, p)$ in which

(1) $(Q, \mathcal{K})$ is a finite knowledge structure;
(2) the mapping $p : \mathcal{K} \to [0,1] : K \mapsto p(K)$ is a probability distribution on $\mathcal{K}$. Thus, for any $K \in \mathcal{K}$, we have $p(K) \geq 0$, and moreover, $\sum_{K \in \mathcal{K}} p(K) = 1$.

A *response function* for a probabilistic knowledge structure $(Q, \mathcal{K}, p)$ is a function $r : (R, K) \mapsto r(R, K)$, defined for all $R \subseteq Q$ and $K \in \mathcal{K}$, and specifying the probability of the *response pattern* R for a subject in state K. Thus, for any $R \in 2^Q$ and $K \in \mathcal{K}$, we have $r(R, K) \geq 0$; moreover, $\sum_{R \subseteq Q} r(R, K) = 1$. A quadruple $(Q, \mathcal{K}, p, r)$, in which $(Q, \mathcal{K}, p)$ is a probabilistic knowledge structure and r its response function will be referred to as a *basic probabilistic model*. This name is justified by the prominent place of this model in this book.

Writing $R \mapsto \rho(R)$ for the resulting probability distribution on the set of all response patterns, we obtain for any $R \subseteq Q$

$$\rho(R) = \sum_{K \in \mathcal{K}} r(R, K) p(K). \qquad (5)$$

The response function r satisfies *local independence* if for each $q \in Q$, there are two constants $\beta_q, \eta_q \in [0,1[$, respectively called *(careless) error probability* and *guessing probability* at q, such that

$$r(R, K) = \left(\prod_{q \in K \setminus R} \beta_q \right) \left(\prod_{q \in K \cap R} (1 - \beta_q) \right) \left(\prod_{q \in R \setminus K} \eta_q \right) \left(\prod_{q \in R \cup K} (1 - \eta_q) \right)$$

$$(6)$$

(with $\overline{R \cup K} = Q \setminus (R \cup K)$ in the last factor). The basic probabilistic model satisfying local independence will be called the *basic local independence model*. These concepts are fundamental, in the sense that all the probabilistic models discussed in this book (in Chapters 8, 10 and 11) will satisfy Equation (5), and most of them also Equation (6). In some models, however, the existence of the probability distribution p on the collection of states $\mathcal{K}$ will not be an axiom. It rather will appear as a consequence of more or less elaborate assumptions regarding the learning process by which a student moves around in the knowledge structure, gradually evolving from the state $\varnothing$ of total ignorance to the state Q of full mastery of the material.

7.3. Remarks. a) In the case of the knowledge space $\mathcal{H}$ of Equation (1), there would thus be $8 + 2 \cdot 5 = 18$ parameters to be estimated from the data: $8 = 9 - 1$ independent probabilities for the states, and two parameters β_q and η_q for each of the five questions in Q. This number of parameters will seem unduly large compared to the 31 degrees of freedom in the data ($31 = 32 - 1$ independent response frequencies of response patterns). There are two ways in which this number of parameters may be reduced. The questions of the test may be designed in such a manner that the probability of finding the correct response by chance or by a sophisticated guess is zero. We have thus $\eta_q = 0$ for all $q \in Q$. (We have in fact made that assumption in our discussion of Example 7.1.) More importantly, the number of independent state probabilities can also be reduced. Later in this chapter, and also in Chapter 8, we investigate a number of learning mechanisms purporting to explain how a subject may evolve from the state $\varnothing$ to the state Q. Typically, such mechanisms set constraints on the state probabilities, effectively reducing the number of parameters. Finally, we point out that this example (which was chosen for expository reasons), is not representative of most empirical situations in that the number of questions is unrealistically small. It is our experience that the ratio of the number of states to the number of possible response patterns decreases dramatically as the number of questions increases (cf. Chapter 12 and, especially, Villano, 1991). The ratio of the number of parameters to the number of response patterns will decrease accordingly.

b) The local independence assumption Equation (6) may be criticized on the grounds that the response probabilities (represented by the parameters

β_q and η_q) are attached to the questions, and do not vary with the knowledge state of the subject. For instance, in Example 7.1, the probability β_c of a careless error to question c is the same in all four states: $\{a, b, c\}$, $\{a, b, c, d\}$, $\{a, b, c, e\}$ and Q. Weakening this assumption would result in a costly increase of the already substantial number of parameters in the model. Such developments will not be considered. The probabilistic models described here are intended for situations in which the core of the model, that is, the knowledge structure with the probability distribution on the collection of states, gives an essentially accurate picture of the data. In other words, the probabilities β_q and η_q of responses not conforming to the subject's state are expected to be small ($\leq .10$). The refined modeling of such relatively minor deviations did not appear to be urgent.

An Empirical Application

The basic local independence model has been successfully applied in a number of situations (Falmagne et al., 1990; Villano, 1991; Lakshminarayan and Gilson, 1993; Taagepera et al., 1997; Lakshminarayan, 1995). For convenience of exposition, we have chosen to present here a simple application to some fictitious data, which will be used repeatedly in this chapter. Lakshminarayan's experiment is described in Chapter 8.

7.4. The data. We consider a situation in which $1,000$ subjects have responded to the five questions of the domain $Q = \{a, b, c, d, e\}$ of Example 7.1. The observed frequencies of the 32 response patterns are contained in Table 7.1. We write $N(R)$ for the observed frequency of the response pattern R. In the example given in Table 7.1, we have thus $N(\emptyset) = 80$, $N\{a\} = 92$, etc. We also denote by

$$N = \sum_{R \subseteq Q} N(R), \tag{7}$$

the number of subjects in the experiment. As in earlier Chapters, we set $|Q| = m$. The basic local independence model (cf. Definition 7.2) depends upon three kinds of parameters: the state probabilities $p(K)$, and the response parameters β_q and η_q entering in Equation (6). Let us denote by θ the vector of all these parameters. In the case of Example 7.1, we set

$$\theta = (p(\emptyset), p(\{a\}), \ldots, p(Q), \beta_a, \ldots, \beta_e, \eta_a, \ldots, \eta_e).$$

R	Obs.	R	Obs.	R	Obs.	R	Obs.
∅	80	ad	18	abc	89	bde	2
a	92	ae	10	abd	89	cde	2
b	89	bc	18	abe	19	abcd	73
c	3	bd	20	acd	16	abce	82
d	2	be	4	ace	16	abde	19
e	1	cd	2	ade	3	acde	15
ab	89	ce	2	bcd	18	bcde	15
ac	16	de	3	bce	16	Q	77

Table 7.1. Frequencies of the response patterns for $1,000$ fictitious subjects.

This parameter θ is a point in a *parameter space* Θ containing all possible vectors of parameter values. The probability $\rho(R)$ of a response pattern R depends upon θ, and we shall make that dependency explicit by writing

$$\rho_\theta(R) = \rho(R). \tag{8}$$

A key component in the application of the model is the estimation of the parameters, that is, the choice of a point of Θ yielding the best fit. We shall briefly review two closely related standard techniques for estimating the parameters and testing the model.

7.5. The Chi-square statistic. Among the most widely used indices for evaluating the goodness-of-fit of a model to data is the *chi-square statistic* defined in this case as the random variable

$$\mathrm{CHI}(\theta; D, N) = \sum_{R \subseteq Q} \frac{\left(N(R) - \rho_\theta(R)N\right)^2}{\rho_\theta(R)N}, \tag{9}$$

in which θ is the vector of parameters, D symbolizes the data (that is, the vector of observed frequencies $N(R)$), and N and $\rho_\theta(R)$ are as in Equations (7) and (8). Thus, the chi-square statistic is a weighted sum of all the square deviations between the observed frequencies $N(R)$ and the predicted frequencies $\rho_\theta(R)N$. The weights $\left(\rho_\theta(R)N\right)^{-1}$ are normalizing factors ensuring that CHI enjoys important asymptotic properties.

Namely, if the model is correct, then the minimum of $\mathrm{CHI}(\theta; D, N)$ for $\theta \in \Theta$ converges (as $N \to \infty$) to a chi-square random variable. This holds under fairly general conditions on the smoothness of ρ_θ as a function of θ, and under the assumption that the response patterns given by different subjects are independent. Specifically, for N large and considering D as a random vector, the random variable $\min_{\theta \in \Theta} \text{CHI}(\theta; D, N)$ is approximately distributed as a chi-square random variable with

$$v = (2^m - 1) - \big((|\mathcal{K}| - 1) + 2m \big) \qquad (10)$$

degrees of freedom. Denoting by $\tilde{\theta}$ a value of θ that gives a global minimum, and by χ_v^2 a chi-square random variable with v degrees of freedom, we recap this result by the formula

$$\min_{\theta \in \Theta} \text{CHI}(\theta; D, N) \;=\; \text{CHI}(\tilde{\theta}; D, N) \overset{d}{\approx} \chi_v^2, \qquad \text{(for } N \text{ large)} \qquad (11)$$

where $\overset{d}{\approx}$ means *approximately distributed as*. The right member of Equation (10) computes the difference between the number of degrees of freedom in the data, and the number of parameters estimated by the minimization. In the case of Example 7.1, we would have $v = (2^5 - 1) - ((9 - 1) + 2 \cdot 5) = 13$. The convergence in Equation (11) is fast. A rule of thumb for the approximation in (11) to be acceptable is that, for every $R \subseteq Q$, the expected number of response patterns equal to R is greater than 5, i.e.

$$\rho_{\tilde{\theta}}(R)N > 5. \qquad (12)$$

The result $c = \text{CHI}(\tilde{\theta}; D, N)$ of such a computation for some observed data D is then compared with a standard table of representative values of the chi-square random variable with the appropriate degrees of freedom. The model is accepted if c lies in the bulk of the distribution, and rejected if c exceeds some critical value.

7.6. Decision procedures. Typical decision procedures are:

(1) **Strong rejection:** $\mathbf{P}(\chi_v^2 > c) \le .01$;
(2) **Rejection:** $\mathbf{P}(\chi_v^2 > c) \le .05$;
(3) **Acceptance:** $\mathbf{P}(\chi_v^2 > c) > .05$.

A value c leading to a rejection (resp. strong rejection) will be called *significant* (resp. *strongly significant*). In practice, a significant or even strongly significant value of a chi-square statistic needs not always lead to a rejection of the model. This is especially true if the model is providing a reasonably simple, detailed description of complex data, and no sensible alternative model is yet available.

7.7. Remarks. a) For large, sparse data tables (as is Table 7.1), the criterion specified by (12) is sometimes judged to be too conservative, and a minimum expected cell size equal to 1 may be appropriate (see Feinberg, 1981). This less demanding criterion will be used in all the analyses reported in this chapter. When this criterion fails, a simple solution is to group the cells with low expected values. Suppose, for instance, that only three cells have low expected frequencies, corresponding to the response patterns R, R', R'', say

$$\rho_{\tilde{\theta}}(R)N = .1, \quad \rho_{\tilde{\theta}}(R')N = .5, \quad \rho_{\tilde{\theta}}(R'')N = .7.$$

These three cells would be grouped into one, with expected frequency 1.3. The chi-square statistic would then be recomputed, replacing the three terms corresponding to the response patterns R, R', R'' by the single term

$$\frac{\left\{N(R) + N(R') + N(R'') - \left(\rho_{\dot{\theta}}(R) + \rho_{\dot{\theta}}(R') + \rho_{\dot{\theta}}(R'')\right)N\right\}^2}{\left(\rho_{\dot{\theta}}(R) + \rho_{\dot{\theta}}(R') + \rho_{\dot{\theta}}(R'')\right)N}.$$

This grouping would result in a loss of two degrees of freedom for the chi-square random variable in the decision procedure. The change of notation from $\tilde{\theta}$ to $\dot{\theta}$ for the estimate of θ in the last equation is meant to suggest that a different global minimum may be obtained after grouping.

b) It sometimes happens that the observed value c of the chi-square statistic lies in the other tail of the distribution, e.g. $\mathbf{P}(\chi_v^2 < c) < .05$. Such a result would suggest that the variability in the data is smaller than should be expected, which is usually indicative of an error (in the computation of the chi-square statistic, the number of degrees of freedom, or some other aspect of the procedure or even the experimental paradigm).

c) The minimum $\mathrm{CHI}(\tilde{\theta}; D, N)$ in Equation (9), often called the *chi-square statistic*, cannot readily be obtained by analytical methods. (Setting equal to zero the derivatives in (9) with respect to each of the parameters results in a non linear system.) Brute force search of the parameter space is possible. A modification, due to Brent (1973), of a conjugate gradient search method of Powell (1964) was used for our optimization problems.

The actual program was PRAXIS (Brent, 1973; Gegenfurtner, 1992)[1]. It should be noted that one application of PRAXIS (or for that matter any other optimizing program of that kind) does not guarantee the achievement of a global minimum. Typically, the procedure is repeated many times, with varying initial values for the parameters, until the researcher is reasonably sure that a global minimum has been reached.

d) All the statistical results and techniques used in this book are well-known, and we shall rarely give supporting references in the text. Listed in increasing order of difficulty, some standard references are Fraser (1958), Lindgren (1968), Brunk (1965), Cramér (1963), Lehman (1959). The last two are classics of the literature.

7.8. Results. The parameter values obtained from the minimization of the chi-square statistic for the data in Table 7.1 are given in Table 7.2. The value of the chi-square statistic $\min_{\theta \in \Theta} \text{CHI}(\theta; D, N)$ was 14.7 for $31 - 18 = 13$ degrees of freedom. According to the decision procedures listed in 7.6, the model should be accepted, pending further results. Some comments on the values obtained for the parameters will be found in the next section.

Response probabilities	
$\beta_a = .17$	$\eta_a = .00$
$\beta_b = .17$	$\eta_b = .09$
$\beta_c = .20$	$\eta_c = .00$
$\beta_d = .46$	$\eta_d = .00$
$\beta_e = .20$	$\eta_e = .03$
State probabilities	
$p(\varnothing) = .05$	$p(\{a, b, c\}) = .04$
$p(\{a\}) = .11$	$p(\{a, b, d\}) = .19$
$p(\{b\}) = .08$	$p(\{a, b, c, d\}) = .19$
$p(\{a, b\}) = .00$	$p(\{a, b, c, e\}) = .03$
$p(\{a, b, c, d, e\}) = .31$	

Table 7.2. Estimated values of the parameters obtained from the minimization of the chi-square statistic of Equations (9)–(11). The value of the chi-square statistic was 14.7 for 31-18 = 13 degrees of freedom.

---

[1] We thank Michel Regenwetter and Yung-Fong Hsu for all the computations reported in this chapter.

The Likelihood Ratio Procedure

Another time-honored method, which will also be used in this book, is the so-called *likelihood ratio procedure.*

7.9. Maximum likelihood estimates. In estimating the parameters of a model, it makes good intuitive and theoretical sense to choose those parameter values which render the observed data most likely. Such estimates are called *maximum likelihood estimates*. In the case of the empirical example discussed in the previous section, the likelihood of the data (for a given parameter point θ) is obtained from the *likelihood function*

$$\prod_{R \subseteq Q} \rho_\theta(R)^{N(R)}. \tag{13}$$

This computation relies on the reasonable assumption that the response patterns given by different subjects are independent. In practice, maximum likelihood estimates of the parameter in the vector θ are obtained by maximizing the logarithm (in base 10 e.g.) of the likelihood function from (13). Let us denote by $\hat{\theta}$ a value of θ maximizing $\sum_{R \subseteq Q} N(R) \log \rho_\theta(R)$, that is

$$\max_{\theta \in \Theta} \sum_{R \subseteq Q} N(R) \log \rho_\theta(R) = \sum_{R \subseteq Q} N(R) \log \rho_{\hat{\theta}}(R).$$

It turns out that, in most cases of interest, the maximum likelihood estimates are asymptotically (that is, for $N \to \infty$) the same as those obtained from the minimization of the chi-square statistic. This means that, if the parameter values in the vector $\hat{\theta}$ are maximum likelihood estimates, then for large N, the chi-square statistic

$$\mathrm{CHI}(\hat{\theta}; D, N) = \sum_{R \subseteq Q} \frac{\left(N(R) - \rho_{\hat{\theta}}(R)N\right)^2}{\rho_{\hat{\theta}}(R)N}, \tag{14}$$

is approximately distributed as a chi-square random variable with

$$v = 2^m - 1 - (|\mathcal{K}| - 1) - 2m$$

degrees of freedom (compare with Equations (9) and (11)).

7.10. The likelihood ratio test. A statistical test of the model can be
obtained from the fact that, for large N,

$$-2\log \frac{\prod_{R\subseteq Q} \rho_{\hat\theta}(R)^{N(R)}}{\prod_{R\subseteq Q} \left(\frac{N(R)}{N}\right)^{N(R)}} \overset{d}{\approx} \chi_v^2, \tag{15}$$

in which v and χ_v^2 have the same meaning as in Equations (10) and (11).
This result holds if ρ is a smooth enough function of the variables in θ,
which is the case in all the models considered here. The decision procedures
are as in 7.6. The statistical test associated with Equation (15) is called
a *likelihood ratio test*. The left member of (15) is referred to as a *log-
likelihood ratio statistic*. The concordance between the likelihood ratio and
the chi-square tests illustrated by Equations (15) and (11) holds in general.
This fact is sometimes expressed by the statement that *the likelihood ratio
test and the chi-square test are asymptotically equivalent*.

Maximum likelihood estimates of the parameters have been computed for
the data of Table 7.1. The results are given in Table 7.3.

Response probabilities	
$\beta_a = .16$	$\eta_a = .04$
$\beta_b = .16$	$\eta_b = .10$
$\beta_c = .19$	$\eta_c = .00$
$\beta_d = .29$	$\eta_d = .00$
$\beta_e = .14$	$\eta_a = .02$
State probabilities	
$p(\varnothing) = .05$	$p(\{a,b,c\}) = .08$
$p(\{a\}) = .10$	$p(\{a,b,d\}) = .15$
$p(\{b\}) = .08$	$p(\{a,b,c,d\}) = .16$
$p(\{a,b\}) = .04$	$p(\{a,b,c,e\}) = .10$
$p(\{a,b,c,d,e\}) = .21$	

Table 7.3. Maximum likelihood estimates of the parameters of the basic local
independence model defined in 7.2. The value of the log-likelihood ratio statistic
was 12.6, for 31-18=13 degrees of freedom.

The value of the log-likelihood ratio statistic was 12.6 for $31 - 18 = 13$ degrees of freedom. As in the case of the chi-square test (and not surprisingly,
in view of the asymptotic equivalence of the two tests), the data supports

the model. A comparison between Tables 7.2 and 7.3 indicates a reasonably good agreement between the estimates of the parameters. In the rest of this book, we shall systematically use likelihood ratio tests. Notice that the estimated values of the η_i parameters in Table 7.3 are quite small, suggesting that the true value of these parameters may be zero. How to test this hypothesis in the framework of likelihood ratio procedures is discussed next (see 7.14).

7.11. Remarks. a) If the statistical analyses reported above were based on real data, the high values obtained for some of the β_q parameters would be a cause of concern. In the case of Item d, for instance, it would be difficult to explain a situation in which a question has been fully mastered, but is nevertheless answered incorrectly in 46% of the cases (according to Table 7.2). However. the data are artificial and the analysis only intended to illustrate some statistical techniques.

b) The estimates obtained for the parameters were actually more unstable than what is suggested by a comparison between the values in Tables 7.2 and 7.3. For some starting values of the parameters used by the PRAXIS routine, the estimated values were sometimes quite different from those in the Tables (for a chi-square value not very different from those reported). This is not atypical for such search procedures. It usually suggests an overabundance of parameters. In such cases, an equally good fit may be obtained by varying only a subset of those parameters. An obvious candidate is the special case of the basic local independence model in which all the η_q values are kept fixed, equal to 0.

c) An examination of Equation (15) suggests that the log-likelihood ratio statistic is based on a comparison of two models, one represented in the denominator, and the other in the numerator. The denominator of (15) computes the likelihood of the data for a very general multinomial model assuming only that there is some probability $F(R)$ associated with each response pattern $R \subseteq Q$. It is well-known that the maximum likelihood estimates of the multinomial parameters $F(R)$ is their relative frequency: we have $\widehat{F(R)} = N(R)/N$. These maximum likelihood estimates are used in the computation of the likelihood in the denominator of (15). In the numerator of (15), we have a similar expression for the basic local independence model, involving also maximum likelihood estimates.

The approximation in Equation (15) is based on a general result which is stated informally below, leaving aside some technical conditions.

7.12. Theorem. *Let Ω be an s-dimensional subset of $\mathbf{R}^s$. Let $f(\omega; D, N)$ be the likelihood of the data D according to some model, where $\omega \in \Omega$ represents a vector of s independent parameters of the model, N represents the number of observations, and f is a smooth function of ω. Let Ω' be a u-dimensional subset of Ω with $0 < u < s$. Suppose that the vector ω_0 of the true values of the parameters lies in Ω'. Then, under fairly general conditions on the sets Ω and Ω', we have, for large N,*

$$-2 \log \frac{\sup_{\omega \in \Omega'} f(\omega; D, N)}{\sup_{\omega \in \Omega} f(\omega; D, N)} \overset{d}{\approx} \chi^2_{s-u}. \tag{16}$$

7.13. Remark. The subset Ω' in Equation (16) specifies a submodel or a special case of the model represented in the denominator. The likelihood ratio statistic in (16) yields a test of the hypothesis $\omega_0 \in \Omega'$ against the general model that $\omega_0 \in \Omega$. The number of degrees of freedom in the chi-square is the difference between the number of estimated parameters in the denominator and in the numerator. In the case of the likelihood ratio test of the basic local independence model, we have $\Omega = \Theta = [0,1]^s$ with $s = 2^m - 1$, and $\Omega' = \Theta'$ is a u-dimensional surface in Θ, with $u = |\mathcal{K}| - 1 + 2m$, specified by Equations (5) and (6). The importance of this theorem from the standpoint of applications is that it justifies a nested sequence of statistical tests of increasing specificity, corresponding to a chain of subsets $\Omega \supset \Omega' \supset \Omega'' \supset \cdots$ of decreasing dimensionality.

7.14. Application. We shall illustrate the use of this theorem to test, for the data of Table 7.1, the hypothesis that $\eta_q = 0$ in the framework of the basic local independence model. This hypothesis specifies the subset Θ'' of Θ' defined by

$$\Theta'' = \{\theta \in \Theta' \,|\, \forall q \in Q : \eta_q = 0\},$$

where θ symbolizes the vectors

$$\theta = (p(\varnothing), p(\{a\}), \ldots, p(Q), \beta_a, \ldots, \beta_e, \eta_a, \ldots, \eta_e).$$

In other words, the model in which all the η_i's are equal to zero is now regarded as a submodel of the basic local independence model. Using Theorem 7.12, we obtain the statistic

$$-2\log\frac{\max_{\theta\in\Theta''}\prod_{R\subseteq Q}\rho_\theta(R)^{N(R)}}{\max_{\theta\in\Theta'}\prod_{R\subseteq Q}\rho_\theta(R)^{N(R)}} \overset{d}{\approx} \chi_5^2, \tag{17}$$

with $5 = 18 - 13$. The denominator of this log-likelihood ratio statistic is the numerator of Equation (15). The value obtained for this chi-square was 1.6, which is nonsignificant (in the terms of 7.6). Accordingly, we temporarily retain, for these data, the special case $\eta_q = 0$ for all $q \in \{a, b, \ldots, e\}$. In the sequel, this model will be referred to as the *basic local independence model with no guessing*.

Learning Models

The number of knowledge states of empirical knowledge structures tend to be quite large. For example, in one experiment reviewed in Chapter 12, the number of states estimated by the QUERY routine, applied to five human experts, ranged from several hundreds to several thousands (for 50 questions). This presents a problem for practical applications of the basic local independence model, because it means that a prohibitively large number of parameters—e.g. the probabilities of all these states in the relevant population—may have to be estimated from the empirical frequencies of the response patterns. Even with substantial data sets, reliable estimates may be hard to obtain[2]. One possible solution to this difficulty is to set constraints on the state probabilities, effectively reducing the number of independent quantities involved.

A natural idea is to postulate some learning mechanism describing the successive transitions of a student, over time, from the state $\varnothing$ of complete ignorance to the state Q of full mastery of the material. Several examples of models will be described. All are based on the following basic principle:

The probability that, at the time of the test, a subject is in a particular state K of the structure is expressed as the probability that this subject

(i) *has successively mastered all the items in the state K, and*
(ii) *has failed to master any item immediately accessible from K.*

---

[2] However, see Villano, 1991.

Here, assuming that the structure is discriminative (cf. Definition 1.4), we consider an item q to be immediately accessible from the state K if $K + \{q\}$ is again a state. In other words, q belongs to the outer fringe of the state K, namely $K^{\mathcal{O}} = \{q \in Q \setminus K \mid K \cup q^* \in \mathcal{K}\}$ (cf. Definition 2.7).

7.15. A simple learning model. The model considered in this section makes strong independence assumptions regarding the learning process. Consider a discriminative knowledge structure $(Q, \mathcal{K})$. For each item q in Q, we introduce one parameter g_q, with $0 \le g_q \le 1$, intended to measure the probability that q is mastered. To define a (plausible) probability distribution p on $\mathcal{K}$, we suppose that, for each state K, all the events in the two classes: 'mastering any item $q \in K$' and 'failing to master any item $q \in K^{\mathcal{O}}$' are (conditionally) independent in the sense of the formula:

$$p(K) = \prod_{q \in K} g_q \prod_{q' \in K^{\mathcal{O}}} (1 - g_{q'}). \tag{18}$$

This formula applies to all states K because of the convention that a product of zero terms equals 1. When p is a probability distribution on $\mathcal{K}$, and a response function r is provided, the quadruple $(Q, \mathcal{K}, p, r)$ is called a *simple learning model*.

This model specifies the state probabilities in terms of only $m = |Q|$ parameters, regardless of the number of states. In the case of our standard example (Figure 7.1), the probabilities of the states are given by the formulas in Table 7.4.

States	Probabilities
$\varnothing$	$(1 - g_a)(1 - g_b)$
$\{a\}$	$g_a(1 - g_b)$
$\{b\}$	$g_b(1 - g_a)$
$\{a, b\}$	$g_a g_b(1 - g_c)(1 - g_d)$
$\{a, b, c\}$	$g_a g_b g_c(1 - g_d)(1 - g_e)$
$\{a, b, d\}$	$g_a g_b g_d(1 - g_c)$
$\{a, b, c, d\}$	$g_a g_b g_c g_d(1 - g_e)$
$\{a, b, c, e\}$	$g_a g_b g_c g_e(1 - g_d)$
$\{a, b, c, d, e\}$	$g_a g_b g_c g_d g_e.$

Table 7.4. Probabilities of the states of the structure $\mathcal{H}$ of Figure 7.1 for the simple learning model.

The probability of the nine states are thus expressed in terms of 5 parameters. It is easy to verify that these particular probabilities add up to one. In general, however, the quantities $p(K)$ defined by Equation (18) do not necessarily specify a probability distribution. Two examples will be given in the next section.

7.16. Test of the simple learning model. This model was tested on the data of Table 7.1. We made the assumption that the response function r was specified by the local independence condition with no guessing. The resulting model is then a special case of the basic local independence model with no guessing, which has already been tested. Accordingly, it makes sense to use a likelihood ratio procedure.

A likelihood ratio test was performed, which yield a value 15.5 for the log-likelihood ratio statistic. The number of degrees of freedom is $3 = 13 - 10$. Indeed, the basic local independence model with no guessing has 13 parameters, while the simple learning model (with the same local independence assumptions) has 10 parameters, and no grouping of the response patterns was necessary. This value of the chi-square statistic is very significant, leading to a strong rejection of the model.

7.17. Remark. This model was introduced to illustrate, in a simple case, how assumptions about the learning process could dramatically reduce the number of parameters. Objections can certainly be made to the hypothesis that the probability of mastering an item q does not depend on the current state K of the subject, provided that the item is learnable from that state, i.e. provided that $q \in K^{\mathcal{O}}$. This seems very strong. The independence assumptions are also difficult to accept. This model can be elaborated by assuming that the probability of mastering an item may depend upon past events, for example upon the last item learned. This idea will not be developed here, however (see Problem 3). A different kind of model is considered later in this chapter, in which a knowledge structure is regarded as the state space of a Markov chain describing the learning process. Before discussing such a model, we return to a question left pending concerning the conditions under which the simple learning model can be defined. We shall ask: under which conditions do the quantities defined by Equation (18) add up to one?

A Combinatorial Result

The simple learning model introduced in 7.15 specifies the values $p(K)$ in terms of the parameters g_q, for all K in $\mathcal{K}$. However, the following two examples show that the real-valued mapping p defined by Equation (18) does not always provide a probability distribution on $\mathcal{K}$. It thus makes sense to search for conditions on a knowledge structure that guarantee that the simple learning model is applicable. In other words, we look for conditions under which Equation (18) defines a genuine probability distribution. Interestingly, the notion of a well-graded space will be crucial (cf. Definitions 1.7 and 2.7).

7.18. Example. Let $\mathcal{G} = \{\varnothing, \{a\}, \{b\}, \{c\}, \{a,b\}, \{a,c\}, \{a,b,c\}\}$ and define $g_q = \frac{1}{2}$ for all $q \in \{a,b,c\}$. Equation (18) gives

$$p(\varnothing) = p(\{a\}) = p(\{a,b\}) = p(\{a,c\}) = p(\{a,b,c\}) = \frac{1}{8},$$

$$p(\{b\}) = p(\{c\}) = \frac{1}{4},$$

yielding $\sum_{K \in \mathcal{G}} p(K) = \frac{9}{8}$. This knowledge structure is well-graded, but is not a knowledge space: we have $\{b\}, \{c\} \in \mathcal{G}$, but $\{b,c\} \notin \mathcal{G}$.

7.19. Example. Let $\mathcal{H} = \{\varnothing, \{c\}, \{a,b\}, \{b,c\}, \{a,b,c\}\}$. Then according to Equation (18),

$$\sum_{K \in \mathcal{H}} p(K) = \big(1 - g_c\big) + \big(g_c(1 - g_b)\big) + \big(g_a g_b (1 - g_c)\big) +$$

$$\big(g_b g_c (1 - g_a)\big) + \big(g_a g_b g_c\big)$$

$$= 1 + g_a g_b (1 - g_c),$$

which gives 1 only for some special values of g_a, g_b, and g_c. It is easily checked that $\mathcal{H}$ is a knowledge space which is not well-graded.

We state a preparatory result.

7.20. Theorem. *Let $(Q, \mathcal{K})$ be a finite (not necessarily discriminative) knowledge structure. Then $(Q, \mathcal{K})$ is a well-graded knowledge space if and only if the following condition is satisfied:*

[U] *Each subset A of Q includes exactly one state $K \subseteq A$ such that $K^{\mathcal{O}} \cap A = \varnothing$.*

PROOF. We establish the sufficiency, and leave the necessity to the reader. Suppose that [U] holds. If $(Q, \mathcal{K})$ is not a space, there are two states L, M in $\mathcal{K}$ with $L \cup M \notin \mathcal{K}$. Take $A = L \cup M$, and let K be the set mentioned in Condition [U]. Thus, by the uniqueness of the set K, either $L^{\mathcal{O}} \cap A \neq \varnothing$ or $M^{\mathcal{O}} \cap A \neq \varnothing$. This implies the existence of some question q with

$$q \in (L \cup M) \setminus L \quad \text{and} \quad L \cup q^* \in \mathcal{K}$$

or

$$q \in (L \cup M) \setminus M \quad \text{and} \quad M \cup q^* \in \mathcal{K},$$

where q^* denotes the notion containing q (cf. Definition 1.4). In the first case, if $L \cup q^* \neq L \cup M$, replace L with $L \cup q^*$. In the second case, if $M \cup q^* \neq L \cup M$, replace M with $M \cup q^*$. Repeating this construction, we will eventually prove $K \cup L \in \mathcal{K}$ and obtain a contradiction. Thus Condition [U] implies that $(Q, \mathcal{K})$ is a space. Condition [U] also implies that $(Q, \mathcal{K})$ is well-graded. Indeed, if this were not true, we would derive by Theorem 2.11(iii) the existence of two states L and M with $L \subseteq M$, and such that $|M^* \setminus L^*| \geq 2$ with no state N satisfying $L \subset N \subset M$. Taking $A = M$, we derive the required contradiction since both L and M can play the rôle of K. □

7.21. Theorem. *When $(Q, \mathcal{K})$ is a finite well-graded knowledge space, the real-valued mapping p defined by Equation (18) specifies a probability distribution on $\mathcal{K}$ for any mapping g from $\mathcal{K}$ to $[0, 1]$. As a partial converse, if $(Q, \mathcal{K})$ is any finite knowledge structure, g is a mapping from Q to $]0, 1[$, and the mapping p defined from Equation (18) is a probability distribution, then $(Q, \mathcal{K})$ is a well-graded knowledge space.*

PROOF[3]. Suppose that $(Q, \mathcal{K})$ is a finite well-graded knowledge space. Thus, by Theorem 7.20, Condition [U] is satisfied. Let g be any function from Q to $[0, 1]$. We define a function $h : 2^Q \to [0, 1]$ by

$$h(A) = \prod_{q \in A} g_q \prod_{q \in Q \setminus A} (1 - g_q).$$

---

[3] Except for the argument that Condition [U] implies wellgradedness, we owe this proof to an anonymous referee of Falmagne (1994).

With $p : \mathcal{K} \to [0,1]$ from Equation (18), using the mapping h just defined, we get

$$
\begin{aligned}
p(K) &= \prod_{q \in K} g_q \prod_{q \in K^\mathcal{O}} (1 - g_q) \\
&= \left(\prod_{q \in K} g_q \prod_{q \in K^\mathcal{O}} (1 - g_q) \right) \cdot \prod_{q \in Q \setminus (K \cup K^\mathcal{O})} (g_q + (1 - g_q)) \\
&= \sum_{A \in 2^Q, \, A \supseteq K, \, A \cap K^\mathcal{O} = \varnothing} h(A).
\end{aligned}
$$

(The last equality follows by distributing the rightmost product). Thus

$$
\sum_{K \in \mathcal{K}} p(K) = \sum_{K \in \mathcal{K}} \left(\sum_{A \in 2^Q, \, A \supseteq K, \, A \cap K^\mathcal{O} = \varnothing} h(A) \right). \tag{19}
$$

On the other hand,

$$
1 = \prod_{q \in Q} (g_q + (1 - g_q)) = \sum_{A \in 2^Q} h(A). \tag{20}
$$

By Condition [U], each subset A of Q includes exactly one state K with $A \cap K^\mathcal{O} = \varnothing$. We conclude that each term $h(A)$, for $A \in 2^Q$, appears exactly once in the right-hand side of Equation (19). Hence, by comparing (19) and (20), we derive $\sum_{K \in \mathcal{K}} p(K) = 1$. Moreover, when g takes its values in $]0, 1[$, the converse also holds (since $h(A) \neq 0$ for all subsets A).   □

Markov Chain Models

The models described in this section also explain the state probabilities of the basic probabilistic model by a learning process. However, these models differ from the simple learning model in that the time is introduced explicitly, taking discrete values $n = 1, 2, \ldots$ The response function r is specified by the parameters β_q and η_q as in the local independence Equation (6). We also assume that the knowledge structure is discriminative, finite and well-graded. All these models are formulated in the framework of Markov chain theory. (For an introduction, see Feller, 1968; Kemeny and Snell, 1960; Parzen, 1962; or Shyryayev, 1996.)

7.22. Markov Chain Model 1. We assume that learning takes place in discrete steps. On any given step, at most one item can be mastered. In the case of our standard example in Figure 7.1, a transition between state $\emptyset$ to state $\{a\}$ or to state $\{b\}$ may occur on step one or later, if neither of the two states $\{a\}$ or $\{b\}$ have been achieved. The probabilities of such transitions are specified by some parameters g_a and g_b. We assume that these probabilities do not depend upon past events. Thus, the probability of a transition from state K to state $K + \{q\}$, with q in the outer fringe of K, is equal to g_q. A sample of subjects tested at a given time is assumed to have accomplished some number n of steps. This number is the same for all subjects and is a parameter, which has to be estimated from the data. If n is large, the probabilities of states containing many items will be large.

In general, this model is a Markov chain having as a state space the knowledge structure $\mathcal{K}$ (thus, the states of the Markov chain coincide with the knowledge states). The transition probabilities, contained in a matrix $M = (m_{KL})$, are defined by

$$m_{KL} = \begin{cases} g_q & \text{if } L = K + \{q\}, \text{ with } q \in K^{\mathcal{O}}, \\ 1 - \sum q \in K^{\mathcal{O}} & \text{if } L = K, \\ 0 & \text{otherwise.} \end{cases} \qquad (21)$$

A case of such a matrix is given in Table 7.5 for our standard example. For simplicity, we represent any knowledge state by a string listing its elements. Also, since the matrix is quite large, we adopt the abbreviations

$$\overline{g}_q = 1 - g_q \quad \text{and} \quad \overline{g}_{qr} = 1 - g_q - g_r. \qquad (22)$$

	$\emptyset$	a	b	ab	abc	abd	$abcd$	$abce$	Q
$\emptyset$	$\overline{g}_{ab}$	g_a	g_b	0	0	0	0	0	0
a	0	$\overline{g}_b$	0	g_b	0	0	0	0	0
b	0	0	$\overline{g}_a$	g_a	0	0	0	0	0
ab	0	0	0	$\overline{g}_{cd}$	g_c	g_d	0	0	0
abc	0	0	0	0	$\overline{g}_{de}$	0	g_d	g_e	0
abd	0	0	0	0	0	$\overline{g}_c$	g_c	0	0
$abcd$	0	0	0	0	0	0	$\overline{g}_e$	0	g_e
$abce$	0	0	0	0	0	0	0	$\overline{g}_d$	g_d
Q	0	0	0	0	0	0	0	0	1

Table 7.5. Transition matrix M of the Markov Model 1, for our standard example.

The process begins with a vector ν_0 specifying the initial probabilities of the states. (Since the states of the Markov chains are confounded with the knowledge states, no clash of terminology can arise.) The probabilities after one step are thus given by the row vector

$$\nu_1 = \nu_0 \, M.$$

If, as may often be sensible, the subjects are assumed to start the learning process in state $\varnothing$, the initial probability vector will take the form

$$\nu_0 = \underbrace{(1, 0, \ldots, 0)}_{|\mathcal{K}| \text{ terms}}.$$

If this assumption is used in the case of our example, the probabilities of the states after the first and the second step would then be

$$\nu_1 = (\overline{g}_{ab}, \, g_a, \, g_b, \, 0, \, 0, \, 0, \, 0, \, 0, \, 0),$$
$$\nu_2 = (\overline{g}^2_{ab}, \, \overline{g}_{ab}g_a + g_a\overline{g}_b, \, \overline{g}_{ab}g_b + g_b\overline{g}_a, \, 2g_ag_b, \, 0, \, 0, \, 0, \, 0, \, 0),$$

that is, the first row of $M^1 = M$ and M^2 respectively. In general, the state probabilities after state n are obtained from

$$\nu_n = \nu_0 \, M^n.$$

Writing $\nu_n(K)$ for the probability of state K at step n, we obtain, for the probability $\rho_n(R)$ of a response pattern R at step n, the prediction

$$\rho_n(R) = \sum_{K \in \mathcal{H}} r(R, K)\nu_n(K). \tag{23}$$

Using standard concepts of the theory of Markov chains[4], it is easy to show that if $g_q > 0$ for all q, then

$$\lim_{n \to \infty} \nu_n(K) = \begin{cases} 1 & \text{if } K = Q \\ 0 & \text{otherwise.} \end{cases}$$

Notice that the number of parameters, including the parameters β_q and η_q entering in the specification of the function r in Equation (6) and the

---

[4] Some basic terminology of Markov chains is recalled in 11.19

step number n representing the time of the test, cannot exceed $3m + 1$, with $m = |Q|$. We shall not develop this model any further here, and no empirical application will be presented.

7.23. Remarks. a) To some extent, this model is vulnerable to the same kind of criticisms as those addressed to the simple learning model. In particular, the probability of learning a new item q depends upon past events only in a trivial way: this probability is equal to g_q if q is learnable from the subject's current state K, that is, if $q \in K^{\mathcal{O}}$, and is equal to 0 otherwise. However, an easy adaptation of this model is available, which is outlined below as Markov Chain Model 2.

b) Another objection lies in the implicit assumption that all the subjects had the same amount of learning, represented by the step number n in Equation (23). The only difference between the subjects' states lies in the chance factors associated with the transition parameters g_q. This model could be generalized by postulating the existence of a probability distribution on the set of positive integers representing the learning step. For instance, we could assume that the learning step is distributed as a negative binomial. These developments will not be pursued here. (However, see Problem 6.)

c) An essential difference between this model and the Simple Learning Model should not be overlooked. Markov Model 1 is capable of predicting the results of data obtained from a sample of subjects having been tested several times. Suppose, for example, that the same test has been given to a sample of subject before and after a special training period. A pair (R, R') of response patterns is thus available for each subject. The required prediction concerns the probabilities $\rho_{n,n+j}(R, R')$ of observing response pattern R at step n and response pattern R' at step $n + j$ (with n and j being parameters). Using standard techniques of Markov chains theory, these predictions could be derived from Markov Model 1 (and also from Markov Model 2). The number of parameters would not exceed $3m + 2$: we would have the same parameters as in the case of a single test, plus one, namely the positive integer j entering in the specification of the step number $n + j$ of the second test. This parameter could be used as a measure of the efficiency of the training period. Such predictions could not be derived from the Simple Learning Model without further elaboration.

7.24. Markov Chain Model 2. The main concepts are as in Markov Chain Model 1, except that the last item learned affects the probabilities of learning a new one. We also have a Markov chain, but we have to keep track of the last item learned. In other words, except for the empty set, the states of the Markov chain take the form of a pair (K, q), with q in the inner fringe $K^{\mathcal{I}} = \{s \in Q \mid K \setminus \{s\} \in \mathcal{K}\}$ of K (cf. Definition 2.7, that we apply in the discriminative case).

To avoid confusion, we shall refer to the states of the Markov chain as *m-states*. The probability of a transition from m-state (K, q), with $q \in K^{\mathcal{I}}$, to some m-state $(K \cup \{s\}, s)$ — this corresponds thus to a transition from the knowledge state K to a knowledge state $K + \{s\}$ — depends only on q and s. We denote it as g_{qs}. For every m-state (K, q), the probabilities g_{qs} satisfy the constraint $\sum_{s \in K^{\mathcal{O}}} g_{qs} \leq 1$.

We also set the probability of remaining in m-state (K, q) equal to $1 - \sum_{s \in K^{\mathcal{O}}} g_{qs}$. Hence the probabilities of a transition from m-state (K, q) to any other m-state (K', s), with $K' \neq K + \{s\}$, are equal to zero. Needless to say, the empty set is also a m-state. The probability of a transition from that state to state $(\varnothing, q)$ is denoted by g_{0q}. All the other details are as in Model 1.

Probabilistic Substructures

In 1.17, we defined a substructure of a parent knowledge structure $(Q, \mathcal{K})$ as a knowledge structure $(A, \mathcal{K}|_A)$ whose states are the traces of the states of $\mathcal{K}$ on some nonempty subset A of Q, thus

$$\mathcal{K}|_A = \{L \in 2^Q \mid \exists K \in \mathcal{K} : L = K \cap A\}.$$

A motivation for this concept is that some practical knowledge structure may be so large that a direct statistical study aimed at estimating the state probabilities may not be feasible, even in the context of learning models such as those described in this chapter. In such a case, partial information on these probabilities can be obtained from studying substructures of the given knowledge structure. Consider, for example, a knowledge structure with a domain containing 400 problems of high school mathematics covering say, the last three grades. The relevant knowledge structure may contain several hundred thousands of knowledge states. The empirical analysis of such a

structure encounters two kinds of difficulties. For one, it will not be practical
to propose a 400 problem test to a large enough number of students. For
the other, even if such a test is given, the analysis of the data in terms
of probabilistic knowledge structures may be very difficult, in particular
because of the large number of parameters to be estimated. However, the
analysis of a number of shorter tests made of problems taken from the same
domain may be manageable, and would reveal useful information on the full
structure.

Notice that the probabilities of the knowledge states of the parent structure have a natural importation into any substructure. This is illustrated
in the Example below.

7.25. Example. Suppose that the states of the knowledge structure

$$\mathcal{H} = \{\varnothing, \{a\}, \{b\}, \{a,b\}, \{a,b,c\}, \{a,b,d\}, \{a,b,c,d\},$$
$$\{a,b,c,e\}, \{a,b,c,d,e\}\}$$

of our standard example occur in a population of reference with the probabilities listed below:

$$
\begin{array}{lll}
p(\varnothing) = .04 & p(\{a,b\}) = .12 & p(\{a,b,c,d\}) = .13 \\
p(\{a\}) = .10 & p(\{a,b,c\}) = .11 & p(\{a,b,c,e\}) = .18 \quad (24) \\
p(\{b\}) = .06 & p(\{a,b,d\}) = .07 & p(\{a,b,c,d,e\}) = .19
\end{array}
$$

If a test consisting only of questions a, d and e is considered, the knowledge
state $\{a,d\}$ of $\mathcal{H}' = \mathcal{H}|_{\{a,d,e\}}$ will occur in the population with a probability

$$p'(\{a,d\}) = p(\{a,b,d\}) + p(\{a,b,c,d\}) = .07 + .13 = .20.$$

Indeed, any student in state $\{a,b,d\}$ or $\{a,b,c,d\}$ of the structure $\mathcal{H}$ will
appear to be in the state $\{a,d\}$ of the structure $\mathcal{H}'$ if only questions a, d
and e are considered.

More generally, the probability $p'(J)$ of any state J of $\mathcal{H}'$ is the sum
of the probabilities $p(K)$ of all the states K of $\mathcal{H}$ having their trace on
$\{a,d,e\}$ equal to J. All the state probabilities p' are therefore as follows:

$$
\begin{array}{lll}
p'(\varnothing) = .10, & p'(\{a\}) = .33, & p'(\{a,d\}) = .20, \\
& p'(\{a,e\}) = .18, & p'(\{a,d,e\}) = .19.
\end{array}
$$

$$(25)$$

Definition 7.27 generalizes this example. We first introduce some useful notation.

7.26. Definition. Let $(Q, \mathcal{K})$ be a knowledge structure, and let $\mathcal{K}' = \mathcal{K}|_{Q'}$ be the substructure induced by $\mathcal{K}$ on some set $Q' \subset Q$. Notice that the mapping $K \mapsto K \cap Q' = J$ from $\mathcal{K}$ onto the substructure $\mathcal{K}|_{Q'}$ defines an equivalence relation on the parent structure $\mathcal{K}$, with equivalence classes

$$[K/Q'] = \{K' \in \mathcal{K} \mid K \cap Q' = K' \cap Q'\}.$$

For any $J \in \mathcal{K}|_{Q'}$ we write $J^\circ = \{K \in \mathcal{K} \mid K \cap Q' = J\}$. The family $J^\circ \subseteq \mathcal{K}$ is called the *parent family* of J, and we have $\cup_{J \in \mathcal{K}'} J^\circ = \mathcal{K}$. The ambiguity that may arise from this compact notation when more than one induced family is considered will always be removable by the context.

7.27. Definition. Suppose that $(Q, \mathcal{K}, p)$ is a probabilistic knowledge structure (cf. 7.2), with $\mathcal{K}'$ and Q' as in Definition 7.26. In this case, the triple $(Q', \mathcal{K}', p')$ is called the *(probabilistic) substructure induced by* $(Q, \mathcal{K}, p)$ on Q' if for all $J \in \mathcal{K}'$

$$p'(J) = \sum_{K \in J^\circ} p(K). \tag{26}$$

Probabilistic subspaces are defined similarly. Since the parent families J° are equivalence classes of a partition of $\mathcal{K}$, we have

$$\sum_{J \in \mathcal{K}'} p'(J) = \sum_{J \in \mathcal{K}'} \sum_{K \in J^\circ} p(K) = \sum_{K \in \mathcal{K}} p(K) = 1.$$

From the standpoint of applications, the inverse case is the important one: the state probabilities of a substructure are known, and one wishes to make some inferences on the state probabilities of the parent structure.

7.28. Example. Let $\mathcal{H}$ and $\mathcal{H}'$ be as Example 7.25, and suppose that only the state probabilities $p'(J)$, for $J \in \mathcal{H}'$, are available, their values being those given in Equation (25). For instance, the only thing we know about the state probabilities $p(\varnothing)$ and $p(\{b\})$ is that they have to satisfy $p(\varnothing) + p(\{b\}) = p'(\varnothing) = .10$. In this case, it may seem reasonable to split the mass of the state $\varnothing$ of $\mathcal{H}'$ into two equal parts, and to set $p(\varnothing) = p(\{b\}) = p'(\varnothing)/2 = .05$.

In general, the idea is to assign, for each $J \in \mathcal{H}'$, the same probability to each state of the equivalence class $J^\diamond$. The resulting probabilities for all the states of $\mathcal{H}$ are as follows

$$p(\varnothing) = .05 \qquad p(\{a,b\}) = .11 \qquad p(\{a,b,c,d\}) = .10$$
$$p(\{a\}) = .11 \qquad p(\{a,b,c\}) = .11 \qquad p(\{a,b,c,e\}) = .18$$
$$p(\{b\}) = .05 \qquad p(\{a,b,d\}) = .10 \qquad p(\{a,b,c,d,e\}) = .19$$

7.29. Definition. Let $\mathcal{K}' = \mathcal{K}|_{Q'}$ be the substructure induced by a knowledge structure $(Q,\mathcal{K})$ on $Q' \subseteq Q$, and suppose that $(Q',\mathcal{K}',p')$ is a probabilistic knowledge structure. Then $(Q,\mathcal{K},p)$ is a *uniform extension* of $(Q',\mathcal{K}',p')$ to $(Q,\mathcal{K})$ if for all $K \in \mathcal{K}$

$$p(K) = \frac{p'(K \cap Q')}{|(K \cap Q')^\diamond|}. \tag{27}$$

Nomenclatures and Classifications

We can also think of combining the information from several probabilistic substructures of a given knowledge structure (on different subsets of its domain) in order to manufacture the probabilistic parent structure. This section contains some preparatory material for such a construction.

7.30. Definition. Let $Q_1, \ldots, Q_k$ be nonempty subsets of the domain of a knowledge structure $(Q,\mathcal{K})$. If $\cup_{i=1}^k Q_i = Q$, then the collection of substructures $\mathcal{K}|_{Q_i}$, $i = 1, \ldots k$ is a *nomenclature* of $(Q,\mathcal{K})$. If $\{Q_1, \ldots, Q_k\}$ is a partition of Q, then the nomenclature is called a *classification*.

A simple result in this connection is as follows.

7.31. Theorem. Let $\{Q_1, \ldots, Q_k\}$ be a finite collection of subsets of the domain Q of a knowledge structure $\mathcal{K}$. Then, for any state K, we have

$$[K/(Q_1 \cup \cdots \cup Q_k)] = [K/Q_1] \cap \cdots \cap [K/Q_k]. \tag{28}$$

Moreover, if the collection of substructures $\mathcal{K}|_{Q_i}$ is a nomenclature, then

$$\{K\} = [K/Q_1] \cap \cdots \cap [K/Q_k]. \tag{29}$$

PROOF. Equation (28) results from the fact that

$$J \cap (Q_1 \cup \cdots \cup Q_k) = K \cap (Q_1 \cup \cdots \cup Q_k) \quad \Longleftrightarrow \quad \wedge_{i=1}^{k}(J \cap Q_i = K \cap Q_i)$$

which is easy to verify. The special case follows immediately.   □

7.32. Remark. A special case of Theorem 7.31 arises when each of the sets Q_i contains a single element. We have then, extending the notation of 1.4 and writing $\mathcal{K}_q$ (resp. $\mathcal{K}_{\bar{q}}$) for the collection of all states containing (resp. not containing) q,

$$[K/\{q\}] = \begin{cases} \mathcal{K}_q & \text{if } q \in K, \\ \mathcal{K}_{\bar{q}} & \text{if } q \notin K. \end{cases}$$

Independent Substructures

Combining the state probabilities of several substructures to construct a probability distribution for the parent structure is feasible, for instance, if the substructures can be considered as 'independent' in a sense made clear in the next definition.

7.33. Definition. Let $(Q, \mathcal{K}, p)$ be a probabilistic knowledge structure (cf. 7.2), and let $\mathbf{P}$ be the induced probability measure on the power set of $\mathcal{K}$. (Thus, for any $\mathcal{F} \subseteq \mathcal{K}$, $\mathbf{P}(\mathcal{F}) = \sum_{K \in \mathcal{F}} p(K)$.) Take two subsets $Q', Q'' \subseteq Q$, and consider the probabilistic substructures $(Q', \mathcal{K}', p')$ and $(Q'', \mathcal{K}'', p'')$ (in the sense of Definition 7.27). Two states $J \in \mathcal{K}'$, $L \in \mathcal{K}''$ are called *independent* if the events J° and L° are independent in the probability space $(\mathcal{K}, 2^{\mathcal{K}}, \mathbf{P})$, that is if

$$\mathbf{P}(J^\circ \cap L^\circ) = \mathbf{P}(J^\circ) \cdot \mathbf{P}(L^\circ) = p'(J) \cdot p''(L).$$

The substructures $(Q', \mathcal{K}', p')$ and $(Q'', \mathcal{K}'', p'')$ are *independent* if any state $K \in \mathcal{K}$ has independent traces on Q' and Q''.

These concepts extend in a natural manner to the case of an arbitrary (finite) number of substructures. Suppose that $\Xi = (Q_i, \mathcal{K}_i, p_i)_{1 \leq i \leq k}$ is a collection of substructures of a probabilistic knowledge structure $(Q, \mathcal{K}, p)$. A collection of traces $J_i \in \mathcal{K}_i$, $i = 1, \ldots, k$ is *independent* if

$$\mathbf{P}(\bigcap_{i=1}^{k} J_i^\circ) = \prod_{i=1}^{k} \mathbf{P}(J_i^\circ) = \prod_{i=1}^{k} p_i(J_i).$$

(It is not sufficient to require that the traces J_i are pairwise independent; see Problem 13.) The collection Ξ is said to be *independent* if any state K of the parent structure $\mathcal{K}$ has an independent collection of traces $K \cap Q_i$. If, in addition Ξ is a nomenclature (cf. 7.30), then it is called an *independent representation* of $(Q, \mathcal{K}, p)$. In this case, we must have for any $K \in \mathcal{K}$, using Equation (29) in Theorem 7.31,

$$P(\{K\}) = P(\cap_{i=1}^{k}[K/Q_i]) = \prod_{i=1}^{k} P([K/Q_i]). \tag{30}$$

7.34. Examples. a) With $\mathcal{H}$ and $\mathcal{H}'$ as in Example 7.25, consider the substructure

$$\mathcal{H}'' = \{\varnothing, \{b\}, \{b, c\}, \{b, c, e\}\},$$

induced by $\mathcal{H}$ on $\{b, c, e\}$. Using Equation (26) we obtain the state probabilities of $\mathcal{H}''$ from those of $\mathcal{H}$ given in Equation (24):

$$p''(\varnothing) = .14, \quad p''\{b\} = .25, \quad p''\{b, c\} = .24, \quad p''\{b, c, e\} = .37.$$

(Notice in passing that $\{\mathcal{H}', \mathcal{H}''\}$ form a nomenclature of $\mathcal{H}$.) Any state H of $\mathcal{H}$ belongs to a class of the partition generated by intersecting the classes of the two partition associated with $\mathcal{H}'$ and $\mathcal{H}''$. We have, for example, for the empty state of $\mathcal{H}$

$$\varnothing \in (\varnothing \cap \{a, d, e\})^\circ \cap (\varnothing \cap \{b, c, e\})^\circ = \{\varnothing, \{b\}\} \cap \{\varnothing, \{a\}\} = \{\varnothing\}.$$

The states $\varnothing \in \mathcal{H}'$ and $\varnothing \in \mathcal{H}''$ are not independent, since we have

$$\begin{aligned}
p'(\varnothing) \times p''(\varnothing) &= \big(p(\varnothing) + p(\{b\})\big) \times \big(p(\varnothing) + p(\{a\})\big) \\
&= (.04 + .06) \times (.04 + .10) \\
&= .014 \\
&\neq .04 = p(\varnothing).
\end{aligned}$$

Thus, the substructure $\mathcal{H}'$ and $\mathcal{H}''$ themselves are not independent (see Problem 8).

b) On the other hand, consider the knowledge structure $\{\varnothing, \{a\}, \{b\}, \{a, b\}\}$, with states probabilities

$$p(\varnothing) = .05, \quad p(\{a\}) = .15, \quad p(\{b\}) = .20, \quad p(\{a, b\}) = .60.$$

It is easy to check that the two probabilistic substructures induced on $\{a\}$ and $\{b\}$ are independent, and form thus a independent representation. (Problem 9). The following result follows immediately from Definition 7.33.

7.35. Theorem. If $(Q_i, \mathcal{K}_i, p_i)$, $i = 1, \ldots, k$ is an independent representation of a knowledge structure $(Q, \mathcal{K}, p)$, then

$$\sum_{K \in \mathcal{K}} \prod_{i=1}^{k} p_i(K \cap Q_i) = 1.$$

PROOF. Writing $\mathbf{P}$ for the probability measure induced on $2^{\mathcal{K}}$, we obtain, using Equation (30),

$$1 = \sum_{K \in \mathcal{K}} p(K) = \sum_{K \in \mathcal{K}} \mathbf{P}(\{K\}) = \sum_{K \in \mathcal{K}} \prod_{i=1}^{k} \mathbf{P}([K/Q_i]) = \sum_{K \in \mathcal{K}} \prod_{i=1}^{k} p_i(K \cap Q_i).$$

$\square$

This suggests the following construction. Consider a collection of substructures $(Q_i, \mathcal{K}_i)$ forming a representation of a basic knowledge structure $(Q, \mathcal{K})$, and suppose that the state probabilities p_i of the substructures are available. If one has reasons to believe that the substructures are approximately independent, then the state probabilities $p(K)$ of the parent could be approximated by the formula

$$p(K) = \frac{\prod_{i=1}^{k} p_i(K \cap Q_i)}{\sum_{L \in \mathcal{K}} \prod_{i=1}^{k} p_i(L \cap Q_i)}.$$

The concept of independence introduced in this section is consistent with other, standard ones. As an example, we consider the correlation between items.

7.36. Definition. Take a probabilistic knowledge structure $(Q, \mathcal{K}, p)$ and suppose that the basic local independence model holds (cf. 7.2). We have thus a collection of parameters $\beta_q, \eta_q \in [0, 1[$, $q \in Q$, specifying the response function r of Equation (6). For any $q \in Q$, define a random variable

$$\mathbf{X}_q = \begin{cases} 1 & \text{if the subject's response is correct,} \\ 0 & \text{otherwise.} \end{cases}$$

The $\mathbf{X}_q$'s will be called *item indicator* random variables.

7.37. Theorem. *Assume the basic probabilistic model holds, and consider the following three conditions, for q, $q' \in Q$ with $q \neq q'$.*

(i) *the item indicator random variables $\mathbf{X}_q$, $\mathbf{X}_{q'}$ are independent;*

(ii) *their covariance vanishes: $\mathrm{Cov}(\mathbf{X}_q, \mathbf{X}_{q'}) = 0$;*

(iii) *the probabilistic substructures $\{\{\varnothing, \{q\}\}\}$ and $\{\varnothing, \{q'\}\}$ induced on $\{q\}$ and $\{q'\}$, respectively, are independent.*

Then $(i) \Leftrightarrow (ii) \Rightarrow (iii)$.

PROOF. We write $\mathcal{K}$ for the knowledge structure, and $\mathbf{P}$ for the probability measure on $2^{\mathcal{K}}$. Let $\beta_q, \eta_q \in [0, 1[$ be the response parameters. Since

$$\mathrm{Cov}(\mathbf{X}_q, \mathbf{X}_{q'}) = \mathrm{E}(\mathbf{X}_q \mathbf{X}_{q'}) - \mathrm{E}(\mathbf{X}_q)\mathrm{E}(\mathbf{X}_{q'})$$
$$= \mathbf{P}(\mathbf{X}_q = 1, \mathbf{X}_{q'} = 1) - \mathbf{P}(\mathbf{X}_q = 1)\mathbf{P}(\mathbf{X}_{q'} = 1), \qquad (31)$$

the equivalence of (i) and (ii) is clear. Developing the right member of Equation (31), using Equations (5) and (6), we obtain

$$\mathbf{P}(\mathcal{K}_q \cap \mathcal{K}_{q'})(1 - \beta_q)(1 - \beta_{q'}) + \mathbf{P}(\mathcal{K}_q \cap \mathcal{K}_{\overline{q'}})(1 - \beta_q)\eta_{q'}$$
$$+ \mathbf{P}(\mathcal{K}_{\overline{q}} \cap \mathcal{K}_{q'})\eta_q(1 - \beta_{q'}) + \mathbf{P}(\mathcal{K}_{\overline{q}} \cap \mathcal{K}_{\overline{q'}})\eta_q \eta_{q'}$$
$$- \big(\mathbf{P}(\mathcal{K}_q)(1 - \beta_q) + \mathbf{P}(\mathcal{K}_{\overline{q}})\eta_q\big)\big(\mathbf{P}(\mathcal{K}_{q'})(1 - \beta_{q'}) + \mathbf{P}(\mathcal{K}_{\overline{q'}})\eta_{q'}\big)$$
$$= \big(\mathbf{P}(\mathcal{K}_q \cap \mathcal{K}_{q'}) - \mathbf{P}(\mathcal{K}_q)\mathbf{P}(\mathcal{K}_{q'})\big)(1 - \beta_q)(1 - \beta_{q'})$$
$$+ \big(\mathbf{P}(\mathcal{K}_q \cap \mathcal{K}_{\overline{q'}}) - \mathbf{P}(\mathcal{K}_q)\mathbf{P}(\mathcal{K}_{\overline{q'}})\big)(1 - \beta_q)\eta_{q'}$$
$$+ \big(\mathbf{P}(\mathcal{K}_{\overline{q}} \cap \mathcal{K}_{q'}) - \mathbf{P}(\mathcal{K}_{\overline{q}})\mathbf{P}(\mathcal{K}_{q'})\big)\eta_q(1 - \beta_{q'})$$
$$+ \big(\mathbf{P}(\mathcal{K}_{\overline{q}} \cap \mathcal{K}_{\overline{q'}}) - \mathbf{P}(\mathcal{K}_{\overline{q}})\mathbf{P}(\mathcal{K}_{\overline{q'}})\big)\eta_q \eta_{q'}. \qquad (32)$$

By definition, the substructures $\{\{\varnothing, \{q\}\}\}$ and $\{\varnothing, \{q'\}\}$ are independent if and only if for every state K of the parent, $[K/\{q\}]$ and $[K/\{q'\}]$ are independent events, in other words if and only if

$$\mathbf{P}(\mathcal{K}_q \cap \mathcal{K}_{q'}) - \mathbf{P}(\mathcal{K}_q)\mathbf{P}(\mathcal{K}_{q'}) = \mathbf{P}(\mathcal{K}_q \cap \mathcal{K}_{\overline{q'}}) - \mathbf{P}(\mathcal{K}_q)\mathbf{P}(\mathcal{K}_{\overline{q'}})$$
$$= \mathbf{P}(\mathcal{K}_{\overline{q}} \cap \mathcal{K}_{q'}) - \mathbf{P}(\mathcal{K}_{\overline{q}})\mathbf{P}(\mathcal{K}_{q'})$$
$$= \mathbf{P}(\mathcal{K}_{\overline{q}} \cap \mathcal{K}_{\overline{q'}}) - \mathbf{P}(\mathcal{K}_{\overline{q}})\mathbf{P}(\mathcal{K}_{\overline{q'}})$$
$$= 0.$$

Substituting in the right member of Equation (32), and noticing that the values of the parameters can be chosen arbitrarily in the interval $[0, 1]$, the implication $(ii) \Rightarrow (iii)$ follows. $\qquad \square$

Original Sources and Related Works

Probabilistic concepts were introduced in knowledge space theory by Falmagne and Doignon (1988a, b; see also Villano et al. 1987, Falmagne 1989a, b).

Several researchers have applied the basic local independence model to real data (Falmagne et al., 1990; Villano, 1991; Lakshminarayan and Gilson, 1993; Taagepera et al., 1997; Lakshminarayan, 1995). Villano (1991)'s work deserves a special mention because of the systematic way he constructed a large knowledge space by testing successive uniform extensions of smaller ones. The method of Cosyn and Thiéry (1999) reviewed in Chapter 12, which relies in part on Villano's techniques but also applies a sophisticated type of QUERY algorithm to question the experts, has been used successfully in the ALEKS system.

The Markov models described in 7.22 and 7.24 were proposed by Falmagne (1994). Some recent empirical tests of these models are described in Fries (1997).

All the Markov chain concepts used in this chapter are standard. For an introduction to Markov chain theory, we have already referred the reader to Feller (1968), Kemeny and Snell (1960) or Parzen (1962).

Problems

1. Modify the local independence assumption in such a manner that the probability of a response pattern R, given a state K varies with the subject. You should obtain an explicit formula for the probability $\rho(R)$ of obtaining a response pattern R in that case.

2. Compute the number of degrees of freedom of the chi-square for the basic probabilistic model, in the case of a 7 item test with open responses. What would this number be in a multiple choice situation, in which 5 alternatives are proposed for each question, and assuming that the options have been designed so as to make all guessing probabilities equal to $\frac{1}{5}$?

3. Generalize the simple learning model by assuming that the probability of mastering an item may depend upon past events, in particular, upon the last item mastered.

4. Suppose that the parent structure satisfies the simple learning model. Does any substructure also satisfies that model? More generally, which of the models discussed in the chapter is preserved under substructures?

5. Could a substructure satisfy the simple learning model, while the parent structure does not satisfy that model?

6. Generalize Markov Chain Model 1, by assuming that subjects of the sample may have different learning step numbers. Specifically, assume that the learning step has a negative binomial distribution, and derive the predictions permitting a test of the model.

7. Develop Markov Chain Model 1 in order to predict the results of a sample of subjects tested at two different times (cf. Remark 7.23(c)).

8. Modify the state probabilities of the knowledge structure of the standard example, in such a manner that the states $\varnothing \in \mathcal{H}'$ and $\varnothing \in \mathcal{H}''$ are independent but the two substructures themselves are not (cf. Example 7.34(a)).

9. In Example 7.34(b), consider the knowledge structure $\{\varnothing, \{a\}, \{b\}, \{a, b\}\}$, with states probabilities $p(\varnothing) = .05$, $p(\{a\}) = .15$, $p(\{b\}) = .20$, $p(\{a, b\}) = .60$. Prove that the two probabilistic substructures induced on $\{a\}$ and $\{b\}$ are independent, and form thus an independent representation.

10. In the Example of Problem 9, the nomenclature forming the representation was a classification (Definition 7.30). Is this condition necessary? Prove your response.

11. If $\Xi = (Q_i, \mathcal{K}_i, p_i)_{1 \leq i \leq k}$ is an independent representation of a probabilistic knowledge structure $(Q, \mathcal{K}, p)$, then the substructures form a nomenclature and are independent in the sense of Definition 7.33. The latter condition means that the traces of any $K \in \mathcal{K}$ are independent. Does this imply that, for any distinct states K, K' of $\mathcal{K}$ and with $i \neq j$, $K \cap Q_i$ and $K' \cap Q_j$ are also independent?

12. Generalize the Simple Learning Model in the case of a knowledge structure which is not (necessarily) well-graded (cf. 7.15).

13. Show by an example that a knowledge structure may have a collection Ξ of pairwise independent substructures, without having Ξ itself being independent in the sense of Definition 7.33.

14. Find an expression for $\mathrm{Cov}(\mathbf{X}_q, \mathbf{X}_{q'})$ when the knowledge structure is a chain, and the parameters $\beta_q, \beta_{q'}, \eta_q, \eta_{q'}$ vanish (cf. 7.33).

Chapter 8

Stochastic Learning Paths

The stochastic theory presented in this chapter is more ambitious than those examined in Chapter 7. The description of the learning process is more complete and takes place in real time, rather than in a sequence of discrete trials. This theory also contains a provision for individual differences. Nevertheless, its basic intuition is similar, in that that any student progresses through some learning path. As time passes, the student successively masters the items—or the associated concepts—encountered along the learning path. The exposition of the theory given here follows closely Falmagne (1993, 1996). As before, we shall illustrate the concepts of the theory in the framework of an example.

A Knowledge Structure in Euclidean Geometry

This empirical application is due to Lakshminarayan (1995), and was alluded to in 0.3. It involves five problems in high school geometry, which are displayed in Figure 8.1. These five items were part of a test given to 959 undergraduate students at the University of California at Irvine. There were two consecutive applications of the test, separated by a short lecture recalling some fundamental facts of Euclidean geometry. Note that the sets of problems in the two applications were equivalent but not identical. (In the terminology introduced in 0.1, the problems in the second application were different instances of the same five items.) The analysis of the data[1], for the five items a, b, c, d and e, resulted in the knowledge structure below, which is represented by the graph in the lower right of Figure 8.2:

$$\mathcal{L} = \{\, \varnothing,\, \{a\}, \{b\},\, \{a,b\},\, \{a,d\},\, \{b,c\},\, \{a,b,c\},$$
$$\{a,b,d\},\, \{b,c,d\},\, \{a,b,c,d\},\, \{a,b,c,d,e,\}\,\}. \qquad (1)$$

---

[1] The details of Lakshminarayan (1995) study are reported later in this chapter.

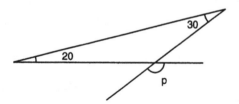

(a) In the figure above, what is the measure
of angle p? Give your answer in degrees.

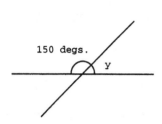

(b) In the figure above, what is the
measure of angle y in degrees?

(c) In the figure above, line L is parallel
to line M. Angle x is 55 degrees. What
is the measure of angle y in degrees?

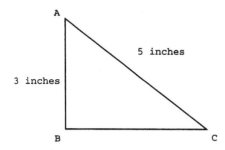

(d) In the triangle shown above, side AB
has the length of 3 inches, and side AC
has the length of 5 inches. Angle ABC is
90 degrees. What is the area of the
triangle?

(e) In the above quadrilateral, side AB
= 1 inch. The angles are as marked in
the figure. The angles marked X are
all equal to each other. What is the
perimeter of the figure ABCD?

Fig. 8.1. The five geometry items in Lakshminarayan's study.

Basic Concepts

The knowledge structure $\mathcal{L}$ is a knowledge space. It has seven learning paths, and all of them happen to be gradations. This is not true for all knowledge spaces. In general, it may happen that the state K' immediately succeeding some state K in a given learning path contains more than one additional item. The theory presented here is suitable for well-graded knowledge structures (cf. 2.4 and 2.9); thus, all learning paths are gradations but closure under union may not be satisfied. Students may differ from each other not only by the particular gradation followed, but also by their learning rate. We assume that, for a given population of subjects, the learning rate is a random variable, which will be denoted by $\mathbf{L}$. A graphical illustration of a density function for this random variable $\mathbf{L}$ is given at the upper left of Figure 8.2. Consider a student equipped with a learning rate $\mathbf{L} = \lambda$, and suppose that this student, for some accidental reasons, is engaged in the gradation

$$\varnothing \subset \{b\} \subset \{b, c\} \subset \{a, b, c\} \subset \{a, b, c, d\} \subset \{a, b, c, d, e\} \qquad (2)$$

marked by the string of arrows in Figure 8.2, starting from the empty box.

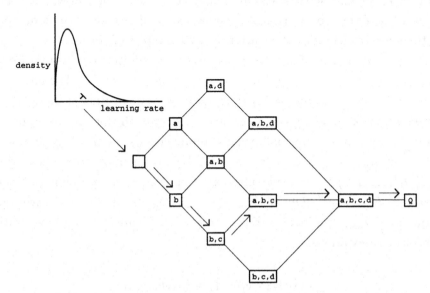

Fig. 8.2. The gradation followed by a certain student in the knowledge structure $\mathcal{L}$.

We also assume that, at the outset, this student is naive. Thus, the first state encountered will be $\varnothing$. Reaching the next state requires mastering the concepts associated with item b. This may or not be laborious, depending on the learning rate of the student, and the difficulty of the item. In the theory developed in this chapter, the time required to reach state $\{b\}$ is a random variable, the distribution of which depends upon these two factors: the difficulty of the item and the student's learning rate, represented here by the number λ. The item difficulty will be measured by a parameter of the theory.

The next state encountered along the gradation in Equation (2) is $\{b, c\}$. Again, the time required for a transition from state $\{b\}$ to state $\{b, c\}$ is a random variable, with a distribution depending upon λ and the difficulty of item c. And so on. If a test takes place, the student will respond according to her current state. However, we do not assume that the responses to items contained in the current state will necessarily be correct. As in Chapter 7, we suppose that careless errors and lucky guesses are always possible. The exact rule specifying the probabilities of both kinds of events will be exactly the same as before (cf. Equation (6) in Chapter 7). If a student is tested n times, at times $t_1, \ldots, t_n$, the n-tuple of observed patterns of responses will depend upon the n knowledge states occupied at the times of testing. Of course, the knowledge states are not directly observable. However, from the axioms of the theory, explicit formulas can be derived for the prediction of the joint probabilities of occurrence of all n-tuples of states.

A formal statement of the theory requires a few additional concepts. Consider a fixed, finite, well-graded knowledge structure $(Q, \mathcal{K})$. For any real number $t \geq 0$, we denote by $\mathbf{K}_t$ the knowledge state at time t; thus, $\mathbf{K}_t$ is a random variable taking its values in $\mathcal{K}$. We write $\mathcal{G}_{\mathcal{K}}$ for the collection of all gradations in $\mathcal{K}$. We shall usually abbreviate $\mathcal{G}_{\mathcal{K}}$ as $\mathcal{G}$ when no ambiguity can arise. Let $K \subset K' \subset \cdots \subset K''$ be any chain of states. We write $\mathcal{G}(K, K', \ldots, K'')$ for the subcollection of all gradations containing all the states $K, K', \ldots, K''$. Obviously, $\mathcal{G}(\varnothing) = \mathcal{G}(Q) = \mathcal{G}$, and in the knowledge structure $(Q, \mathcal{L})$ specified by Equation (1) (cf. Figure 8.2), we have, with obvious abbreviations,

$$\mathcal{G}(\{a\}, \{a, b, c, d\}) = \{\nu_1, \nu_2, \nu_3\},$$
$$\nu_1 = adbce, \quad \nu_2 = abdce, \quad \nu_3 = abcde.$$

The gradation taken by a subject will also be regarded as a random variable. We write $\mathbf{C} = \nu$ to signify that ν is the gradation followed by the student. Thus, the random variable $\mathbf{C}$ takes its values in $\mathcal{G}$. The existence of a probability distribution on the set of all gradations is a fairly general hypothesis. It involves, as possible special cases, various mechanisms describing the gradation followed by the subject as resulting from a succession of choices along the way. We shall go back to this issue later in this chapter.

For convenient reference, all the basic concepts are recalled in a glossary.

8.1. Glossary.

Q	a finite, nonempty set, called the *domain*;
$\mathcal{K}$	a well-graded knowledge structure on Q;
$\mathbf{K}_t = K \in \mathcal{K}$	the subject is in state K at time $t \geq 0$;
$\mathbf{R}_t = R \in 2^Q$	specifies the set R of correct responses given by the subject at time $t \geq 0$;
$\mathbf{L} = \lambda$	the learning rate of the subject is equal to $\lambda \geq 0$;
$\mathcal{G}$	the collection of all gradations;
$\mathcal{G}(K, K', \ldots)$	the collection of all gradations containing $K, K', \ldots$;
$\mathbf{C} = \nu \in \mathcal{G}$	the gradation of the subject is ν.

8.2. General Axioms. We begin by formulating four axioms specifying the general stochastic structure of the theory. Precise hypotheses concerning the distributions of the random variables $\mathbf{L}$ and $\mathbf{C}$ and the conditional probabilities of the response patterns, given the states, will be made in later sections of this chapter. Each axiom is followed by a paraphrase in plain language. These axioms define, up to some functional relations, the probability measure associated with the triple

$$((\mathbf{R}_t, \mathbf{K}_t)_{t \geq 0}, \mathbf{C}, \mathbf{L}).$$

The selection of a subject in the population corresponds to a choice of values $\mathbf{L} = \lambda$ and $\mathbf{C} = \nu$ for the learning rate and the gradation of the subject, respectively. In turn, these values specify a stochastic process

$$((\mathbf{R}_t, \mathbf{K}_t)_{t \geq 0}, \nu, \lambda)$$

describing the progress of the particular subject through the material, along gradation ν.

[B] Beginning State. For all $\nu \in \mathcal{G}$ and $\lambda \in \mathbf{R}$,

$$\mathbf{P}(\mathbf{K}_0 = \varnothing \,|\, \mathbf{L} = \lambda, \mathbf{C} = \nu) = 1.$$

(*With probability one, the initial state of the subject is the empty state, independently of the gradation and of the learning rate*[2].)

[I] Independence of Gradation and Learning Rate. The random variables $\mathbf{L}$ and $\mathbf{C}$ are independent. That is, for all $\lambda \in \mathbf{R}$ and $\nu \in \mathcal{G}$,

$$\mathbf{P}(\mathbf{L} \le \lambda, \mathbf{C} = \nu) = \mathbf{P}(\mathbf{L} \le \lambda) \cdot \mathbf{P}(\mathbf{C} = \nu).$$

(*The gradation followed is independent of the learning rate.*)

[R] Response Rule. There is a function $r : 2^Q \times \mathcal{K} \to [0, 1]$ such that, for all $\nu \in \mathcal{G}$, all positive integers n, all real numbers λ and $t_n > t_{n-1} > \cdots > t_1 \ge 0$, and any event $\mathcal{E}$ determined only by

$$\big((\mathbf{R}_{t_{n-1}}, \mathbf{K}_{t_{n-1}}), (\mathbf{R}_{t_{n-2}}, \mathbf{K}_{t_{n-2}}), \ldots, (\mathbf{R}_{t_1}, \mathbf{K}_{t_1})\big),$$

we have

$$\begin{aligned}
\mathbf{P}(\mathbf{R}_{t_n} = R_{t_n} \,|\, \mathbf{K}_{t_n} = K_n, \mathcal{E}, \mathbf{L} = \lambda, \mathbf{C} = \nu) &= \mathbf{P}(\mathbf{R}_{t_n} = R_{t_n} \,|\, \mathbf{K}_{t_n} = K_n) \\
&= r(R_n, K_n).
\end{aligned}$$

(*If the state at a given time is known, the probability of a response pattern at that time only depends upon that state through the function r. It is independent of the learning rate, the gradation, and any sequence of past states and responses.*)

[L] Learning Rule. There is a function $\ell : \mathcal{K} \times \mathcal{K} \times \mathbf{R}^2 \times \mathcal{G} \to [0, 1]$, such that, for all $\nu \in \mathcal{G}$, all positive integers n, all real numbers λ and $t_{n+1} > t_n > \cdots > t_1 \ge 0$, and any event $\mathcal{E}$ determined only by

$$\big((\mathbf{R}_{t_{n-1}}, \mathbf{K}_{t_{n-1}}), (\mathbf{R}_{t_{n-2}}, \mathbf{K}_{t_{n-2}}), \ldots, (\mathbf{R}_{t_1}, \mathbf{K}_{t_1})\big),$$

---

[2] As $\mathbf{L}$ is a real-valued random variable, the event $\mathbf{L} = \lambda$ may have measure 0. As usual, the conditional probability in Axiom [B] and in other similar expressions in this chapter is defined by a limiting operation (cf. Parzen, 1960, page 335).

we have

$$\mathbf{P}(\mathbf{K}_{t_{n+1}} = K_{n+1} \,|\, \mathbf{K}_{t_n} = K_n, \mathbf{R}_n = R_n, \mathcal{E}, \mathbf{L} = \lambda, \mathbf{C} = \nu)$$
$$= \mathbf{P}(\mathbf{K}_{t_{n+1}} = K_{n+1} \,|\, \mathbf{K}_{t_n} = K_n, \mathbf{L} = \lambda, \mathbf{C} = \nu)$$
$$= \ell(K_n, K_{n+1}, t_{n+1} - t_n, \lambda, \nu).$$

Moreover, the function ℓ is assumed to satisfy the following two conditions:
For any $K, K' \in \mathcal{K}$, $\nu, \nu' \in \mathcal{G}$, $\delta > 0$ and any $\lambda \in \mathbf{R}$,

$$\ell(K, K', \delta, \lambda, \nu) = 0 \quad \text{if} \quad \nu \notin \mathcal{G}(K, K'); \tag{L1}$$
$$\ell(K, K', \delta, \lambda, \nu) = \ell(K, K', \delta, \lambda, \nu') \quad \text{if} \quad \nu, \nu' \in \mathcal{G}(K, K'). \tag{L2}$$

(*The probability of a state at a given time only depends upon the last state recorded, the time elapsed, the learning rate and the gradation. It does not depend upon previous states and responses. This probability is the same for all gradations containing both the last recorded state and the new state, and vanishes otherwise.*)

8.3. Definition. A triple $\mathcal{S} = ((\mathbf{K}_t, \mathbf{R}_t)_{t \geq 0}, \mathbf{C}, \mathbf{L})$ satisfying the four Axioms [B], [I], [R] and [L] is a *system of stochastic learning paths*, or more briefly, a *SLP system*. The functions r and ℓ of Axioms [R] and [L] will be called the *response rule* and the *learning rule* of the system $\mathcal{S}$, respectively.

Of special interest is another condition on the learning rule ℓ which essentially states that there is no forgetting. When ℓ also satisfies

$$\ell(K, K', \delta, \lambda, \nu) = 0 \quad \text{if} \quad K \not\subseteq K', \tag{L3}$$

for all $K, K' \in \mathcal{K}$, $\nu \in \mathcal{G}$, $\delta > 0$ and $\lambda \in \mathbf{R}$, we shall say that $\mathcal{S}$ is *progressive*. Only progressive SLP systems will be considered in this chapter (see Problem 1, however).

8.4. Remarks. a) One may object to Axiom [B] that when a subject is tested, the time elapsed since the beginning of learning—the index t in $\mathbf{R}_t$ and $\mathbf{K}_t$—is not always known exactly. We may perhaps assume, for example, that the learning of mathematics begins roughly at age 2 or 3, but this may not be precise enough. In some situations, there is no difficulty in considering at least some of these indexed times as parameters to be estimated from the data.

b) Axiom [I], which concerns the independence of the learning rate and the learning path, was hard to avoid—however unrealistic it may perhaps appear. Its strong appeal is that it renders the derivations relatively straightforward. For the time being, there does not seem to be any obvious alternative. Our hope is that the predictions of the model will be robust to that assumption.

c) Axiom [R] seems reasonable enough. This axiom will be specialized in a later section as the local independence assumption already encountered in Chapter 7 (see 7.2).

d) On the other hand, the assumptions embedded in [L] require a thorough examination. To begin with, a two or k-dimensional version of the learning rate may be required in some cases. The number λ in the argument of the function ℓ would then be replaced by a real vector, the components of which would measure different aspect of the learning process. The theory could be elaborated along such lines if the need arises.

Also, one might be tempted to formulate, instead of Axiom [L], the much stronger Markovian condition formalized by the equation

$$\mathbf{P}(\mathbf{K}_{t_{n+1}} = K_{n+1} \mid \mathbf{K}_{t_n} = K_n, \mathbf{R}_{t_n} = R_n, \mathcal{E}) = \mathbf{P}(\mathbf{K}_{t_{n+1}} = K_{n+1} \mid \mathbf{K}_{t_n} = K_n),$$

(with $\mathcal{E}$ as in Axiom [L]). Thus, the probability of a state at time t_{n+1} would only depend upon the last recorded state, and possibly, the time elapsed since that observation. This assumption is inappropriate because the more detailed history embedded in the event $\mathcal{E}$ in the left member of the above equation may provide information on the learning rate and the learning path, which in turn, may modify the probabilities of the states at time t_{n+1}. In Problem 2, we ask the reader to provide a numerical example falsifying this equation, based on the domain Q and the knowledge structure $\mathcal{L}$ of Equation (1) and Figure 8.2.

Much more can be said about Axiom [L]. It turns out that, in the framework of the other Axioms [B], [I] and [R], this axiom considerably restricts the distributional form of the latencies (see 8.13). We postpone further discussion of Axiom [L] to derive a few results which only depend upon Axioms [B], [I], [R] and [L]. In other words, no assumptions are made regarding the functional form of r or ℓ.

Basic Results

We begin with a preparatory result.

8.5. Theorem. *For all integer $n > 0$, all real numbers $t_n > \cdots > t_1 \geq 0$, and any event $\mathcal{E}$ determined only by $(\mathbf{R}_{t_{n-1}}, \mathbf{R}_{t_{n-2}}, \dots, \mathbf{R}_{t_1})$, we have*

$$\mathbf{P}(\mathbf{K}_{t_n} = K_n \,|\, \mathbf{K}_{t_{n-1}} = K_{n-1}, \dots, \mathbf{K}_{t_1} = K_1, \mathcal{E})$$
$$= \mathbf{P}(\mathbf{K}_{t_n} = K_n \,|\, \mathbf{K}_{t_{n-1}} = K_{n-1}, \dots, \mathbf{K}_{t_1} = K_1).$$

8.6. Convention. To lighten the notation, we occasionally use the abbreviations:

$$\kappa_n = \cap_{i=1}^n [\mathbf{K}_{t_i} = K_i],$$
$$\rho_n = \cap_{i=1}^n [\mathbf{R}_{t_i} = R_i].$$

Notice that the choice of the times $t_n > \cdots > t_1 \geq 0$ is implicit in this notation. We also write p_ν for $\mathbf{P}(\mathbf{C} = \nu)$. Our proof of Theorem 8.5 requires a preparatory result.

8.7. Lemma. *For any real number λ and any learning path ν, and with $\mathcal{E}$ only depending on ρ_n, we have*

$$\mathbf{P}(\mathcal{E} \,|\, \kappa_n, \mathbf{L} = \lambda, \mathbf{C} = \nu) = \mathbf{P}(\mathcal{E} \,|\, \kappa_n).$$

PROOF: By induction, using Axioms [R] and [L] in alternation (Problem 11). $\square$

8.8. Proof of Theorem 8.5. By Lemma 8.7, for any positive integer n,

$$\frac{\mathbf{P}(\kappa_n, \mathcal{E} \,|\, \mathbf{L} = \lambda, \mathbf{C} = \nu)}{\mathbf{P}(\kappa_n, \mathcal{E})} = \frac{\mathbf{P}(\mathcal{E} \,|\, \kappa_n, \mathbf{L} = \lambda, \mathbf{C} = \nu)\mathbf{P}(\kappa_n \,|\, \mathbf{L} = \lambda, \mathbf{C} = \nu)}{\mathbf{P}(\mathcal{E} \,|\, \kappa_n)\mathbf{P}(\kappa_n)}$$
$$= \frac{\mathbf{P}(\mathcal{E} \,|\, \kappa_n)\mathbf{P}(\kappa_n \,|\, \mathbf{L} = \lambda, \mathbf{C} = \nu)}{\mathbf{P}(\mathcal{E} \,|\, \kappa_n)\mathbf{P}(\kappa_n)}$$
$$= \frac{\mathbf{P}(\kappa_n \,|\, \mathbf{L} = \lambda, \mathbf{C} = \nu)}{\mathbf{P}(\kappa_n)} = g(\kappa_n, \lambda, \nu),$$

the last equality defining the function g. Setting $p_\nu = \mathbf{P}(\mathbf{C} = \nu)$, we have successively

$$\mathbf{P}(\mathbf{K}_{t_n} = K_n \mid \kappa_{n-1}, \mathcal{E}) = \int_{-\infty}^{\infty} \sum_{\nu \in \mathcal{G}} \mathbf{P}(\mathbf{K}_{t_n} = K_n \mid \kappa_{n-1}, \mathcal{E}, \mathbf{L} = \lambda, \mathbf{C} = \nu)$$

$$\times \frac{\mathbf{P}(\kappa_{n-1}, \mathcal{E} \mid \mathbf{L} = \lambda, \mathbf{C} = \nu)}{\mathbf{P}(\kappa_{n-1}, \mathcal{E})} \, p_\nu \, d\mathbf{P}(\mathbf{L} \le \lambda)$$

$$= \int_{-\infty}^{\infty} \sum_{\nu \in \mathcal{G}} \mathbf{P}(\mathbf{K}_{t_n} = K_n \mid \kappa_{n-1}, \mathbf{L} = \lambda, \mathbf{C} = \nu) \, g(\kappa_{n-1}, \lambda, \nu) \, p_\nu \, d\mathbf{P}(\mathbf{L} \le \lambda)$$

$$= \mathbf{P}(\mathbf{K}_{t_n} = K_n \mid \kappa_{n-1}). \qquad \square$$

As indicated, the results concerning the observable patterns of responses are our prime concern. However, an examination of Axiom [R] suggests that these results could be derived from a study of the process $(\mathbf{K}_t, \mathbf{L}, \mathbf{C})_{t \ge 0}$. The next Theorem makes this idea precise.

8.9. Theorem. *For any positive integer* n, *any response patterns* $R_i \in 2^Q$, $1 \le i \le n$, *and any real numbers* $t_n > t_{n-1} > \cdots > t_1 \ge 0$,

$$\mathbf{P}\big(\cap_{i=1}^n [\mathbf{R}_{t_i} = R_i] \big) = \sum_{K_1 \in \mathcal{K}} \cdots \sum_{K_n \in \mathcal{K}} \Big(\prod_{i=1}^n r(R_i, K_i) \Big) \mathbf{P}\big(\cap_{i=1}^n [\mathbf{K}_{t_i} = K_i] \big).$$

PROOF: We have

$$\mathbf{P}\big(\cap_{i=1}^n [\mathbf{R}_{t_i} = R_i] \big) = \mathbf{P}(\rho_n) = \sum_{(\kappa_n)} \mathbf{P}(\rho_n, \kappa_n). \qquad (3)$$

Developing the general term, we obtain successively, using Axiom [R] and Theorem 8.5,

$$\mathbf{P}(\rho_n, \kappa_n) =$$
$$\mathbf{P}(\mathbf{R}_{t_n} = R_n \mid \rho_{n-1}, \kappa_n) \mathbf{P}(\mathbf{K}_{t_n} = K_n \mid \rho_{n-1}, \kappa_{n-1})$$
$$\times \mathbf{P}(\mathbf{R}_{t_{n-1}} = R_{n-1} \mid \rho_{n-2}, \kappa_{n-1}) \mathbf{P}(\mathbf{K}_{t_{n-1}} = K_{n-1} \mid \rho_{n-2}, \kappa_{n-2}) \times \cdots$$
$$\times \mathbf{P}(\mathbf{R}_{t_2} = R_2 \mid \mathbf{K}_{t_1} = K_1, \mathbf{R}_{t_1} = R_1, \mathbf{K}_{t_2} = K_2)$$
$$\times \mathbf{P}(\mathbf{K}_{t_2} = K_2 \mid \mathbf{K}_{t_1} = K_1, \mathbf{R}_{t_1} = R_1)$$
$$\times \mathbf{P}(\mathbf{R}_{t_1} = R_1 \mid \mathbf{K}_{t_1} = K_1) \mathbf{P}(\mathbf{K}_{t_1} = K_1)$$
$$= \big(r(R_n, K_n) r(R_{n-1}, K_{n-1}) \cdots r(R_1, K_1) \big) \mathbf{P}(\mathbf{K}_{t_n} = K_n \mid \kappa_{n-1})$$
$$\times \mathbf{P}(\mathbf{K}_{t_{n-1}} = K_{n-1} \mid \kappa_{n-2}) \cdots \mathbf{P}(\mathbf{K}_{t_2} = K_2 \mid \mathbf{K}_{t_1} = K_1) \mathbf{P}(\mathbf{K}_{t_1} = K_1)$$

$$= \left(\prod_{i=1}^{n} r(R_i, K_i)\right) \mathbf{P}(\kappa_n). \tag{4}$$

The result follows from (3) and (4). $\qquad\qquad\qquad\qquad\qquad\qquad$ $\square$

Thus, the joint probabilities of the response patterns can be obtained from the joint probabilities of the states, and from the conditional probabilities captured by the function r. We now turn to a study of the process $(\mathbf{K}_t, \mathbf{L}, \mathbf{C})_{t \geq 0}$.

8.10. Theorem. *For all real numbers $\lambda, t > 0$, and all $\nu \in \mathcal{G}$,*

$$\mathbf{P}(\mathbf{K}_t = K \mid \mathbf{L} = \lambda, \mathbf{C} = \nu) = \ell(\varnothing, K, t, \lambda, \nu).$$

PROOF: By Axiom [B], $\mathbf{P}(\mathbf{K}_0 = K' \mid \mathbf{L} = \lambda, \mathbf{C} = \nu) = 0$ for any $K' \neq \varnothing$. We obtain:

$$\mathbf{P}(\mathbf{K}_t = K \mid \mathbf{L} = \lambda, \mathbf{C} = \nu)$$
$$= \sum_{K' \in \mathcal{K}} \mathbf{P}(\mathbf{K}_t = K \mid \mathbf{K}_0 = K', \mathbf{L} = \lambda, \mathbf{C} = \nu) \mathbf{P}(\mathbf{K}_0 = K' \mid \mathbf{L} = \lambda, \mathbf{C} = \nu)$$
$$= \mathbf{P}(\mathbf{K}_t = K \mid \mathbf{K}_0 = \varnothing, \mathbf{L} = \lambda, \mathbf{C} = \nu)$$
$$= \ell(\varnothing, K, t, \lambda, \nu). \qquad\qquad\qquad\qquad\qquad\qquad\qquad\qquad \square$$

8.11. Theorem. *For all integers $n > 0$ and times $t_n > \cdots > t_1 > t_0 = 0$, and with $K_0 = \varnothing$,*

$$\mathbf{P}(\mathbf{K}_{t_1} = K_1, \ldots, \mathbf{K}_{t_n} = K_n)$$
$$= \int_{-\infty}^{\infty} \sum_{\nu \in \mathcal{G}} \left(\prod_{i=0}^{n-1} \ell(K_i, K_{i+1}, t_{i+1} - t_i, \lambda, \nu)\right) p_\nu \, d\mathbf{P}(\mathbf{L} \leq \lambda).$$

PROOF: With the notation κ_n as in Convention 8.6, we have

$$\mathbf{P}(\kappa_n) = \int_{-\infty}^{\infty} \sum_{\nu \in \mathcal{G}} \mathbf{P}(\kappa_n \mid \mathbf{L} = \lambda, \mathbf{C} = \nu) p_\nu \, d\mathbf{P}(\mathbf{L} \leq \lambda). \tag{5}$$

Using Axiom [L] and Theorem 8.10,

$$P(\kappa_n \,|\, \mathbf{L} = \lambda, \mathbf{C} = \nu)$$
$$= P(\mathbf{K}_{t_n} = K_{t_n} \,|\, \kappa_{n-1}, \mathbf{L} = \lambda, \mathbf{C} = \nu)\, P(\mathbf{K}_{t_{n-1}} = K_{t_{n-1}} \,|\, \kappa_{n-2}, \mathbf{L} = \lambda, \mathbf{C} = \nu)$$
$$\cdots P(\mathbf{K}_{t_2} = K_2 \,|\, \mathbf{K}_{t_1} = K_1, \mathbf{L} = \lambda, \mathbf{C} = \nu)\, P(\mathbf{K}_{t_1} = K_1 \,|\, \mathbf{L} = \lambda, \mathbf{C} = \nu)$$
$$= \prod_{i=0}^{n-1} \ell(K_i, K_{i+1}, t_{i+1} - t_i, \lambda, \nu).$$

The result obtains after substituting in (5).   □

Assumptions on Distributions

This theory is of limited practical use without making specific hypotheses concerning the distributions of the random variable $\mathbf{L}$ measuring the learning rate, and the random variables implicit in the expression of the learning rule ℓ of Axiom [L], which govern the time required to master the items. We also need to specify the response rule r of Axiom [R], representing the probabilities of the response patterns, conditional to the states. We formulate here two axioms specifying the response rule and the distribution of the learning rate. The distributions of the learning latencies are discussed in the next section. It is fair to say that our choice of axioms result from a compromise between realism and applicability. These axioms are by no means the only feasible ones. Different compromises, still in the framework of Axioms [B], [I], [R] and [L], could be adopted.

8.12. Axioms on r and $\mathbf{L}$. The first axiom embodies the standard "local independence" condition of psychometric theory (Lord and Novick, 1974), and has been used in Chapter 7.

[N] **Local Independence.** For each item $q \in Q$, there is a parameter β_q, $0 \le \beta_q < 1$, representing the probability of a careless error in responding to this item if it is present in the current knowledge state. There is also a collection of parameters η_q representing the probability of a lucky guess[3] for

---

[3] We recall that the terms 'lucky guess' and 'careless error' have no cognitive interpretation. They always refer to a current knowledge state. By convention, a correct response to an item not belonging to the knowledge state of reference is a lucky guess, and an incorrect response to an item belonging to that state is a careless error.

a response to an item $q \in Q$ not present in the current learning state. These parameters specify the function r of Axiom [R], that is, the probability of a response set R, conditional to a knowledge state K, according to the following formula:

$$r(R,K) = \left(\prod_{q \in K \setminus R} \beta_q \right) \left(\prod_{q \in K \cap R} (1 - \beta_q) \right) \left(\prod_{q \in R \setminus K} \eta_q \right) \left(\prod_{q \in \overline{R \cup K}} (1 - \eta_q) \right),$$

in which the complement $\overline{R \cup K}$ in the last factor is taken with respect to the domain Q.

[A] Learning Ability. The random variable $\mathbf{L}$ measuring the learning rate is continuous, with a density function f, and a mass concentrated on the positive reals. Specifically, it is assumed that $\mathbf{L}$ is distributed gamma, with parameters $\alpha > 0$ and $\xi > 0$; that is:

$$f(\lambda) = \begin{cases} \frac{\xi^\alpha}{\Gamma(\alpha)} \lambda^{\alpha - 1} e^{-\xi \lambda} & \text{for } \lambda > 0 \\ 0 & \text{for } \lambda \leq 0. \end{cases}$$

Thus, $E(\mathbf{L}) = \frac{\alpha}{\xi}$ and $Var(\mathbf{L}) = \frac{\alpha}{\xi^2}$.

The Learning Latencies

The four axioms [B], [R], [I] and [L] put severe constraints on the functional form of the learning latencies. For example, we may not simply assume that these latencies are distributed gamma[4] because we show here that the latency distributions are in fact forced to be exponential.

8.13. Remarks. Let us write $\mathbf{T}_{q,K,\lambda}$ for the random variable measuring the time required to master some new item q, for a subject with a learning rate λ, this subject being in some state K from which item q is learnable, that is, $K \cup \{q\}$ is a state in the structure. More specifically, for any gradation ν containing both K and $K \cup \{q\}$ and any $\tau > 0$, we infer from Axiom [L] that

$$\mathbf{P}(\mathbf{T}_{q,K,\lambda} \leq \tau) = \ell(K, K \cup \{q\}, \tau, \lambda, \nu),$$

---

[4] This assumption was made by Falmagne (1993). As argued by Stern and Lakshminarayan (1995), it is incorrect. The history of this issue is summarized in the last section of this chapter.

or equivalently

$$P(\mathbf{T}_{q,K,\lambda} > \tau) = \ell(K, K, \tau, \lambda, \nu). \tag{6}$$

For any $t, \delta, \delta', \lambda > 0$, any state K in $\mathcal{K}$, and any gradation ν containing both K and $K \cup \{q\}$, we have

$$
\begin{aligned}
&P(\mathbf{K}_{t+\delta+\delta'} = K \,|\, \mathbf{K}_t = K, \mathbf{L} = \lambda, \mathbf{C} = \nu) \\
&= P(\mathbf{K}_{t+\delta+\delta'} = K \,|\, \mathbf{K}_{t+\delta} = K, \mathbf{K}_t = K, \mathbf{L} = \lambda, \mathbf{C} = \nu) \\
&\qquad\qquad\qquad \times P(\mathbf{K}_{t+\delta} = K \,|\, \mathbf{K}_t = K, \mathbf{L} = \lambda, \mathbf{C} = \nu) \\
&+ P(\mathbf{K}_{t+\delta+\delta'} = K \,|\, \mathbf{K}_{t+\delta} \neq K, \mathbf{K}_t = K, \mathbf{L} = \lambda, \mathbf{C} = \nu) \\
&\qquad\qquad\qquad \times P(\mathbf{K}_{t+\delta} \neq K \,|\, \mathbf{K}_t = K, \mathbf{L} = \lambda, \mathbf{C} = \nu).
\end{aligned}
$$

By Axiom [L], the second term vanishes and the equation simplifies into

$$
\begin{aligned}
&P(\mathbf{K}_{t+\delta+\delta'} = K \,|\, \mathbf{K}_t = K, \mathbf{L} = \lambda, \mathbf{C} = \nu) \\
&= P(\mathbf{K}_{t+\delta+\delta'} = K \,|\, \mathbf{K}_{t+\delta} = K, \mathbf{L} = \lambda, \mathbf{C} = \nu) \\
&\qquad\qquad\qquad \times P(\mathbf{K}_{t+\delta} = K \,|\, \mathbf{K}_t = K, \mathbf{L} = \lambda, \mathbf{C} = \nu);
\end{aligned}
$$

that is, in terms of the function ℓ

$$\ell(K, K, \delta + \delta', \lambda, \nu) = \ell(K, K, \delta', \lambda, \nu)\, \ell(K, K, \delta, \lambda, \nu).$$

In turn, this can be rewritten as

$$P(\mathbf{T}_{q,K,\lambda} > \delta + \delta') = P(\mathbf{T}_{q,K,\lambda} > \delta')\, P(\mathbf{T}_{q,K,\lambda} > \delta). \tag{7}$$

Fixing q, K and λ and writing $g(\tau) = P(\mathbf{T}_{q,K,\lambda} > \tau)$ for $\tau > 0$, we get

$$g(\delta + \delta') = g(\delta')g(\delta) \tag{8}$$

for any $\delta, \delta' > 0$. The function g is defined on $]0, \infty[$ and decreasing (because $1 - g(\tau)$ is a distribution function). Standard functional equations apply (cf. Aczél, 1966), yielding for any $\tau > 0$ and some $\theta > 0$

$$g(\tau) = e^{-\theta\tau}.$$

The constant θ may of course depend upon q, K and λ. We obtain

$$P(\mathbf{T}_{q,K,\lambda} \leq \tau) = 1 - e^{-\theta(q,K,\lambda)\tau}. \tag{9}$$

Note that $E(\mathbf{T}_{q,K,\lambda}) = 1/\theta(q, K, \lambda)$.

The above argument includes the possibility that the distribution of the learning latencies may depend on the state through the parameters θ. For the rest of this section, however, we shall assume that $\theta(q, K, \lambda)$ is independent of K for all states K such that q is learnable from K. In other words, K can be dropped in Equations (7) and (9). (We discuss the validity of this assumption in a later section of this chapter; see 8.22.)

8.14. Remark. It is reasonable to require that a subject having twice the learning rate μ of some other subject would master, on the average, a given item in half the time:

$$E(\mathbf{T}_{q,2\mu}) = \frac{1}{\theta(q, 2\mu)} = \frac{E(\mathbf{T}_{q,\mu})}{2} = \frac{1}{2\theta(q, \mu)}.$$

Generalizing this observation leads to

$$\frac{\theta(q, \lambda\mu)}{\lambda} = \theta(q, \mu) \tag{10}$$

for all $\lambda > 0$. Setting $\mu = 1$ and $\gamma_q = 1/\theta(q, 1)$ in Equation (10), we obtain: $\theta(q, \lambda) = \lambda/\gamma_q$. The distribution function of the learning latencies is thus

$$\mathbf{P}(\mathbf{T}_{q,\lambda} \leq \tau) = 1 - e^{-(\lambda/\gamma_q)\tau}, \tag{11}$$

with $E(\mathbf{T}_{q,\lambda}) = \gamma_q/\lambda$. The form of this expectation is appealing. It entails that the difficulties of the items encountered along a learning path are additive in the following sense. Suppose that a subject with learning rate λ successively solves items $q_1, \ldots, q_n$. The total time to master all of these items has expectation

$$E(\mathbf{T}_{q_1,\lambda} + \cdots + \mathbf{T}_{q_n,\lambda}) = \frac{\sum_{i=1}^{n} \gamma_{q_i}}{\lambda}. \tag{12}$$

In words: the average time required to solve successively a number of items is the ratio of the sum of their difficulties, to the learning rate of the subject.

At the beginning of this section, we stated that the four Axioms [B], [I], [R] and [L] implicitly specified the functional form of the learning latencies. In fact, for a variety of reasons, we cannot simply derive the form of the latency distributions as a theorem. For example, we could not deduce from the stated axioms that the random variables $\mathbf{T}_{q,\lambda}$ do not depend upon K and that Equation (11) holds for $\mathbf{T}_{q,K,\lambda} = \mathbf{T}_{q,\lambda}$, which is what we want, and (because of Equation (12)) will make the theory manageable. Accordingly, we formulate below a new axiom specifying the form of these latency distributions. The following notation will be useful.

8.15. Definition. Let ν be a gradation containing a state $K \neq Q$. We write K^ν for the state immediately following state K in gradation ν. We have thus $K \subset K^\nu \in \nu$ with $|K^\nu \setminus K| = 1$, and for any $S \in \nu$ with $K \subset S$, we must have $K^\nu \subseteq S$.

[T] **Learning Times.** We assume that, for a subject with learning rate λ, the time required for the mastery of item q (this item being accessible from that subject's current state) is a random variable $\mathbf{T}_{q,\lambda}$, which has an exponential distribution with parameter λ/γ_q, where $\gamma_q > 0$ is an index measuring the difficulty of item q. Thus, $E(\mathbf{T}_{q,\lambda}) = \frac{\gamma_q}{\lambda}$ and $Var(\mathbf{T}_{q,\lambda}) = (\frac{\gamma_q}{\lambda})^2$. All these random variables (for all q and λ) are assumed to be independent. The items are mastered successively. This means that the total time to master successively items $q, q', \ldots$ encountered along some gradation ν is the sum $\mathbf{T}_{q,\lambda} + \mathbf{T}_{q',\lambda} + \cdots$ of exponentially distributed random variables, with parameters λ/γ_q, $\lambda/\gamma_{q'}, \ldots$ Formally, for any positive real numbers δ and λ, any learning path $\nu \in \mathcal{G}$, and any two states $K, K' \in \nu$ with $K \subseteq K'$, we have

$$
\ell(K, K', \delta, \lambda, \nu)
$$
$$
= \begin{cases}
\mathbf{P}(\mathbf{T}_{q,\lambda} > \delta) \text{ with } \{q\} = K^\nu \setminus K & \text{if } K = K' \neq Q, \\
\mathbf{P}(\sum_{q \in K' \setminus K} \mathbf{T}_{q,\lambda} \leq \delta) - \mathbf{P}(\sum_{q \in K'^\nu \setminus K} \mathbf{T}_{q,\lambda} \leq \delta) & \text{if } K \subset K' \neq Q, \\
\mathbf{P}(\sum_{q \in K' \setminus K} \mathbf{T}_{q,\lambda} \leq \delta) & \text{if } K \subset K' = Q, \\
1 & \text{if } K = K' = Q, \\
0 & \text{in all other cases.}
\end{cases}
$$

Notice that the total time required to solve all the items in a state K is the random variable $\sum_{q \in K} \mathbf{T}_{q,\lambda}$ which is distributed as a sum of independent exponential random variables.

Empirical Predictions

In this section and the next one, we suppose that we have a SLP system $S = ((\mathbf{K}_t, \mathbf{R}_t)_{t \geq 0}, \mathbf{C}, \mathbf{L})$ with the distributions satisfying axioms [N], [A] and [T]. The predictions given below were derived by Stern and Lakhsminarayan (1995) and Lakhsminarayan (1995). We begin with a well-known result (Adke and Manjunath, 1984).

8.16. Theorem. *Let $\mathbf{T}_1, \mathbf{T}_2, \ldots, \mathbf{T}_n$ be jointly distributed independent exponential random variables with parameters $\lambda_1, \lambda_1, \ldots, \lambda_n$ respectively, and suppose that $\lambda_i \neq \lambda_j$ for $i \neq j$. Then the density function $h_{\mathbf{T}}$ and the associated distribution function $H_{\mathbf{T}}$ of the sum $\mathbf{T} = \mathbf{T}_1 + \cdots + \mathbf{T}_n$ are given, for $t \geq 0$ and $\delta \geq 0$, by*

$$h_{\mathbf{T}}(t) = \left(\prod_{i=1}^{n} \lambda_i \right) \sum_{j=1}^{n} e^{-\lambda_j t} \left(\prod_{\substack{k=1 \\ k \neq j}}^{n} (\lambda_k - \lambda_j) \right)^{-1}, \tag{14}$$

$$H_{\mathbf{T}}(\delta) = \int_0^{\delta} h_{\mathbf{T}}(t)\, dt = \left(\prod_{i=1}^{n} \lambda_i \right) \sum_{j=1}^{n} \frac{1 - e^{-\lambda_j \delta}}{\lambda_j \prod_{\substack{k=1 \\ k \neq j}}^{n} (\lambda_k - \lambda_j)}\, dt. \tag{15}$$

Thus, the results in the right members of (14) and (15) do not depend upon the order assigned by the index i. For example, with $n = 3$, we get

$$h_{\mathbf{T}}(t) = \lambda_1 \lambda_2 \lambda_3 \left(\frac{e^{-\lambda_1 t}}{(\lambda_2 - \lambda_1)(\lambda_3 - \lambda_1)} + \right.$$
$$\left. \frac{e^{-\lambda_2 t}}{(\lambda_1 - \lambda_2)(\lambda_3 - \lambda_2)} + \frac{e^{-\lambda_3 t}}{(\lambda_1 - \lambda_3)(\lambda_2 - \lambda_3)} \right). \tag{16}$$

Since none of the above references contains a proof, one is included below.

PROOF: The MGF (moment generating function) $M_{\mathbf{T}_i}(\theta)$ of an exponential random variable $\mathbf{T}_i$ with parameter λ_i is given by $M_{\mathbf{T}_i}(\theta) = \lambda_i (\lambda_i - \theta)^{-1}$. Accordingly, the MGF of the sum of n independent exponential random variables $\mathbf{T}_i$ with parameter λ_i, $1 \leq i \leq n$, is

$$M_{\mathbf{T}}(\theta) = \prod_{i=1}^{n} \frac{\lambda_i}{\lambda_i - \theta}. \tag{17}$$

The proof proceeds by showing that the MGF of the random variable with the density function specified by Equation (14) is that given by (17). The result follows from the fact that the MGF of a random variable uniquely determines its distribution. In passing, this would establish that the right members of (14) and (15) do not depend upon the order assigned by the index i. Multiplying by $e^{\theta t}$ in both members of (14) and integrating from 0 to ∞ over t (assuming that $\theta < \lambda_j$ for all indices j) yields

$$E(e^{\theta \mathbf{T}}) = \left(\prod_{i=1}^{n} \lambda_i\right) \sum_{j=1}^{n} \left((\lambda_j - \theta) \prod_{\substack{k=1 \\ k \neq j}}^{n} (\lambda_k - \lambda_j)\right)^{-1} = \left(\prod_{i=1}^{n} \frac{\lambda_i}{\lambda_i - \theta}\right) \times D$$

with (see Problem 4)

$$D = \frac{\sum_{j=1}^{n} (-1)^{n-j} \left[\left(\prod_{i \neq j}(\lambda_i - \theta)\right)\left(\prod_{\substack{k \neq j, l \neq j \\ k < l}}(\lambda_k - \lambda_l)\right)\right]}{\prod_{i<j}(\lambda_i - \lambda_j)}. \qquad (18)$$

It suffices to show that $D = 1$ for all values of λ_i, $1 \leq i \leq n$ satisfying the conditions. The numerator in Equation (18) is a polynomial in θ of degree $n-1$. For $i = 1, 2, \ldots, n$, it is easy to check that the value of this polynomial evaluated at $\theta = \lambda_i$ is equal to the denominator of (18) (see Problem 5). Thus $D = 1$ for all values of θ, which completes the proof of Equation (14). Equation (15) follows easily. $\square$

A similar result can be obtained in the case where some of the λ_i's have identical values. We shall not treat this case here.

8.17. Convention. For the remainder of this section, we suppose that all the difficulty parameters γ_q associated with the items have different values. Accordingly, the parameters λ/γ_q of the exponential learning latencies have different values.

8.18. Definition. For any $S \subseteq Q$ and $\lambda > 0$, we denote by $g_{S,\lambda}$ the density of the sum $\mathbf{T}_{S,\lambda} = \sum_{q \in S} \mathbf{T}_{q,\lambda}$, where each $\mathbf{T}_{q,\lambda}$ is an exponential random variable with parameter γ_q/λ and the random variables $\mathbf{T}_{q,\lambda}$ are pairwise independent. We also denote by $G_{S,\lambda}$ the corresponding distribution function. Thus, $g_{S,\lambda}$ and $G_{S,\lambda}$ are specified by Equations (14) and (15) up to the values of λ and the γ_q's. This notation is justified because any permutation of the index values in the right member of Equation (14) yield the same density function, and hence the same distribution function.

We will use Theorem 8.16 to obtain an explicit expression for the function ℓ in terms of the postulated exponential distributions. The next theorem is a restatement of Axiom [T] following immediately from Theorem 8.16 and Definition 8.18.

8.19. Theorem. *For any positive real numbers δ and λ, any learning path $\nu \in \mathcal{G}$, and any two states $K, K' \in \nu$, we have*

$$\ell(K, K', \delta, \lambda, \nu) = \begin{cases} e^{-\frac{\lambda}{7q}\delta} & \text{with } \{q\} = K^\nu \setminus K & \text{if } K = K' \neq Q, \\ G_{K'\setminus K,\lambda}(\delta) - G_{K'^\nu\setminus K,\lambda}(\delta) & \text{if } K \subset K' \neq Q, \\ G_{K'\setminus K,\lambda}(\delta) & \text{if } K \subset K' = Q, \\ 1 & \text{if } K = K' = Q, \\ 0 & \text{in all other cases.} \end{cases}$$

We are now equipped to derive explicit predictions for the joint probabilities of the states occurring at successive times. The next theorem contains one example. As a convention, we set $G_{\varnothing,\lambda}(\delta) = 1$ and $G_{Q^\nu\setminus S,\lambda}(\delta) = 0$. Notice that $G_{S,\mathbf{L}}(\delta)$ is a random variable, with

$$E\left[G_{S,\mathbf{L}}(\delta)\right] = \int_0^\infty G_{S,\lambda}(\delta) f(\lambda) d\lambda,$$

where f comes from 8.12, Axiom [A].

8.20. Theorem. *For all integers $n > 0$, all states $K_1 \subseteq \cdots \subseteq K_n$, and all real numbers $t_n > \cdots > t_1 \geq 0$, we have*

$$\mathbf{P}(\mathbf{K}_{t_1} = K_1, \ldots, \mathbf{K}_{t_n} = K_n)$$

$$= \sum_\nu p_\nu \left\{ \int_0^\infty \prod_{i=0}^{n-1} \left[G_{K_{i+1}\setminus K_i,\lambda}(t_{i+1} - t_i) - G_{K_{i+1}^\nu\setminus K_i,\lambda}(t_{i+1} - t_i) \right] f(\lambda) d\lambda \right\}$$

$$= \sum_\nu p_\nu E \left\{ \prod_{i=0}^{n-1} \left[G_{K_{i+1}\setminus K_i,\mathbf{L}}(t_{i+1} - t_i) - G_{K_{i+1}^\nu\setminus K_i,\mathbf{L}}(t_{i+1} - t_i) \right] \right\},$$

with $t_0 = 0$, $K_0 = \varnothing$ and the sum extending over all $\nu \in \mathcal{G}(K_1, \ldots, K_n)$.

The predictions for the Case $n = 2$ are useful for the application described at the end of the chapter (Lakshminarayan, 1995). We have clearly:

8.21. Theorem. *For any pair of response patterns $(R_2, R_1) \in 2^Q \times 2^Q$, and any real numbers $t_2 > t_1 \geq 0$,*

$$\mathbf{P}(\mathbf{R}_{t_1} = R_1, \mathbf{R}_{t_2} = R_2)$$

$$= \sum_{K_1 \in \mathcal{K}} \sum_{K_2 \in \mathcal{K}} r(R_1, K_1) r(R_2, K_2) \, \mathbf{P}(\mathbf{K}_{t_1} = K_1, \mathbf{K}_{t_2} = K_2) \qquad (19)$$

with $r(R_1, K_1)$ and $r(R_2, K_2)$ specified by Axiom [N] in terms of the parameters β_q and η_q, and $\mathbf{P}(\mathbf{K}_{t_1} = K_1, \mathbf{K}_{t_2} = K_2)$ given by Theorem 8.20.

An explicit expression of the joint probabilities $\mathbf{P}(\mathbf{R}_{t_1} = R_1, \mathbf{R}_{t_2} = R_2)$ is now easy to obtain. To begin with, we can replace the values $r(R_i, K_i)$ of the response rule by their expressions in terms of the probabilities of the careless errors and of the lucky guesses given by Axiom [N]. Next, the values of the joint probabilities of the state $\mathbf{P}(\mathbf{K}_{t_1} = K_1, \mathbf{K}_{t_2} = K_2)$ in terms of the distributions of the learning rate and the learning latencies can be obtained by routine integration via Axioms [A] and Theorems 8.21, 8.16, 8.19 and 8.20. We shall not spell out these results here (see Lakshminarayan, 1995).

Limitations of this Theory

The assumption that the time required to master some item q does not depend upon the current state K —provided that item q can be learned from K—was made for the sake of simplicity, but is not immune from criticisms. One can easily imagine situations in which it might fail. Intuitively, if an item q can be learned from both K and K', a student in state K may conceivably be better prepared for the learning of q than a student in state K'. The argument in the discussion and the example below makes this idea concrete[5].

8.22. Remarks. We consider a knowledge space $\mathcal{K}$ which is a substructure of some large, idealized knowledge structure $\widehat{\mathcal{K}}$ containing all the items in a given field of information, with domain $\widehat{Q}$. Let us denote by $\mathcal{Q}$ the surmise relation of $\widehat{\mathcal{K}}$ (cf. Definition 1.43); thus:

$$q'\mathcal{Q}q \iff (\forall K \in \widehat{\mathcal{K}} : q \in K \Rightarrow q' \in K).$$

---

[5] This argument was suggested by Lakshminarayan (personal communication; see also Stern and Lakshminarayan, 1995). The details are as in Falmagne (1996).

We also define, for any q in $\hat{Q}$ and any subset S of $\hat{Q}$,

$$Q^{-1}(q) = \{r \in \hat{Q} \,|\, rQq\}$$
$$Q^{-1}(S) = \{r \in \hat{Q} \,|\, rQq, \text{ for some } q \in S\} = \cup_{q \in S}\, Q^{-1}(q).$$

Consider a state K in $\mathcal{K}$ and suppose that items a and b can be learned from K. Specifically, both $K \cup \{a\}$ and $K \cup \{b\}$ are states of $\mathcal{K}$; moreover, since $\mathcal{K}$ is a knowledge space, $K \cup \{a, b\}$ is also a state. This means that item a, for example, can be learned from either state K or state $K \cup \{b\}$. It is reasonable to suppose that the difficulty of mastering item a from state K must depend on the items from $\hat{Q} \setminus K$ that must be mastered before mastering a; that is, it depends on the set

$$S(a, K) = Q^{-1}(a) \setminus Q^{-1}(K).$$

Similarly, the difficulty of mastering a from the state $K \cup \{b\}$ depends on the set

$$S(a, K \cup \{b\}) = Q^{-1}(a) \setminus Q^{-1}(K \cup \{b\}).$$

The assumption that the difficulty of an item does not depend upon the state of the subject leads one to require that $S(a, K) = S(a, K \cup \{b\})$. In fact, we have by definition $S(a, K \cup \{b\}) \subseteq S(a, K)$, but the equality holds only in special circumstances. Indeed, some manipulation yields (cf. Problem 6):

$$S(a, K) \setminus S(a, K \cup \{b\}) = Q^{-1}(a) \cap \overline{Q^{-1}(K)} \cap Q^{-1}(K \cup \{b\}).$$

Supposing that the intersection in the right member is empty means that there is no item q in $\hat{Q}$ preceding both a and b, and not preceding at least one item in K. Such an assumption does not hold for general knowledge structures. A counterexample is easy to find.

8.23. Example. Consider the knowledge space

$$\hat{\mathcal{K}} = \{\varnothing, \{c\}, \{d\}, \{c, d\}, \{a, c, d\}, \{b, c, d\}, \{a, b, c, d\}\}.$$

Thus $\hat{Q} = \{a, b, c, d\}$. The subspace of $\hat{\mathcal{K}}$ defined by $Q = \{a, b, c\}$ is

$$\mathcal{K} = \{\varnothing, \{c\}, \{a, c\}, \{b, c\}, \{a, b, c\}\}.$$

If we take $K = \{c\} \in \mathcal{K}$, we get

$$Q^{-1}(a) = \{a, c, d\}, \quad Q^{-1}(K) = \{c\},$$
$$\overline{Q^{-1}(K)} = \{a, b, d\}, \quad Q^{-1}(K \cup \{b\}) = \{b, c, d\}$$

with

$$Q^{-1}(a) \cap \overline{Q^{-1}(K)} \cap Q^{-1}(K \cup \{b\}) = \{d\}.$$

A cure is easy but costly. As argued earlier, we can replace all the difficulty parameters γ_q by parameters $\gamma_{q,K}$ explicitly depending on the current
state K of the subject. Less prohibitive solutions would be preferable, of
course. In practice, it may turn out that the dependence of the difficulty of
mastering some item q on the state from which q is accessed, while theoretically justified, is in fact mild, either because the estimates of the parameters
$\gamma_{q,K}$ and $\gamma_{q,K'}$ do not differ much, or because this dependence only affects
a small proportion of the states and items. This question was investigated
empirically by Lakshminarayan (1995) who showed that a model with exponentially distributed learning latencies and with difficulty parameters γ_q
that did not depend on the state was able to fit the data quite well.

In any event, our discussion opens the following problem: Under wich
conditions on the knowledge structure $(\widehat{Q}, \widehat{\mathcal{K}})$ are all the set differences
$S(a, K) \setminus S(a, K \cup \{b\})$ empty, for all substructures $(Q, \mathcal{K})$? It is clear that
a sufficient condition is that $\widehat{\mathcal{K}}$ is a chain. This condition is not necessary,
however.

The above discussion concerning a possible lack of invariance of the difficulty parameters was organized around the concept of a substructure $(Q, \mathcal{K})$
of a parent structure $(\widehat{Q}, \widehat{\mathcal{K}})$. In short, it was argued that the difficulty of
acquiring a new item a accessible from two states K and $K \cup \{b\}$ in $\mathcal{K}$,
could differ because the implicit paths in $\widehat{\mathcal{K}}$ leading from K to $K \cup \{a\}$ and
from $K \cup \{b\}$ to $K \cup \{a, b\}$ could required the mastery of different items
in $\widehat{Q} \setminus Q$. The same type of argument can be used to show that, from the
standpoint of the learning latencies, it cannot be the case that the axioms
of the model holds for both $\mathcal{K}$ and $\widehat{\mathcal{K}}$. Specifically, if the learning latencies
are exponentially distributed in a model assumed to hold for the parent
knowledge structure $(Q, \widehat{\mathcal{K}})$, then these latencies cannot, in general, be exponentially distributed for the substructure $(Q, \mathcal{K})$. (Rather, they have to
be sums of exponential random variables, or even mixtures of such sums.)

Consequently, because the exponential form is necessary, the model cannot in principle hold for $(Q, \mathcal{K})$. For a more detailed discussion on this point, see Stern and Lakshminarayan (1995).

Simplifying Assumptions

Another type of objection that may be raised against this theory is that the number of gradations may be very large. Because a probability is attached to each gradation, the number of parameters to be estimated from the data may be prohibitive. However, some fairly natural simplifying assumptions can be made which would result in a substantial decrease in the number of parameters attached to the gradations. We discuss an example.

8.24. Markovian Learning Paths. Suppose that the probability of a gradation

$$\varnothing \subset \{a\} \subset \{a, b\} \subset \{a, b, c\} \subset \{a, b, c, d\} \subset Q \qquad (20)$$

in some well-graded knowledge structure with domain $Q = \{a, b, c, d, e\}$ can be obtained by multiplying the successive transition probabilities from state to state along the gradation. Let us denote by $p_{\varnothing, \{a\}}$ the conditional probability that $\{a\}$ is the first state visited after the initial empty state. We also write $p_{\{a\}, \{a, b\}}$ for the conditional probability of a transition from state $\{a\}$ to state $\{a, b\}$, etc. Note that these transition probabilities can be computed from the gradation probabilities. For example, we have (using the notation of 8.1 for the subsets of gradations)

$$p_{\{a\}, \{a, b\}} = \frac{\mathbf{P}(\mathbf{C} \in \mathcal{G}(\{a\}, \{a, b\}))}{\mathbf{P}(\mathbf{C} \in \mathcal{G}(\{a\}))}.$$

Our simplifying assumption is that the probability of the gradation in (20) is given by the product

$$p_{\varnothing, \{a\}} \cdot p_{\{a\}, \{a, b\}} \cdot p_{\{a, b\}, \{a, b, c\}} \cdot p_{\{a, b, c\}, \{a, b, c, d\}}.$$

(Note that the conditional probability of a transition from state $\{a, b, c, d\}$ into state Q, is equal to one.)

Let us generalize this example. We extend a notation introduced earlier. For any state $K \neq \varnothing$ and any gradation $\nu \in \mathcal{G}(K)$, we write ${}^\nu K$ for the

state immediately preceding state K in ν. The probability of any learning
path ν is then obtained from the product of the transition probabilities

$$p_\nu = p_{\varnothing,\varnothing^\nu} \cdot p_{\varnothing^\nu,(\varnothing^\nu)^\nu} \cdot \ldots \cdot p_{\nu Q,Q}, \tag{21}$$

some of which may be equal to one. (For instance, we have $p_{\nu Q,Q} = 1$, for
all learning paths ν.) It is our experience that, for large typical well-graded
knowledge structures, this assumption will dramatically reduce the number
of parameters in the model.

Clearly, more extreme simplifying assumptions can be tested. We could
assume, for instance, that p_{K,K^ν} only depends upon $K^\nu \backslash K$. In the case of a
well-graded knowledge structure, this would reduce the number of transition
probabilities to at most $|Q|$.

Remarks on Application and Use of the Theory

Suitable data for this theory consists in the frequencies of n-tuples of response patterns $R_1, R_2, \ldots, R_n$, observed at times $t_1 < t_2 < \cdots < t_n$. Let
us consider the case $n = 2$. Thus, a sample of subjects has been selected,
and these subjects have been tested twice, at times t and $t + \delta$. We denote
by $N(R, R')$, with $R, R' \subseteq Q$, the number of subjects having produced the
two patterns of responses R and R' at times t and $t + \delta$.

8.25. A maximum likelihood procedure. The parameters may be estimated by maximizing the loglikelihood function

$$\sum_{R,R' \subseteq Q} N(R,R') \log \mathbf{P}(\mathbf{R}_t = R, \mathbf{R}_{t+\delta} = R') \tag{22}$$

in terms of the various parameters of the theory, namely, the response parameters β_q and η_q, the parameters α and ξ of the distribution of the
learning rates, the item difficulty parameters γ_q, and the probabilities p_ν
of the gradations. In some cases, it is also possible to consider the times t
and $t + \delta$ as parameters. For example, the time t elapsed since the beginning of learning may be difficult to assess accurately. Applying a theory of
such a complexity raises a number of issues, which we now address.

8.26. Remarks. a) To begin with, some readers may cringe at this plethora
of parameters, and wonder whether an application of this theory is a realistic prospect. Actually, as indicated earlier in this chapter, the theory

has been successfully applied to several sets of data (Falmagne et al. 1990; Taagepera et al., 1997; Lakshminarayan, 1995). Notice that, when a test is administered several times, the number of parameters does not increase, whereas the number of response categories—i.e. the number of degrees of freedom in the data—increases exponentially. If the number of gradations is not prohibitive, or if some Markovian assumptions in the spirit of the last section are satisfied, the complexity of the theory will remain well beneath that of the data to be explained. For example, under the simplest Markovian assumption mentioned at the end of the last section, the number of parameters of the theory is of the order of $|Q|$, while the number of response categories is of the order of $2^{n|Q|}$ (with n applications of the test to the same sample of subjects), which is a good return. The maximization of the loglikelihood function in Equation (22) may be achieved by a procedure such as the Conjugate Gradient Search algorithm by Powell (1964) which is available in form of the C subroutine PRAXIS (cf. Gegenfurtner, 1992). In practice, the procedure is applied many times[6], with different starting values for the parameters, to ensure that the final estimates do not correspond to a local maximum.

b) The first two applications mentioned above (Falmagne et al. 1990; Taagepera et al., 1992) were performed under the hypothesis that the learning latencies had general gamma (rather than the correct exponential) distributions. In 8.13, the gamma assumption was shown to be inconsistent with the other axioms. However, because this inconsistency does not propagate to the predictions (in the sense that Equation (19), for example, defines a genuine distribution on the set of all pairs (R_1, R_2) of patterns of responses), it does not preclude a good fit to the data. Moreover, these learning latencies affect the predictions only indirectly, after smearing by the learning rate random variable. This suggest that the predictions of the model may be robust to the particular assumptions made on the learning latencies.

c) Another issue concerns the knowledge structure $\mathcal{K}$ at the core of the theory. For the theory to be applicable, a knowledge structure must be assumed. Methods to obtain such a knowledge structure have been discussed in Chapter 5 (cf. Koppen, 1989; Koppen and Doignon, 1990; Falmagne et al., 1990; Kambouri et al., 1993; Müller, 1989; Dowling, 1991, 1993a; Villano,

---

[6] Several hundred times.

1991; Cosyn and Thiéry, 1999). A reasonable procedure is to start the analysis with a tentative knowledge structure, presumed to contain all the right states, and possibly some subsets of questions which are not states. If the application of the model proves to be successful, this starting knowledge structure can be progressively refined by assuming that some learning paths have probability zero, thereby eliminating some states. This method is exemplified in Falmagne et al. (1990), with real data, for the case of a single test. Chapter 12 summarizes the work of Kambouri et al. (1993) and Cosyn and Thiéry (1999) which are devoted to this question.

An Application of the Theory to the Case $n = 2$

The most ambitious application of the theory described in this Chapter is due to Lakshminarayan (1995), and we summarize it here. Only the main lines of the analysis and of the results will be reported.

8.27. The items. The domain contains the five problems in high school geometry displayed in Figure 8.1, which deal with basic concepts such as angles, parallel lines, triangles and the Pythagorean Theorem. These problems are labeled a, b, c, d and e. Two versions of each problem were generated, forming two equivalent sets V1 and V2 of five problems which were applied at different times[7]. The differences between the two versions only concerned the particular numbers used as measures of angles or length.

8.28. Procedure. The experiment had three phases:

(1) **Pretest.** The subjects were presented with one version, and asked to solve the problems. Approximately half of the subjects were given version V1, and the rest, version V2.

(2) **Lesson.** After completion of Phase 1, the test sheets were removed and a 9-page booklet containing a lesson in high school geometry was distributed to the subjects, who were required to study it. The lesson was directly concerned with the problems to be solved.

(3) **Posttest.** The subjects were given the other version of the test, and were asked to solve the problems.

There was no delay between the phases, and the subjects were allowed

---

[7] V1 and V2 were in fact embedded into two larger equivalent sets containing 14 problems each.

to spend as much time as they wanted on each problem. Typically, the students spent between 25 to 55 minutes to complete the three phases of the experiment.

The data consist in the observed frequencies of each of the $2^5 \times 2^5 = 1024$ possible pairs of response patterns. The subjects were undergraduate students of the University of California at Irvine; 959 subjects participated in the experiment.

8.29. Parameter estimation. The overall data set provided by the 959 subjects was split into two unequal part. Those of 159 subjects were set aside, and kept for testing the predictions of the model. Those of the remaining 800 subjects were used to find out the knowledge structure and to estimate parameters. A tentative knowledge structure was initially postulated, based on the content of the items. This structure was then gradually refined on the basis of goodness-of-fit (likelihood ratio) calculation.

All the problems have open responses. Accordingly, all the guessing parameters η_q were set equal to zero. The remaining parameters were estimated by a maximum likelihood method.

8.30. Results. The final knowledge structure was a knowledge space included in that displayed in Figure 8.2. Two of the original gradations, namely *abcde* and *adbce* were dropped in the course of the analysis because their estimated probabilities were not significantly different from zero. As a consequence, the state $\{a, d\}$ was removed. The resulting space is

$$\{ \varnothing, \{a\}, \{b\}, \{a,b\}, \{b,c\}, \{a,b,c\},$$
$$\{a,b,d\}, \{b,c,d\}, \{a,b,c,d\}, \{a,b,c,d,e,\} \}.$$

This space has six gradations, five of which were assigned a non zero probability. All but one the remaining gradations had small probabilities ($< .10$). Gradation *bcade* occurred with an estimated probability of .728. The estimated probabilities of the five gradations are given in Table 8.1, together with the estimates of the other parameters of the model. Note that t and $t + \delta$ (the two times of testing) are regarded as parameters because the time elapsed since the beginning of learning geometry could not be assessed accurately. The unit of t and δ is the same as that of ξ and is arbitrary.

The fit of the model based of the value of the log-likelihood statistic was good. In general, the estimated values of the parameter seem reasonable.

Parameters	Estimates
$\mathbf{P}(\mathbf{C} = abcde)$	.047
$\mathbf{P}(\mathbf{C} = bcdae)$	.059
$\mathbf{P}(\mathbf{C} = bcade)$	.728
$\mathbf{P}(\mathbf{C} = bacde)$	.087
$\mathbf{P}(\mathbf{C} = badce)$	.079
γ_a	11.787
γ_b	25.777
γ_c	11.542
γ_d	34.135
γ_e	90.529
α	113.972
t	11.088
δ	2.448
ξ	13.153
β_a	.085
β_b	.043
β_c	.082
β_d	.199
β_e	.169

Table 8.1. Estimates of the parameters. We recall that all the parameters η_q have been set equal to 0 a priori. The quantities t and $t+\delta$ representing the two times of testing are regarded as parameters in this application because the time t elapsed since the beginning of learning geometry could not be assessed accurately.

In particular, the values obtained for the β_q are small, which is consistent with the interpretation of these parameters as careless error probabilities. The value of δ is overly large compared to that of t. For example if for the sake of illustration we set the unit of t as equal to one year, then the estimated total time of learning until the first test amounts to eleven years and one month, while the time between the first and the second test is estimated to be approximately two years and five months, which may seem absurd. The most likely explanation is that—relative to the two years and five months—the eleven years and one month is an overestimation reflecting the forgetting that took place for most students between the end of their learning geometry in high school and the time of the test. If the first test had occurred while the students were learning geometry in high school, the estimates of t and δ would presumably have been more consistent.

Another subset of 4 items was analyzed by the same methods by Lakshminarayan (1995), which yielded a much less satisfactory fit. At this point, deriving negative conclusions from these other results would be premature.

As we mentioned earlier, fitting a model of that kind, with so many parameters, is a complex procedure which is not guaranteed to lead automatically to the best knowledge structure and the best set of estimates of the parameters for the data.

Original Sources and Related Works

In this chapter, we formalize the learning occurring in the framework of a well-graded knowledge structure, as involving the choice of a gradation, paired with a stochastic process describing the progressive acquisition of the items along that gradation. This concept is natural enough and was already exploited to some extent in the models discussed in Chapter 7. The model discussed here is to date the most elaborate implementation of this idea. As made clear by our presentation, this model attempts to give a realistic picture of learning by making an explicit provision for individual differences and by describing learning as a real time stochastic process. Thus, the target data for the model is made of n-tuples $(R_1, \ldots, R_n)$ of responses patterns observed at arbitrary times $t_1, \ldots, t_n$. This attempt was only partly successful. The potential drawbacks of the current model are summarized below together with the relevant literature.

A first pass at constructing such a model was made by Falmagne (1989a). Even though most of the ideas of this chapter were already used in that earlier attempt, the 1989 model was not fully satisfactory because it was not formulated explicitly as a stochastic process. The 1989 model was tested against a standard unidimensional psychometric model by Lakshminarayan and Gilson (1993), with encouraging results. In his 1993 paper published in the Journal of Mathematical Psychology, Falmagne developed what is essentially the model of this chapter, except that the learning latencies were assumed to be distributed as general gammas. In the notation of Equation (11), we had thus

$$P(\mathbf{T}_{q,\lambda} \leq \delta) = \int_0^\delta \frac{\tau^{\gamma_q} \lambda^{\gamma_q - 1} e^{-\lambda \tau}}{\Gamma(\gamma_q)} d\tau, \tag{23}$$

instead of

$$P(\mathbf{T}_{q,\lambda} \leq \delta) = 1 - e^{-(\lambda/\gamma_q)\delta} \tag{24}$$

as assumed in this chapter, the interpretation of the parameters being the same in both cases. A simulation study of the 1993 model was made by Falmagne and Lakshminarayan (1993). This model was successfully applied to real data by Taagepera et al. (1997). Around May 1994, Stern and Lakshminarayan discovered that the assumption that the learning latencies are distributed gamma was inconsistent with the axioms of the theory as stated here, and surmised that the appropriate distributions for these latencies had to be exponential (Stern and Lakshminarayan, 1995; Lakshminarayan, 1995; Falmagne, 1996).

Problems

1. Formulate an axiom concerning the learning rule ℓ resulting in a SLP system that is not progressive (cf. Definition 8.3). (Hint: Allow for the possibility of forgetting.)

2. Argue that the Markovian assumption

$$\mathbf{P}(\mathbf{K}_{t_{n+1}} = K_{n+1} \,|\, \mathbf{K}_{t_n} = K_n, \mathbf{R}_{t_n} = R_n, \mathcal{E}) = \mathbf{P}(\mathbf{K}_{t_{n+1}} = K_{n+1} \,|\, \mathbf{K}_{t_n} = K_n)$$

(with $\mathcal{E}$ as in Axiom [L]) discussed in Remark 8.4(d) is inappropriate. (Hint: Show by an example that the more detailed history embedded in the event $\mathcal{E}$ in the left member may provide information on the learning rate and/or the learning path, which in turn, may modify the probabilities of the states at time t_{n+1}.)

3. Investigate a possible reformulation of the learning rule that would not imply, via Equations (7) and (8), that the learning latencies are distributed exponentially.

4. Verify Equation (18).

5. Establish all the assertions following Equation (18) in the proof of Theorem 8.16.

6. Verify the computation of $S(a, K) \setminus S(a, K \cup \{b\})$ in Remark 8.22.

7. Following up on the discussion of Remark 8.22, find a counterexample to the Condition $Q^{-1}(a) \cap \overline{Q^{-1}(K)} \cap Q^{-1}(K \cup \{b\}) = \varnothing$, different from the one in Example 8.23.

8. In relation with Example 8.23 and Remark 8.22, find necessary and sufficient conditions on the knowledge structure $(\widehat{Q}, \widehat{\mathcal{K}})$ such that, for all substructures $(Q, \mathcal{K})$, all state $K \in \mathcal{K}$ and all pair of items a and b, the set difference $S(a, K) \setminus S(a, K \cup \{b\})$ is empty. The result may evoke some ironical reflexions. Check this impulse. Instead, work on Problem 9.

9. Investigate realistic cures for the difficulty pointed out by Example 8.23.

10. For any $t > 0$ and any item q, let $N_{t,q}$ be the number of subjects having provided a correct response to item q in a sample of N subjects. Let $\overline{N}_{t,q}$ denote the number of subjects having provided an incorrect response to that item. Thus $N_{t,q} + \overline{N}_{t,q} = N$. Consider a situation in which a test has been applied twice to a sample of N subjects, at times t and $t + \delta$. Investigate the statistical properties of $\overline{N}_{t+\delta,q}/N_{t,q}$ as an estimator of the careless error probability β_q. Is this estimator unbiased, that is, do we have $E(\overline{N}_{t+\delta,q}/N_{t,q}) = \beta_q$?

11. Prove Lemma 8.7

Chapter 9

Descriptive and Assessment Languages⋆

How can we economically describe a state in a knowledge structure? The question is inescapable because realistic states will typically be quite large. In such cases, it is impractical to describe a state by giving the full list of items that it contains. It is also unnecessary: because of the redundancy in many real-life knowledge structures[1], a state will often be characterizable by a relatively small set of features. This idea is not new. In Chapter 2, we proved that any state in a well-graded knowledge structure could be fully described by simply listing its inner and outer fringes (cf. Theorem 2.9 and Remark 2.10(a)). Here, we consider this issue more systematically. This chapter is somewhat eccentric to the rest of this book and can be skipped without harm at first reading. We begin by illustrating the main ideas in the context of a simple example encountered before (see e.g. 3.1).

Languages and Decision Trees

9.1. Example. Consider the knowledge structure (which is in fact a discriminative knowledge space, cf. Example 3.1)

$$\mathcal{L} = \{\, \varnothing, \{a\}, \{b, d\}, \{a, b, c\}, \{b, c, e\}, \{a, b, d\},$$
$$\{a, b, c, d\}, \{a, b, c, e\}, \{b, c, d, e\}, \{a, b, c, d, e\} \,\} \qquad (1)$$

on the domain $Q = \{a, b, c, d, e\}$. The state $\{a, b, c, e\}$ is the only state of $\mathcal{L}$ containing a, e and not d. It can be characterized by stating that $\{a, b, c, e\}$ is a particular state K of $\mathcal{L}$ satisfying

$$a \in K, \quad d \notin K \quad \text{and} \quad e \in K. \qquad (2)$$

Similarly, the state $M = \{b, c, d, e\}$ is specified by the statement

$$a \notin M \text{ and } d, e \in M. \qquad (3)$$

---

[1] For example, the knowledge structure for arithmetic used by the **Aleks** system (cf. Chapter 0) and intended to cover all the elementary concepts from 2-digit addition to negative exponents has around 100 items. Yet, the number of states used by the assessment engine of the system is only around 38,000.

A compact notation will be adopted. We shall represent (2) and (3) respectively by the strings $a\bar{d}e$ and $\bar{a}de$. Extending these concepts, all the states of $\mathcal{L}$ can be represented, for example, by the strings listed in Table 9.1.

States	Strings
$\{a, b, c, d, e\}$	ade
$\{a, b, c, d\}$	$acd\bar{e}$
$\{a, b, c, e\}$	$a\bar{d}e$
$\{a, b, c\}$	$ac\bar{d}\bar{e}$
$\{a, b, d\}$	$a\bar{c}d$
$\{a\}$	$a\bar{b}$
$\{b, c, d, e\}$	$\bar{a}de$
$\{b, c, e\}$	$\bar{a}\bar{d}e$
$\{b, d\}$	$\bar{a}\bar{c}d$
$\varnothing$	$\bar{a}\bar{b}$

Table 9.1. The states of $\mathcal{L}$ and their representing strings.

Such strings are referred to as 'words' and the set of words

$$L_1 = \{ade, acd\bar{e}, a\bar{d}e, ac\bar{d}\bar{e}, a\bar{c}d, a\bar{b}, \bar{a}de, \bar{a}\bar{d}e, \bar{a}\bar{c}d, \bar{a}\bar{b}\}$$

is called a 'descriptive' language (for $\mathcal{L}$). Some descriptive languages are of particular interest for the subject of this monograph because they represent sequential procedures for recognizing the states. Such languages can in principle be used in assessing the knowledge states of individuals. For example, the language

$$L_2 = \{acde, acd\bar{e}, ac\bar{d}e, ac\bar{d}\bar{e}, a\bar{c}d, a\bar{c}\bar{d}, \bar{a}ed, \bar{a}e\bar{d}, \bar{a}\bar{e}b, \bar{a}\bar{e}\bar{b}\}$$

also specifies the states of $\mathcal{L}$, but with words satisfying certain rules. Notice that every word of L_2 begins with a or $\bar{a}$. Also, in any word beginning with a, the next symbol is either c or $\bar{c}$. Similarly, if a word begins with $\bar{a}$, the next symbol is either e or $\bar{e}$; etc. This illustrates a general pattern represented in Figure 9.1 in the form of a decision tree. The words of L_2 are read from the tree by following the branches from left to right. Each leaf corresponds to a word specifying a state. Such a tree represents a possible knowledge assessment procedure.

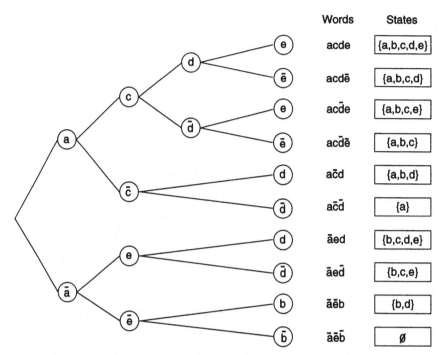

Fig. 9.1. Sequential decision tree corresponding to the language L_2 for the knowledge structure $\mathcal{L}$.

For example, beginning at the extreme left node of the tree, one could first check whether the knowledge state of a subject contains a, by proposing an instance of that item to the subject. Suppose that the subject solves a. We already know then that the subject's state is in

$$\mathcal{L}_a = \{ \, \{a\}, \{a,b,c\}, \{a,b,d\}, \{a,b,c,d\}, \{a,b,c,e\}, \{a,b,c,d,e\} \, \}.$$

(We recall that if $\mathcal{K}$ is a knowledge structure and q is any item, then $\mathcal{K}_q = \{K \in \mathcal{K} \,|\, q \in K\}$ and $\mathcal{K}_{\bar q} = \{K \in \mathcal{K} \,|\, q \notin K\}$; cf. 1.4 and 7.32. Thus $\mathcal{K} = \mathcal{K}_q \cup \mathcal{K}_{\bar q}$.) The next item proposed is c. Suppose that c is not solved, but that the subject then solves d. The complete questioning sequence in this case corresponds to the word $a\bar{c}d$ which identifies the state $\{a,b,d\}$ because

$$\mathcal{L}_a \cap \mathcal{L}_{\bar c} \cap \mathcal{L}_d = \{ \, \{a,b,d\} \, \}.$$

A language which is representable by a decision tree, such as L_2, will be called an 'assessment language.' (An exact definition is given in 9.4.)

There are two difficulties with this deterministic approach to the assessment of knowledge in real-life cases. One is that human behavior in testing situations is often unreliable: subjects may temporarily forget the solution of a problem that they know well, or make careless errors in responding; they could also in some situations guess the correct response to a problem not already mastered. This feature suggests that a straightforward decision tree is inadequate as an knowledge assessment device.

The other difficulty is that we evidently wish to minimize (in some sense) the number of questions asked, which raises the issue of 'optimality' of a decision tree. At least two parameters could be minimized. The first one is the largest number of questions to be asked for specifying any state. In Figure 9.1, this *worst case number* or *depth of the tree*, as it is also called, equals 4 (and is attained for four states). A second parameter often used is the average number of questions, which is here $3.4 = (4 \times 4 + 6 \times 3)/10$. An *optimal* tree is one that minimizes either the depth or the average number of internal nodes, depending on the context. The design of an optimal decision tree for the task at hand is known from theoretical computer science to be a hard problem—where 'hard' is meant in its technical sense, namely the corresponding decision problem is 'NP-complete.' For a good introduction to complexity theory, we refer the reader to Garey and Johnson (1979); for the particular problem at hand here, our reference is Hyafill and Rivest (1976) (see in particular the proof of the main result). We shall not discuss here the various results obtained regarding the construction of optimal decision trees. The reason is that in our setting, deterministic procedures (encoded as binary decision trees) rely on the untenable assumptions that student's answers truly reflect the student's knowledge.

In Chapters 10 and 11, we describe probabilistic assessment procedures which are more robust than those based on decision trees, and are capable of uncovering the knowledge state even when the subject's behavior is somewhat erratic.

There are nevertheless some theoretical questions worth studying in a deterministic framework, and which are also relevant to knowledge assessment. For instance, imagine that we observe a teacher conducting an oral examination of students. Idealizing the situation, suppose that this observation is taking place over a long period, and that we manage to collect all the sequences of questions asked by the teacher. Would we then be able to infer the knowledge structure relied upon by the teacher? We assume here that all potential sequences are revealed during the observation, and that the teacher is correct in assuming that the students' responses reflect their true knowledge state. An example of the results presented in this chapter is that if a knowledge structure is known to be ordinal (in the sense of 1.47), then it can be uncovered from any of its assessment languages (cf. Corollary 9.14). In general, an arbitrary structure cannot be reconstructed on the basis of a single assessment language. Suppose however that we have observed many teachers for a long time, and that all the assessment languages have been observed. Another result is that any knowledge structure can be uncovered from the set of all its feasible assessment languages (cf. 9.17).

To spell out this and other relevant results with a satisfactory precision, we need to introduce a basic terminology concerning 'words', 'languages' and related concepts.

Terminology

9.2. Definition. We start from a finite set Q which we call *alphabet*. Any element in Q is called a *positive literal*. The *negative literals* are the elements of Q marked with an overbar. Thus for any $q \in Q$, we have the two literals q and $\bar{q}$. A *string* over the alphabet Q is any finite sequence of literals written as $\alpha_1 \alpha_2 \ldots \alpha_n$, for some natural number n and literals α_1, α_2, ..., α_n. Denoting by Σ the set of all literals, we equip the set Σ^* of all strings with the operation of concatenation

$$(\rho, \rho') \longmapsto \rho\rho' \in \Sigma^*$$

and hence get a semigroup (because this concatenation is associative), having the empty string 1 as its neutral element. Any subset L of Σ^* is called a *language* over Q. An element ω of a language L is called a *word* of that language.

A string ρ of Σ^* is a *prefix* (resp. *suffix*) of the language L if $\rho\rho'$ (resp. $\rho'\rho$) is a word of L for some string ρ' of Σ^*. A prefix (resp. suffix) ρ is *proper* if there exists a nonempty string π such that $\rho\pi$ (resp. $\pi\rho$) is a word. For any string ρ and any language L, we denote by ρL the language containing all words of the form $\rho\omega$, for $\omega \in L$.

Strings, words, prefixes and suffixes are *positive* (resp. *negative*) when they are formed with positive (resp. negative) literals only. If $\rho = \alpha_1\alpha_2\ldots\alpha_n$ is a string over the alphabet Q, we set $\bar\rho = \bar\alpha_1\bar\alpha_2\ldots\bar\alpha_n$, with the convention $\bar{\bar\alpha} = \alpha$ for any literal α.

9.3. Example. For $Q = \{a, b, c, d, e\}$, we have ten literals. Consider the language consisting of all words of length at most 2. This language consists of $1 + 10 + 10^2$ words. Every word is both a prefix and a suffix. However, there are only $1+10$ proper prefixes, and the same number of proper suffixes. The language has $1 + 5 + 5^2$ positive words (coinciding with the positive prefixes) and $1 + 5$ positive, proper prefixes.

We recall from Definition 1.11 that a collection $\mathcal{K}$ on a domain Q is a collection $\mathcal{K}$ of subsets of Q. We also write $(Q, \mathcal{K})$ to denote the collection and call states the elements of $\mathcal{K}$. Thus, a knowledge structure $(Q, \mathcal{K})$ is a collection $\mathcal{K}$ which contains both $\varnothing$ and Q. Notice that a collection may be empty. The trace of a structure $(Q, \mathcal{K})$ on a nonempty subset A of Q is defined as $\mathcal{K}|_A = \{K \cap A \,|\, K \in \mathcal{K}\}$ (cf. Definition 1.17). This concept of trace of $(Q, \mathcal{K})$ on a subset A of Q extends to collections. The meaning of $\mathcal{K}_q$ and $\mathcal{K}_{\bar q}$ is the same for collections and for structures (cf. 1.4 and 7.32).

In this chapter, we assume that the domain Q is finite.

9.4. Definition. An *assessment language* for the collection $(Q, \mathcal{K})$ is a language L over the alphabet Q that is empty if $|\mathcal{K}| = 0$, has only the word 1 if $|\mathcal{K}| = 1$, and otherwise satisfies $L = qL_1 \cup \bar qL_2$, for some q in Q, where

(1) L_1 is an assessment language for the trace $\mathcal{K}_q|_{Q\setminus\{q\}}$;

(2) L_2 is an assessment language for the collection $\mathcal{K}_{\bar q}$ with domain $Q \setminus \{q\}$.

It is easily verified that the words of an assessment language L for $\mathcal{K}$ are in a one-to-one correspondence with the states of $\mathcal{K}$ (see Problem 2). Figure 9.1 lists the words of an assessment language for the knowledge structure of Equation (1).

We can also characterize assessment languages nonrecursively, by specific properties. The following concept is instrumental in that respect.

9.5. Definition. A *binary classification language* over a finite alphabet Q is a language L which either consists of the empty word alone or satisfies the two following conditions:

(1) a letter may not appear more than once in a word;

(2) if π is a proper prefix of L, then there exist exactly two prefixes of the form $\pi\alpha$ and $\pi\beta$, where α and β are literals; moreover $\bar{\alpha} = \beta$.

Condition (1) implies that L does not contain any word of the form $\pi x \rho x \sigma$ or $\pi x \rho \bar{x} \sigma$ with x a literal and π, ρ, σ strings. Consequently, a binary classification language L is finite.

9.6. Theorem. *For any proper prefix ρ of a nonempty binary classification language L, there exist a unique positive suffix σ and a unique negative suffix τ such that $\rho\sigma$, $\rho\tau \in L$. In particular, L has exactly one positive and one negative word.*

The proof is left as Problem 4. We now show that the conditions in Definition 9.5 provide us a nonrecursive characterization of assessment languages.

9.7. Theorem. *Any assessment language is a binary classification language. Conversely, any binary classification language is an assessment language for some collection.*

PROOF. It is easily proved that an assessment language L for the knowledge structure $(Q, \mathcal{K})$ is a binary classification language over the alphabet Q. Conversely, assume L is a binary classification language over the alphabet Q that contains more than one word. Then 1 is a proper prefix, and by Condition (2) of Definition 9.5, there is a letter q such that all words are of the form $q\pi$ or $\bar{q}\sigma$ for various strings π and σ. Notice that $L_1 = \{\pi \mid q\pi \in L\}$ is again a binary classification language, this time over the alphabet $Q \setminus \{q\}$. Proceeding by induction, we infer the existence of a collection $\mathcal{K}_1$ on $Q \setminus \{q\}$ for which L_1 is an assessment language. Similarly, $L_2 = \{\pi \mid \bar{q}\pi \in L\}$ is an assessment language for some collection $\mathcal{K}_2$ on $Q \setminus \{q\}$. It is easily verified that L is an assessment language for the collection $\{K \cup \{q\} \mid K \in \mathcal{K}_1\} \cup \mathcal{K}_2$. □

A less restrictive class of languages than the assessment languages, or equivalently the binary classification languages, will be used in the next section. We proceed to define it.

9.8. Definition. Let $(Q, \mathcal{K})$ be a finite collection and let L be a language over Q. We write $\alpha \vdash \omega$ to mean that α is a literal of the word ω. A word ω *describes* a state K iff K is the only state that satisfies

$$\text{for any } x \in Q: (x \vdash \omega \Rightarrow x \in K) \text{ and } (\bar{x} \vdash \omega \Rightarrow x \notin K).$$

The language L is a *descriptive language* for the collection $\mathcal{K}$ when the two following conditions hold:

(1) any word of L describes a unique state in K;
(2) any state in $\mathcal{K}$ is described by at least one word of L.

Then we also say that L *describes* $\mathcal{K}$.

While the words of an assessment language for a collection $\mathcal{K}$ are in a one-to-one correspondence with the states of $\mathcal{K}$, we only have a surjective mapping from a descriptive language for $\mathcal{K}$ onto $\mathcal{K}$.

We denote by $\mathrm{ASL}(\mathcal{K})$ and $\mathrm{DEL}(\mathcal{K})$, respectively, the collections of all assessment languages and descriptive languages for $\mathcal{K}$. We also write $(Q, \overline{\mathcal{K}})$ for the *dual* collection of a collection $(Q, \mathcal{K})$, with $\overline{\mathcal{K}} = \{Q \setminus K \mid K \in \mathcal{K}\}$. (This extends the notation and terminology introduced in Definition 1.7 in the case of a structure.) If L is a language, we denote by $\bar{L}$ the language obtained by replacing in each word of L any literal α with $\bar{\alpha}$. We then have (see Problem 6) the two equivalences

$$L \in \mathrm{ASL}(\mathcal{K}) \quad \text{iff} \quad \bar{L} \in \mathrm{ASL}(\overline{\mathcal{K}}),$$
$$L \in \mathrm{DEL}(\mathcal{K}) \quad \text{iff} \quad \bar{L} \in \mathrm{DEL}(\overline{\mathcal{K}}).$$

Recovering Ordinal Knowledge Structures

We prove here that any (partially) ordinal space $(Q, \mathcal{K})$ can be recovered from any of its descriptive languages, and thus also from any of its assessment languages. In this section, we will denote by $\mathcal{P}$ a partial order on Q from which $\mathcal{K}$ is derived (in the sense of Definition 1.50). Maximality and minimality of elements of Q are understood with respect to $\mathcal{P}$. Also, we use $\mathcal{K}(q) = \cap \mathcal{K}_q$ to denote the smallest state that contains item q. Moreover, $\mathcal{H}$ will be the covering relation or Hasse diagram of $\mathcal{P}$ (cf. 0.21). Finally, L will be a descriptive language for $\mathcal{K}$.

9.9. Lemma. *Let ω be a word of L describing the state K from $\mathcal{K}$;*

(i) *if x is a maximal element of K, then $x \vdash \omega$;*

(ii) *if y is a minimal element of $Q \setminus K$, then $\bar{y} \vdash \omega$.*

PROOF. If x is maximal in K, it follows that $K \setminus \{x\}$ is also a state of $\mathcal{K}$. Since ω distinguishes between K and $K \setminus \{x\}$, we conclude $x \vdash \omega$. The second assertion obtains similarly, since $K \cup \{y\}$ is also a state. □

9.10. Corollary. *For each q in Q, the language L uses both literals q and $\bar{q}$.*

PROOF. The smallest state $\mathcal{K}(q)$ containing q has q as a maximal element. Moreover, $\mathcal{K}(q) \setminus \{q\}$ is a state whose complement has q as a minimal element. □

9.11. Theorem. *Define a relation $\mathcal{S}$ on Q by declaring $q\mathcal{S}r$ to hold when*

(1) *there exists some word ω of L such that $q \vdash \omega$ and $\bar{r} \vdash \omega$;*

(2) *there is no word ρ of L such that both $\bar{q} \vdash \rho$ and $r \vdash \rho$.*

Let $\widehat{\mathcal{P}}$ denote the strict partial order obtained from $\mathcal{P}$ by deleting all loops. Then necessarily

$$\mathcal{H} \subseteq \mathcal{S} \subseteq \widehat{\mathcal{P}}. \tag{4}$$

PROOF. Assume $q\mathcal{H}r$ and set $K = \mathcal{K}(r) \setminus \{r\}$. Then K is a state having q as a maximal element; also, r is a minimal element in $Q \setminus K$. By Lemma 9.9, any word ω describing K makes Condition (1) true. Moreover, Condition (2) also holds since from $q\mathcal{H}r$ we deduce that any state containing r also contains q. This establishes $\mathcal{H} \subseteq \mathcal{S}$.

Assume now $q\mathcal{S}r$. Take any state K described by the word ω whose existence is asserted in Condition (1). As $q \in K$ and $r \notin K$, we cannot have $r\mathcal{P}q$. Thus $q \neq r$. It only remains to show that q and r are comparable with respect to $\mathcal{P}$. If this were not the case, $(\mathcal{K}(q) \setminus \{q\}) \cup \mathcal{K}(r)$ would be a state K' having r as a maximal element; moreover, q would be minimal in $Q \setminus K'$. Any word describing K' would then contradict Condition (2). □

Each of the two inclusions in Equation (4) can be either strict or an equality, as shown by the following two examples.

9.12. Example. Consider the set $Q = \{a, b, c\}$ equipped with the alphabetical order $\mathcal{P}$. Figure 9.2 specifies an assessment language for the corresponding ordinal structure $\mathcal{K}$; here we have $\mathcal{H} \subset \mathcal{S} = \widehat{\mathcal{P}}$.

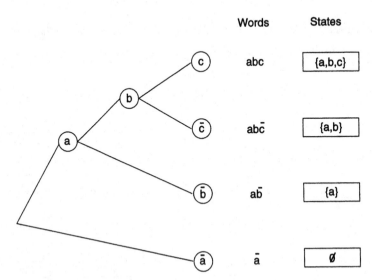

Fig. 9.2. Decision tree, words and states from Example 9.12.

9.13. Example. Figure 9.3 describes another assessment language for the same ordinal knowledge structure as in previous example. We have this time $\mathcal{H} = \mathcal{S} \subset \widehat{\mathcal{P}}$.

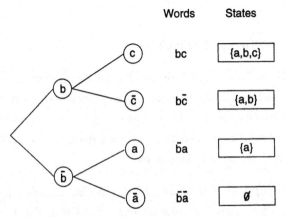

Fig. 9.3. Decision tree, words and states from Example 9.13.

9.14. Corollary. *If a finite knowledge structure is known to be ordinal, it can be recovered from any of its descriptive languages.*

Corollary 9.14 is the main result of this section. It readily follows from Theorem 9.11: $\mathcal{P}$ is always the transitive closure of $\mathcal{S}$.

9.15. Remarks. A language can describe two distinct knowledge structures only one of which is ordinal. Such a case is described by Figure 9.4.

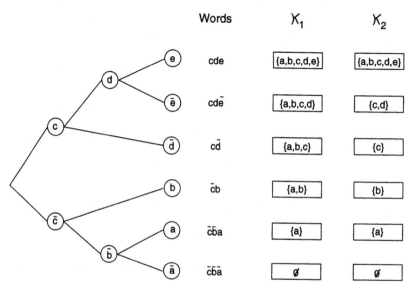

Fig. 9.4. Decision tree, words and states from two distinct knowledge structures, each one described by the language (cf. Remark 9.15). Only $\mathcal{K}_1$ is ordinal.

Recovering Knowledge Structures

As the example in Figure 9.4 indicates, a knowledge structure cannot be recovered from just one of its assessment languages (except if it is known to be ordinal, cf. Corollary 9.13). We will prove by induction on the number of items that any structure $\mathcal{K}$ can be recovered from the complete collection ASL($\mathcal{K}$) of all its assessment languages.

9.16. Lemma. *Suppose $A \subseteq Q$ and let L_A be an assessment language for the trace $\mathcal{K}|_A$. There exists an assessment language L for $\mathcal{K}$ such that*

   (1) *in each word of L, the letters from A precede the letters from $Q \setminus A$;*

(2) *truncating the words from* L *to their literals from* A *give all the words in* L_A *(possibly with repetitions).*

Omitting the proof, we derive from 9.16 the main result of this section.

9.17. Theorem. *For any two knowledge structures* $(Q, \mathcal{K})$ *and* $(Q', \mathcal{K}')$, *the following statements are equivalent:*

   (i) $ASL(\mathcal{K}) = ASL(\mathcal{K}')$;
   (ii) $\mathcal{K} = \mathcal{K}'$.

PROOF. (i) implies (ii). Assume Condition (i) holds. If some item q belongs to Q but not to Q', we can use it as the root of a decision tree for $\mathcal{K}$; the corresponding assessment language would of course not belong to $ASL(\mathcal{K}')$. This shows $Q \subseteq Q'$, and by symmetry $Q' = Q$.

We now proceed by induction on $|Q|$ to prove that the negation of Condition (ii) implies a contradiction. As Condition (ii) does not hold, select in $\mathcal{K} \bigtriangleup \mathcal{K}'$ a maximal element, say K in $\mathcal{K} \setminus \mathcal{K}'$. Then $K \neq Q$. From Lemma 9.16, we have $ASL(\mathcal{K}|_K) = ASL(\mathcal{K}'|_K)$, and thus also by the induction hypothesis $\mathcal{K}|_K = \mathcal{K}'|_K$. Again using Lemma 9.16, we construct an assessment language L for $\mathcal{K}$ by taking letters from K systematically before letters from $Q \setminus K$. The word ω in L describing K must then be of the form

$$k_1 k_2 \ldots k_m \bar{y}_1 \bar{y}_2 \ldots \bar{y}_n$$

with $k_i \in K$ and $y_j \in Q \setminus K$. By our assumption (i), L is also an assessment language for $\mathcal{K}'$. The word ω describes some state K' in $\mathcal{K}'$. In view of the prefix $k_1 k_2 \ldots k_m$ and of $\mathcal{K}|_K = \mathcal{K}'|_K$, we see $K \subseteq K'$. Then from the maximality of K we derive $K' \in \mathcal{K}$. Hence the word ω of L describes two distinct states of $\mathcal{K}$, namely K and K', a contradiction.   □

9.18. Corollary. *The two following statements are equivalent:*

   (1) $DEL(\mathcal{K}) = DEL(\mathcal{K}')$;
   (2) $\mathcal{K} = \mathcal{K}'$.

Original Sources and Related Works

The chapter closely follows Degreef, Doignon, Ducamp and Falmagne (1986). This paper contains some additional, open questions about how languages allow to recover knowledge structures.

Problems

1. Verify all the numbers in Example 9.3.

2. Prove that for any collection $\mathcal{K}$ and any assessment language L for $\mathcal{K}$, there is a one-to-one correspondence between the states of $\mathcal{K}$ and the words of L (cf. Definition 9.4).

3. Describe an optimal decision tree for the ordinal knowledge structure derived from a linear order.

4. Give a proof of Theorem 9.6.

5. Give a proof of the following two equivalences (see after Definition 9.8)
$$L \in \mathrm{ASL}(\mathcal{K}) \text{ iff } \bar{L} \in \mathrm{ASL}(\bar{\mathcal{K}}), \qquad L \in \mathrm{DEL}(\mathcal{K}) \text{ iff } \bar{L} \in \mathrm{DEL}(\bar{\mathcal{K}}).$$

6. Prove that if L is a descriptive language for $\mathcal{K}$, then any language formed by arbitrarily changing the orders of the litterals in the words of L is also a descriptive language for $\mathcal{K}$.

Chapter 10

Uncovering the State of an Individual: A Continuous Markov Procedure

Suppose that, having applied the techniques described in the preceding chapters, we have obtained a particular knowledge structure. We now ask: how can we uncover, by appropriate questioning, the knowledge state of a particular individual? Two broad classes of stochastic assessment procedures are described in this chapter and the next one.

A Deterministic Algorithm

By way of introduction, we first consider a simple algorithm in the spirit of those discussed in Chapter 9 and which is suitable when there are no errors of any kind.

10.1. Example. Our discussion will be illustrated by the knowledge structure

$$\mathcal{K} = \{ \varnothing, \{a\}, \{c\}, \{a,b\}, \{a,c\}, \{b,c\},$$
$$\{a,b,c\}, \{a,b,c,d\}, \{a,b,c,d,e\} \}, \tag{1}$$

which has domain $Q = \{a, b, c, d, e\}$, and is also displayed in Figure 10.1.

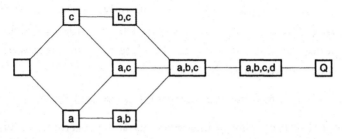

Fig. 10.1. The well-graded knowledge space $\mathcal{K}$ of Equation (1).

This knowledge structure is a well-graded knowledge space, with four gradations. These special features will play no rôle in the assessment, however.

We shall assume that a subject's response never results from a lucky guess or a careless error. (The impact of this assumption is discussed later in this chapter.) Consider an assessment in which the first question asked is Question b. If an incorrect response is obtained, all states containing b must be discarded. We indicate this conclusion by marking with the symbol $\sqrt{}$ the remaining possible states in the second column of Table 10.1; they form the collection $\mathcal{K}_{\bar{b}}$ (cf. 1.4 and 7.32 for this notation).

Problems	b	a	c
Responses	0	1	1
$\varnothing$	$\sqrt{}$		
$\{a\}$	$\sqrt{}$	$\sqrt{}$	
$\{c\}$	$\sqrt{}$		
$\{a,b\}$			
$\{a,c\}$	$\sqrt{}$	$\sqrt{}$	$\sqrt{}$
$\{b,c\}$			
$\{a,b,c\}$			
$\{a,b,c,d\}$			
$\{a,b,c,d,e\}$			

Table 10.1. Inferences from the successive responses of a subject for the knowledge structure $\mathcal{K}$ of Figure 10.1. The states remaining after each response are indicated by the symbol '$\sqrt{}$'.

Next, Problem a is presented, and a correct response is recorded, eliminating the two states $\varnothing$ and $\{c\}$. Thus, the only feasible states after two questions are $\{a\}$ and $\{a,c\}$. The last problem asked is c, which elicits a correct response, eliminating $\{a\}$. In this deterministic framework, $\{a,c\}$ is the only state consistent with the data:

$$(b, \text{false}), (a, \text{correct}), (c, \text{correct}).$$

Clearly, all the states can be uncovered by this procedure, which can be represented by a binary decision tree (see Figure 10.2). The procedure is certainly economical. For instance, if the states are equiprobable, a state can be uncovered by asking an average of $3\frac{2}{9}$ questions out of 5 questions.

This representation was investigated from a formal viewpoint by Degreef, Doignon, Ducamp and Falmagne (1986) and was reviewed in Chapter 9.

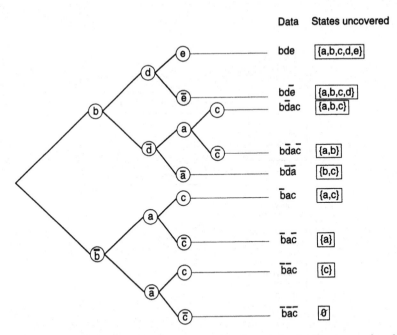

Fig. 10.2. Binary decision tree for uncovering the states in the example of Figure 10.1.

A serious difficulty for this type of algorithm is that it cannot effectively deal with a possible intrinsic randomness, or even instability, of the subject performance. Obvious examples of randomness are the careless errors and lucky guesses formalized in the models developed in Chapter 7 and 8. A case of instability may arise if the subject's state varies in the course of the questioning. This might happen, for example, if the problems of the test cover concepts learned by the subject a long time earlier. The first few questions asked may jolt the subject's memory, and facilitate the retrieval of some material relevant to the last part of the test. In any event, more robust procedures are needed that are capable of uncovering a subject's state, or at least approaching it closely, despite noisy data.

Outline of a Markovian Stochastic Process

The two Markov procedures described in this Chapter and in Chapter 11 enter into a general framework illustrated by Figure 10.3.

At the beginning of step n of the procedure, all the information gath-

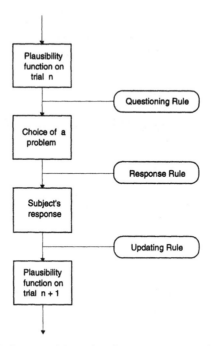

Fig. 10.3. Diagram of the transitions for the two classes of Markovian procedures.

ered from step 1 to step $n - 1$ is summarized in a 'plausibility function' which assigns a positive real number—a 'plausibility value'—to each of the knowledge states in the structure. This plausibility function is used by the procedure to select the next question to ask. The mechanism of this selection lies in a 'questioning rule', an operator applied to the plausibility function, whose output is the question chosen. The subject's response is then observed, and it is assumed that it is generated by the subject's knowledge state, through a 'response rule.' In the simplest case, it is assumed that the response is correct if the question belongs to the subject's state, and false otherwise. Careless error and lucky guess parameters may also be introduced at this stage. (These will play only a minor role in this chapter; however, see Remark 10.27.) Finally, the plausibility function is recomputed, through an 'updating rule', based on the question asked and the subject's observed response.

In this chapter, we consider a case in which the plausibility function is a likelihood function on the set of states. (Our presentation follows closely Falmagne and Doignon, 1988a.) As in Chapters 7 and 8, we consider a probabilistic knowledge structure $(Q, \mathcal{K}, L)$. Thus, for every state K, we represent by $L(K)$ the probability of state K in the population of reference. Note that we assume $0 < L < 1$. For concreteness, we consider the knowledge structure $\mathcal{K}$ of Example 10.1 and Figure 10.1 and we suppose that the knowledge states have the probabilities represented by the histogram of Figure 10.4A. In each of the three graphs of Figure 10.4, the knowledge states are represented by squares, and the items by circles. A link between a square and a circle means that the corresponding state contains the corresponding item. For instance, the square at the extreme left of Figure 10.4A represents the knowledge state $\{a\}$. We have $L(\{a\}) = .10$, $L(\{a, b\}) = .05$, etc.

The probability distribution L will be regarded as the a priori distribution representing the uncertainty of the assessor at the beginning of assessment. We set $\mathbf{L}_1 = L$ to denote the likelihood on step 1 of the procedure. As before, suppose that item b is presented to the subject, who fails to respond correctly. This information will induce a transformation of the likelihood $\mathbf{L}_1$ by an operator, which will decrease the probabilities of all the states containing b, and increase the probabilities of all the states not containing that item. The resulting distribution $\mathbf{L}_2$ is pictured by the histogram in Figure 10.4B. Next, items a and c e.g. are presented successively, eliciting two correct responses. The accumulated effect of these events on the likelihood is depicted by the histogram in Figure 10.4C representing $\mathbf{L}_4$, in which the probability of state $\{a, c\}$ is shown as much higher than that of any other state. This result is similar to what we had obtained, with the same sequence of events, from the deterministic algorithm represented in Figure 10.2, but much less brutal: a knowledge state is not suddenly eliminated; rather, its likelihood decreases. Needless to say, there are many ways of implementing this idea; several implementations will be considered in this chapter.

A possible source of noise in the data obviously lies in the response mechanisms, which we have formalized by the Local Independence Axiom [N] of Chapter 8. For most of this chapter, we shall assume that such factors play a minor role and can be neglected. In other words, all the parameters β_q and η_q take value zero during the main phase of the assessment (thus there are no lucky guesses and no careless errors). Once the assessment algorithm has terminated, the result of the assessment can be refined by reviving these

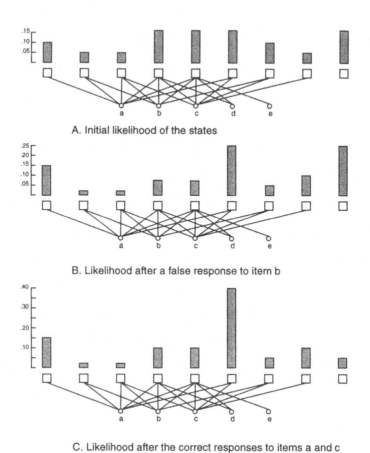

A. Initial likelihood of the states

B. Likelihood after a false response to item b

C. Likelihood after the correct responses to items a and c

Fig. 10.4. Successive transformations of the likelihood induced by the events: (item b, false), (item a, correct), (item c, correct).

response mechanisms (see Remark 10.27). In the meantime, the response rule will be simple. Suppose that the subject is in some knowledge state K_0 and that some question q is asked. The subject's response will be correct with probability one if $q \in K_0$, and incorrect with probability one in the other case.

Notation and Basic Concepts

10.2. Definition. Let $(Q, \mathcal{K}, L)$ be some arbitrary probabilistic knowledge structure with a finite domain Q. The set of all probability distributions on

$\mathcal{K}$ is denoted by Λ. (We have thus $L \in \Lambda$.) We suppose that the subject is, with probability one, in some unknown knowledge state K_0 which will be called *latent* and has to be uncovered.

Any application of an assessment procedure in the sense of this chapter is a realization of a stochastic process. The first step or *trial*, numbered $n = 1$, of this process begins with the initial likelihood $\mathbf{L}_1 = L \in \Lambda$. Any further trial $n > 1$ begins with an updated estimate $\mathbf{L}_n \in \Lambda$ of that likelihood. In the framework of the stochastic process, $\mathbf{L}_n$ is a random vector in Λ. For any state $K \in \mathcal{K}$, we denote by $\mathbf{L}_n(K)$ the likelihood, for the observer, that the subject under examination is in the state K at the beginning of trial n (no confusion should arise from the fact that both $\mathbf{L}_n$ and $\mathbf{L}_n(K)$, being a random vector and a random variable, are also functions of elementary events). More generally, we write for any $\mathcal{F} \subseteq \mathcal{K}$,

$$\mathbf{L}_n(\mathcal{F}) = \sum_{K \in \mathcal{F}} \mathbf{L}_n(K). \tag{2}$$

The second feature of a trial involves the question asked. We write $\mathbf{Q}_n$ for the question asked on trial $n \geq 1$. Thus, $\mathbf{Q}_n$ is a random variable taking its values in the set Q of all items. In general, the choice of a question is governed by a function $(q, \mathbf{L}_n) \mapsto \Psi(q, \mathbf{L}_n)$ mapping $Q \times \Lambda$ into the interval $[0, 1]$ and specifying the probability that $\mathbf{Q}_n = q$. The function Ψ will be called the *questioning rule*. Two special cases of this function are analyzed in this chapter (see Definitions 10.13 and 10.14 below).

The response observed on trial n is denoted by $\mathbf{R}_n$. Only two cases are considered: (i) a correct response, which is coded as $\mathbf{R}_n = 1$; (ii) an error, which is coded as $\mathbf{R}_n = 0$. The probability of a correct response to question q is equal to one if $q \in K_0$, and to zero otherwise.

At the core of the procedure is a Markovian transition rule stating that, with probability one, the likelihood function $\mathbf{L}_{n+1}$ on trial $n + 1$ only depends on the likelihood function $\mathbf{L}_n$ on trial n, the question asked on that trial, and the observed response. This transition rule is formalized by the equation

$$\mathbf{L}_{n+1} \overset{\text{a.s.}}{=} u(\mathbf{R}_n, \mathbf{Q}_n, \mathbf{L}_n), \tag{3}$$

in which u is a function mapping $\{0, 1\} \times Q \times \Lambda$ into Λ. (The specification 'a.s.' is an abbreviation of 'almost surely' and means that the equality holds with probability one. A similar remark holds for the convergence $\overset{\text{a.s.}}{\longrightarrow}$

further down on this page.) The function u is referred to as the *updating rule*. Two cases of this function will be considered (see later Definitions 10.8 and 10.10). The diagram below summarizes these transitions:

$$(\mathbf{L}_n \to \mathbf{Q}_n \to \mathbf{R}_n) \to \mathbf{L}_{n+1}. \tag{4}$$

The Cartesian product $\Gamma = \{0,1\} \times Q \times \Lambda$ is thus the state space of the process, each trial n being characterized by a triple $(\mathbf{R}_n, \mathbf{Q}_n, \mathbf{L}_n)$. We denote by Ω the sample space, that is, the set of all sequences of points in Γ. The complete history of the process from trial 1 to trial n is denoted by

$$\mathbf{W}_n = ((\mathbf{R}_n, \mathbf{Q}_n, \mathbf{L}_n), \dots, (\mathbf{R}_1, \mathbf{Q}_1, \mathbf{L}_1)).$$

The notation $\mathbf{W}_0$ stands for the empty history.

In general, the assessment problem consists in uncovering the latent state K_0. This quest has a natural formalization in terms of the condition

$$\mathbf{L}_n(K_o) \xrightarrow{\text{a.s.}} 1.$$

When this condition holds for some particular assessment procedure, we shall sometimes say that K_0 is *uncoverable* (by that procedure).

We recall that for any subset A of a fixed set S, the *indicator* of A is a function $x \mapsto \iota_A(x)$ is defined on S by

$$\iota_A(x) = \begin{cases} 1 & \text{if } x \in A \\ 0 & \text{if } x \in S \setminus A. \end{cases}$$

For convenience, the main concepts are listed in the glossary below (where "r.v." stands for random variable).

10.3. Glossary.

$(Q, \mathcal{K}, L)$	a finite probabilistic knowledge structure;
Λ	the set of all probability distributions on $\mathcal{K}$;
Γ	the state space of the process $(\mathbf{R}_n, \mathbf{Q}_n, \mathbf{L}_n)_{n \in \mathbb{N}}$;
K_0	the latent knowledge state of the subject;
$\mathbf{L}_1 = L$	the initial likelihood of the states, $0 < L < 1$;
$\mathbf{L}_n(K)$	a r.v. representing the likelihood of state K on trial n;
$\mathbf{Q}_n$	a r.v. representing the question asked on trial n;

$\mathbf{R}_n$	a r.v. representing the response given on trial n;
Ψ	$(q, \mathbf{L}_n) \mapsto \Psi(q, \mathbf{L}_n)$, the questioning rule;
u	$(\mathbf{R}_n, \mathbf{Q}_n, \mathbf{L}_n) \mapsto u(\mathbf{R}_n, \mathbf{Q}_n, \mathbf{L}_n)$, the updating rule;
$\mathbf{W}_n$	random history of the process from trial 1 to trial n;
ι_A	the indicator of a set A.

10.4. General Axioms. The axioms concern a probabilistic knowledge structure $(Q, \mathcal{K}, L)$, the distinguished latent state K_0 of the subject, and the sequence of random vectors $(\mathbf{R}_n, \mathbf{Q}_n, \mathbf{L}_n)$.

[U]. **Updating Rule.** We have $\mathbf{P}(\mathbf{L}_1 = L) = 1$, and for any positive integer n and all measurable sets $B \subseteq \Lambda$,

$$\mathbf{P}(\mathbf{L}_{n+1} \in B \,|\, \mathbf{W}_n) = \iota_B\big(u(\mathbf{R}_n, \mathbf{Q}_n, \mathbf{L}_n)\big),$$

where u is a function mapping the interior of Γ into the interior of Λ. Writing u_K for the coordinate of u associated with the knowledge state K, we have thus

$$\mathbf{L}_{n+1}(K) \stackrel{a.s.}{=} u_K(\mathbf{R}_n, \mathbf{Q}_n, \mathbf{L}_n).$$

Moreover, the function u satisfies the following condition:

$$u_K(\mathbf{R}_n, \mathbf{Q}_n, \mathbf{L}_n) \begin{cases} > \mathbf{L}_n(K) & \text{if } \iota_K(\mathbf{Q}_n) = \mathbf{R}_n, \\ < \mathbf{L}_n(K) & \text{if } \iota_K(\mathbf{Q}_n) \neq \mathbf{R}_n. \end{cases}$$

[Q]. **Questioning Rule.** For all $q \in Q$ and all positive integers n,

$$\mathbf{P}(\mathbf{Q}_n = q \,|\, \mathbf{L}_n, \mathbf{W}_{n-1}) = \Psi(q, \mathbf{L}_n)$$

where Ψ is a function mapping $Q \times \Lambda$ into the interval $[0, 1]$.

[R]. **Response Rule.** For all positive integers n,

$$\mathbf{P}(\mathbf{R}_n = \iota_{K_0}(q) \,|\, \mathbf{Q}_n = q, \mathbf{L}_n, \mathbf{W}_{n-1}) = 1$$

where K_0 is the latent state.

Note that, as a knowledge structure, $\mathcal{K}$ contains at least two states, namely $\varnothing$ and Q.

10.5. Definition. We shall refer to a process $(\mathbf{R}_n, \mathbf{Q}_n, \mathbf{L}_n)$ satisfying the Axioms [U], [Q] and [R] in 10.4 as a *stochastic assessment procedure for*

$(Q, \mathcal{K}, L)$, *parametrized by* u, Ψ *and* K_0. The functions u, Ψ and ι_{K_0} are called the *updating rule*, the *questioning rule*, and the *response rule*, respectively.

10.6. Remarks. Axiom [R] is straightforward. Axioms [Q] and [U] govern, respectively, the choice of a question $\mathbf{Q}_n$ to be asked, and the reallocation of the mass of $\mathbf{L}_n$ on trial $n + 1$ depending on the values of $\mathbf{Q}_n$ and $\mathbf{R}_n$. Axiom [U] states that if $\mathbf{L}_n = \ell$, $\mathbf{Q}_n = q$ and $\mathbf{R}_n = r$ then $\mathbf{L}_{n+1}$ is almost surely equal to $u(r, q, p)$. As a general scheme, this seems reasonable, since we want our procedure to specify the likelihood of each of the knowledge states on each trial. This axiom ensures that no knowledge state will ever have a likelihood of zero, and that the likelihood of any state K will increase whenever we observe either a correct response to a question $q \in K$, or an incorrect response to a question $q \notin K$, and decrease in the two remaining cases. Notice that the first two axioms pertain to the assessment procedure *per se*, while the third describes some hypothetical mechanism governing a student's response.

It is easily shown that each of $(\mathbf{R}_n, \mathbf{Q}_n, \mathbf{L}_n)$, $(\mathbf{Q}_n, \mathbf{L}_n)$, and $(\mathbf{L}_n)$ is a Markov process (see Theorem 10.16). An important question concerns general conditions on the functions u and Ψ under which $\mathbf{L}_n$ converges to some random vector $\mathbf{L}$ independent of the initial distribution L. This aspect of the process will not be investigated here, however. Rather, we shall focus on the problem of defining useful procedures capable of uncovering the subject's knowledge state. Such procedures will be discussed in the section entitled "Uncovering the latent scale."

The class of processes defined by Axioms [U], [Q] and [R] is very large. Useful special cases can be obtain by specializing the questioning rule and the updating rule.

Special Cases

The initial likelihood L may be estimated, for example, by testing a representative sample of subjects from the population, using one of the models discussed in Chapters 7 and 8. In the absence of information on that initial likelihood, we may reasonably set

$$\mathbf{L}_1(K) = \frac{1}{|\mathcal{K}|}.$$

10.7. Two examples of the updating rule u. Suppose that some question q is presented on trial n, and that the subject's response is correct; thus, $\mathbf{Q}_n = q$ and $\mathbf{R}_n = 1$. Axiom [U] requires that the likelihood of all those states containing q should almost surely increase, and the likelihood of all those states not containing q should almost surely decrease. If the response is incorrect, the opposite result should obtain. Some questions may be judged more informative than others. For instance, it may be argued that, since a correct response to a multiple choice question may be due to a lucky guess, it should not be given as much weight as, say, a correct numerical response resulting from a computation. Moreover, the response itself may be taken into account: a correct numerical response may signify the mastery of a question, but an error does not necessarily imply complete ignorance. These considerations will be implemented into two rather different exemplary updating rules, in which the reallocation of the mass of $\mathbf{L}_n$ on trial $n+1$ will be governed by a parameter which may depend upon the question asked and on the response given on trial n.

10.8. Definition. The updating rule u of Axiom [U] will be called *convex with parameters* $\theta_{q,r}$, where $0 < \theta_{q,r} < 1$ for $q \in Q, r = 0, 1$, if the function u of Axiom [U] satisfies the following condition:
    For all $K \in \mathcal{K}$ and with $\mathbf{R}_n = r$, $\mathbf{Q}_n = q$

$$u_K(r, q, \mathbf{L}_n) = (1 - \theta_{q,r})\mathbf{L}_n(K) + \theta_{q,r}g_K(r, q, \mathbf{L}_n) \qquad (5)$$

where

$$g_K(r, q, \mathbf{L}_n) = \begin{cases} r\frac{\mathbf{L}_n(K)}{\mathbf{L}_n(\mathcal{K}_q)}, & \text{if } K \in \mathcal{K}_q \\ (1 - r)\frac{\mathbf{L}_n(K)}{\mathbf{L}_n(\mathcal{K}_{\bar{q}})}, & \text{if } K \in \mathcal{K}_{\bar{q}}. \end{cases}$$

Thus, the right member of Equation (5) specifies a convex combination between the current likelihood $\mathbf{L}_n$ and a conditional one, obtained from discarding all the knowledge states inconsistent with the observed response. The updating rule is *convex with constant parameter* θ if Equation (5) holds with $\theta_{q,r} = \theta$ for all $q \in Q$ and $r \in \{0, 1\}$.

One objection to this particular form of the updating rule is that it is not 'commutative.' One could require that the likelihood on trial $n+1$ should not depend, as it does in Equation (5) (cf. Problem 1), on the order of the pairs of questions and responses up to that trial. Consider the two cases

(1) $(\mathbf{Q}_{n-1} = q, \mathbf{R}_{n-1} = r)$, $(\mathbf{Q}_n = q', \mathbf{R}_n = r')$,
(2) $(\mathbf{Q}_{n-1} = q', \mathbf{R}_{n-1} = r')$, $(\mathbf{Q}_n = q, \mathbf{R}_n = r)$.

It could be argued that, for a given value of the likelihood $\mathbf{L}_{n-1} = \ell$, the likelihood on trial $n+1$ should be the same in these two cases because they convey the same information. Slightly changing our notation by setting $\xi = (q, r)$, $\xi' = (q', r')$ and $F(\ell, \xi) = u(r, q, \ell)$, this translates into the condition

$$F\big(F(\ell, \xi), \xi'\big) = F\big(F(\ell, \xi'), \xi\big). \tag{6}$$

In the functional equation literature, an operator F satisfying (6) is called *permutable* (Aczél, 1966, p. 270). In some special cases, permutability greatly reduces the possible form of an operator. However, the side conditions used by Aczél (1966) are too strong for our purpose (see Luce, 1964, and Marley, 1967, in this connection). In any event, this concept is of obvious interest

10.9. Definition. We shall call *permutable* an updating rule u with an operator F satisfying Equation (6).

An example of a permutable updating rule is given below.

10.10. Definition. The updating rule is called *multiplicative with parameters* $\zeta_{q,r}$, where $1 < \zeta_{q,r}$ for $q \in Q, r = 0, 1$, if the function u of Axiom [U] satisfies the condition: with $\mathbf{Q}_n = q$, $\mathbf{R}_n = r$ and

$$\zeta_{q,r}^K = \begin{cases} 1 & \text{if } \iota_K(q) \neq r, \\ \zeta_{q,r} & \text{if } \iota_K(q) = r \end{cases} \tag{7}$$

we have

$$u_K(r, q, \mathbf{L}_n) = \frac{\zeta_{q,r}^K \mathbf{L}_n(K)}{\sum_{K' \in \mathcal{K}} \zeta_{q,r}^{K'} \mathbf{L}_n(K')}. \tag{8}$$

It is easy to verify that this multiplicative rule is permutable (Problem 2).

Other updating rules applicable in different, but similar situations are reviewed by Landy and Hummel (1986). The two examples of updating rules introduced in 10.8 and 10.10 have been inspired by some operators used in mathematical learning theory (see the Sources Section).

10.11. Remarks. It was pointed out to us by Mathieu Koppen (personal communication) that the multiplicative updating rule can be seen as a Bayesian updating. The latter occurs when the values of $\zeta_{q,r}^K$ are linked in a specific manner with the probabilities of respectively a careless error and a lucky guess in the answer for item q (for the introduction of these probabilities, see 7.2). Fixing question q, and slightly changing our notation, we write

$P_q(K)$ for the a priori probability of state K,

$P_q(K \mid r)$ for the a posteriori probability of state K

after having observed response r,

with a similar interpretation for $P_q(r \mid K)$. From Bayes Theorem, we have

$$P_q(K \mid r) = \frac{P_q(r \mid K)P_q(K)}{\sum_{K' \in \mathcal{K}} P_q(r \mid K')P_q(K')}. \tag{9}$$

We see that Equations (8) and (9) have the same form, but $\zeta_{q,r}^K$ cannot be regarded as a conditional probability. In particular, we do not generally have

$$\zeta_{q,0}^K + \zeta_{q,1}^K = 1.$$

However, let us compare the multiplicative updating rule given in Equation (8) with the explicit form of the Baeysian rule. Equation (8) requires that the values $\mathbf{L}_{n+1}(K)$ are proportional to the values $\zeta_{q,r}^K \mathbf{L}_n(K)$ (for a fixed question q and answer r), with

$$\zeta_{q,r}^K = \begin{cases} \zeta_{q,1} & \text{if } q \in K, r = 1, \\ 1 & \text{if } q \notin K, r = 1, \\ 1 & \text{if } q \in K, r = 0, \\ \zeta_{q,0} & \text{if } q \notin K, r = 0. \end{cases}$$

Similarly, Baeysian updating (Equation (9)) makes the values $\mathbf{L}_{n+1}(K)$ proportional to the values $Z_{q,r}^K \mathbf{L}_n(K)$, where the real numbers $Z_{q,r}^K$ are specified by

$$Z_{q,r}^K = \begin{cases} 1 - \beta_q & \text{if } q \in K, r = 1, \\ \gamma_q & \text{if } q \notin K, r = 1, \\ \beta_q & \text{if } q \in K, r = 0, \\ 1 - \gamma_q & \text{if } q \notin K, r = 0. \end{cases}$$

Thus, the multiplicative updating rule coincides with Baeysian updating iff for all items q

$$\frac{\zeta_{q,1}}{1 - \beta_q} = \frac{1}{\gamma_q} \quad \text{and} \quad \frac{1}{\beta_q} = \frac{\zeta_{q,0}}{1 - \gamma_q}.$$

These equations can be rewritten as

$$\zeta_{q,1} = \frac{1 - \beta_q}{\gamma_q} \quad \text{and} \quad \zeta_{q,0} = \frac{1 - \gamma_q}{\beta_q}$$

or as

$$\beta_q = \frac{\zeta_{q,1} - 1}{\zeta_{q,1}\zeta_{q,0} - 1} \quad \text{and} \quad \gamma_q = \frac{\zeta_{q,0} - 1}{\zeta_{q,1}\zeta_{q,0} - 1}.$$

10.12. Two examples of questioning rule. A simple idea for the questioning rule is to select, on any trial n, a question q that partitions the set $\mathcal{K}$ of all the states into two subsets $\mathcal{K}_q$ and $\mathcal{K}_{\bar{q}}$ with a mass as equal as possible; that is, such that $\mathbf{L}_n(\mathcal{K}_q)$ is as close as possible to $\mathbf{L}_n(\mathcal{K}_{\bar{q}}) = 1 - \mathbf{L}_n(\mathcal{K}_q)$. Note in this connection that any likelihood $\mathbf{L}_n$ defines a set $S(\mathbf{L}_n) \subseteq Q$ containing all those questions q minimizing

$$|2\mathbf{L}_n(\mathcal{K}_q) - 1|.$$

Under this questioning rule, we must have $\mathbf{Q}_n \in S(\mathbf{L}_n)$ with a probability equal to one. The questions in the set $S(\mathbf{L}_n)$ are then chosen with equal probability.

10.13. Definition. The questioning rule will be called *half-split* when

$$\Psi(q, \mathbf{L}_n) = \frac{\iota_{S(\mathbf{L}_n)}(q)}{|S(\mathbf{L}_n)|}. \tag{10}$$

Another method may be used, which is computationally more demanding and may seem at first blush more exact. The uncertainty of the examiner on trial n of the procedure may be evaluated by the entropy of the likelihood on that trial, that is, by the quantity

$$H(\mathbf{L}_n) = - \sum_{K \in \mathcal{K}} \mathbf{L}_n(K) \log_2 \mathbf{L}_n(K).$$

It seems sensible to choose a question so as to reduce that entropy as much as possible. For $\mathbf{Q}_n = q$ and $\mathbf{L}_n = \ell$ the expected value of the entropy on trial $n + 1$ is given by

$$\mathbf{P}(\mathbf{R}_n = 1 \mid \mathbf{Q}_n = q)H\big(u(1, q, \ell)\big) + \mathbf{P}(\mathbf{R}_n = 0 \mid \mathbf{Q}_n = q)H\big(u(0, q, \ell)\big). \tag{11}$$

But the conditional probability $\mathbf{P}(\mathbf{R}_n = 1 \,|\, \mathbf{Q}_n = q)$ of a correct response to question q is unknown, since it depends on the latent state K_0. Thus, (11) cannot be computed. However, it makes sense to replace, in the evaluation of (11), the conditional probability $\mathbf{P}(\mathbf{R}_n = 1 \,|\, \mathbf{Q}_n = q)$ by the likelihood $L_n(K_q)$ of a correct response to question q. The idea is thus to minimize the quantity

$$\tilde{H}(q, \ell) = \ell(K_q) H\big(u(1, q, \ell)\big) + \ell(K_{\bar{q}}) H\big(u(0, q, \ell)\big), \tag{12}$$

over all possible $q \in Q$. Let $J(\ell) \subseteq Q$ be the set of questions q minimizing $\tilde{H}(q, \ell)$. The question asked on trial $n+1$ is then chosen in the set $J(\ell)$.

10.14. Definition. This particular form of the questioning rule, which is specified by the equation

$$\Psi(q, \ell) = \frac{\iota_{J(\ell)}(q)}{|J(\ell)|}, \tag{13}$$

will be referred to as *informative*.

Note that, in Equation (13), the choice of a question varies with the updating rule. This is not the case for Equation (10). Surprisingly, for the convex updating rule with a constant parameter θ, the half-split and the informative questioning rule induce the same drawing of questions. We shall postpone the proof of this fact for the moment, however (see 10.23 and 10.28).

General Results

10.15. Convention. In the rest of this chapter, we shall assume that $(\mathbf{R}_n, \mathbf{Q}_n, \mathbf{L}_n)$ is a stochastic assessment procedure for a probabilistic knowledge structure (Q, K, L) with $L > 0$, parametrized by u, Ψ and K_0. Special cases of this procedure will be specified whenever appropriate.

10.16. Theorem. *The process* $(\mathbf{L}_n)$ *is Markovian. That is, for any positive integer* n *and any measurable set* $B \subseteq \Lambda$

$$\mathbf{P}(\mathbf{L}_{n+1} \in B \,|\, \mathbf{L}_n, \ldots, \mathbf{L}_1) = \mathbf{P}(\mathbf{L}_{n+1} \in B \,|\, \mathbf{L}_n). \tag{14}$$

A similar property holds for the processes $(\mathbf{R}_n, \mathbf{Q}_n, \mathbf{L}_n)$ *and* $(\mathbf{Q}_n, \mathbf{L}_n)$.

PROOF. Using successively Axioms [U], [R] and [L], we have

$$
\begin{aligned}
&\mathbf{P}(\mathbf{L}_{n+1} \in B \,|\, \mathbf{L}_n, \ldots, \mathbf{L}_1) \\
&= \sum_{(\mathbf{R}_n, \mathbf{Q}_n)} \mathbf{P}(\mathbf{L}_{n+1} \in B \,|\, \mathbf{R}_n, \mathbf{Q}_n, \mathbf{L}_n, \ldots, \mathbf{L}_1) \mathbf{P}(\mathbf{R}_n, \mathbf{Q}_n \,|\, \mathbf{L}_n, \ldots, \mathbf{L}_1) \\
&= \sum_{(\mathbf{R}_n, \mathbf{Q}_n)} \iota_B\big(u(\mathbf{R}_n, \mathbf{Q}_n, \mathbf{L}_n)\big) \mathbf{P}(\mathbf{R}_n, \mathbf{Q}_n \,|\, \mathbf{L}_n, \ldots, \mathbf{L}_1) \\
&= \sum_{(\mathbf{R}_n, \mathbf{Q}_n)} \iota_B\big(u(\mathbf{R}_n, \mathbf{Q}_n, \mathbf{L}_n)\big) \mathbf{P}(\mathbf{R}_n \,|\, \mathbf{Q}_n, \mathbf{L}_n \ldots \mathbf{L}_1) \mathbf{P}(\mathbf{Q}_n \,|\, \mathbf{L}_n, \ldots, \mathbf{L}_1) \\
&= \sum_{(\mathbf{R}_n, \mathbf{Q}_n)} \iota_B\big(u(\mathbf{R}_n, \mathbf{Q}_n, \mathbf{L}_n)\big) \iota_{K_0}(\mathbf{Q}_n) \Psi(\mathbf{Q}_n, \mathbf{L}_n)
\end{aligned}
$$

which only depends on the set B and $\mathbf{L}_n$. We leave the two other cases to the reader (Problems 3 and 4). $\qquad\square$

In general, a stochastic assessment procedure is not necessarily capable of uncovering a latent state K_0. The next theorem gathers some simple, but very general results in this connection. We recall that $\triangle$ denotes the symmetric difference between sets (cf. 0.23).

10.17. Theorem. *If the latent state is K_0, then for all positive integers n, all real numbers ϵ with $0 < \epsilon < 1$ and all states $K \neq K_0$,*

$$\mathbf{P}\big(\mathbf{L}_{n+1}(K_0) > \mathbf{L}_n(K_0)\big) = 1; \tag{15}$$

$$\mathbf{P}\big(\mathbf{L}_{n+1}(K_0) \geq 1 - \epsilon\big) \geq \mathbf{P}\big(\mathbf{L}_n(K_0) \geq 1 - \epsilon\big); \tag{16}$$

$$\mathbf{P}\big(\mathbf{L}_{n+1}(K) < \mathbf{L}_n(K)\big) = \mathbf{P}(\mathbf{Q}_n \in K \triangle K_0). \tag{17}$$

Moreover

$$\lim_{n \to \infty} \mathbf{P}\big(\mathbf{L}_{n+1}(K_0) \geq 1 - \epsilon > \mathbf{L}_n(K_0)\big) = 0. \tag{18}$$

Equation (16) implies that the sequence $\mathbf{P}\big(\mathbf{L}_n(K_0) \geq 1 - \epsilon\big)$ converges.

PROOF. Equation (15) is an immediate consequence of Axioms [U] and [R]. It implies

$$\mathbf{P}\big(\mathbf{L}_{n+1}(K_0) \geq 1 - \epsilon \,|\, \mathbf{L}_n(K_0) \geq 1 - \epsilon\big) = 1. \tag{19}$$

Consequently,

$$P(L_{n+1}(K_0) \geq 1 - \epsilon)$$
$$= P(L_{n+1}(K_0) \geq 1 - \epsilon \,|\, L_n(K_0) \geq 1 - \epsilon)\, P(L_n(K_0) \geq 1 - \epsilon)$$
$$+ P(L_{n+1}(K_0) \geq 1 - \epsilon \,|\, L_n(K_0) < 1 - \epsilon)\, P(L_n(K_0) < 1 - \epsilon)$$
$$\geq P(L_n(K_0) \geq 1 - \epsilon). \qquad (20)$$

This establishes Equation (16). Writing $\mu(\epsilon) = \lim_{n \to \infty} P(L_n(K_0) \geq 1-\epsilon)$, and taking limits on both sides of Equation (20), we obtain using (19) again

$$\mu(\epsilon) = \mu(\epsilon) + \lim_{n \to \infty} P(L_{n+1}(K_0) \geq 1 - \epsilon > L_n(K_0)),$$

yielding (18). The left member of (17) can be decomposed into

$$P(L_{n+1}(K) < L_n(K) \,|\, Q_n \in K \,\triangle\, K_0)\, P(Q_n \in K \,\triangle\, K_0)$$
$$+ P(L_{n+1}(K) < L_n(K) \,|\, Q_n \in \overline{K \,\triangle\, K_0})\, P(Q_n \in \overline{K \,\triangle\, K_0}).$$

By Axioms [R] and [U], the factor $P(L_{n+1}(K) < L_n(K) \,|\, Q_n \in K \,\triangle\, K_0)$ in the first term is equal to one, and the last term vanishes. Thus, Equation (17) follows. $\qquad \square$

Uncovering the Latent State

We show that, under some fairly general conditions on the updating and the questioning rules, the latent state K_0 can be uncovered. These general conditions include the cases in which the updating rule is convex or multiplicative, and the questioning rule is half-split. We first consider an example using a convex updating rule with a constant parameter θ and the half-split questioning rule.

10.18. Example. Let $Q = \{a, b, c\}$ and

$$\mathcal{K} = \{\, \varnothing,\, \{a\},\, \{b, c\},\, \{a, c\},\, \{a, b, c\} \,\},$$

with the latent state $K_0 = \{b, c\}$ and $L_1(K) = .2$ for all $K \in \mathcal{K}$. Since the questioning rule is half-split, and

$$|2L_1(\mathcal{K}_q) - 1| = .2$$

for all $q \in Q$, we have $S(\mathbf{L}_1) = \{a, b, c\}$ (in the notation of 10.12), i.e. on trial one, the questions are selected in $S(\mathbf{L}_1)$ with equal probabilities. Notice that

$$\mathbf{L}_1(\mathcal{K}_{\bar{a}}) = \mathbf{L}_1(\mathcal{K}_{\bar{b}}) = .4, \quad \text{while} \quad \mathbf{L}_1(\mathcal{K}_{\bar{c}}) = .6.$$

For the likelihood of the state $K_0 = \{b, c\}$ on trial two, we obtain thus, by the convex updating rule

$$\mathbf{L}_2(K_0) = \begin{cases} (1-\theta).2 + \theta\frac{.2}{.4} & \text{with prob. } \frac{1}{3} \quad (a \text{ is chosen}); \\ (1-\theta).2 + \theta\frac{.2}{.4} & \text{with prob. } \frac{1}{3} \quad (b \text{ is chosen}); \\ (1-\theta).2 + \theta\frac{.2}{.6} & \text{with prob. } \frac{1}{3} \quad (c \text{ is chosen}). \end{cases}$$

In accordance with 10.17, Equation(15), this implies

$$\mathbf{P}\big(\mathbf{L}_2(K_0) > \mathbf{L}_1(K_0)\big) = 1.$$

In fact, Theorem 10.24 shows that K_0 is uncoverable.

We now turn to a general result of convergence, based on some strengthening of the conditions defining a stochastic assessment procedure.

10.19. Definition. An updating rule u is called *regular* if there is a nonincreasing function $v :]0, 1[\to \mathbf{R}$ such that

(1) $v(t) > 1$ for all $t \in]0, 1[$;
(2) $u_K(r, q, \ell) \geq v\big(\ell(\mathcal{K}_q)\big) \ell(K)$, if $\iota_K(q) = r = 1$;
(3) $u_K(r, q, \ell) \geq v\big(\ell(\mathcal{K}_{\bar{q}})\big) \ell(K)$, if $\iota_K(q) = r = 0$.

10.20. Theorem. *Both the convex and the multiplicative updating rules are regular.*

PROOF. For the convex updating rule, if $\iota_K(q) = r = 1$, Equation (5) can be rewritten as

$$u_K(r, q, \ell) = \left(1 + \theta_{q,r}\left(\frac{1}{\ell(\mathcal{K}_q)} - 1\right)\right)\ell(K).$$

Setting $\theta = \min\{\theta_{q,r} \mid q \in Q, r \in \{0, 1\}\}$, we get

$$u_K(r, q, \ell) \geq \left(1 + \theta\left(\frac{1}{\ell(\mathcal{K}_q)} - 1\right)\right)\ell(K).$$

In the case $\iota_K(q) = r = 0$, we obtain similarly

$$u_K(r, q, \ell) = \left(1 + \theta_{q,r}\left(\frac{1}{\ell(K_{\bar{q}})} - 1\right)\right)\ell(K)$$

$$\geq \left(1 + \theta\left(\frac{1}{\ell(K_{\bar{q}})} - 1\right)\right)\ell(K).$$

Thus, (1)–(3) in Definition 10.19 are satisfied with

$$v(t) = 1 + \theta\left(\frac{1}{t} - 1\right), \quad \text{for } t \in]0, 1[.$$

For the multiplicative updating rule, in the case $\iota_K(q) = r = 1$, we have

$$u_K(r, q, \ell) = \frac{\zeta_{q,r}}{\zeta_{q,r}\ell(K_q) + \ell(K_{\bar{q}})}\ell(K)$$

$$= \frac{\zeta_{q,r}}{1 + (\zeta_{q,r} - 1)\ell(K_q)}\ell(K),$$

and similarly, if $\iota_K(q) = r = 0$,

$$u_K(r, q, \ell) = \frac{\zeta_{q,r}}{\ell(K_q) + \zeta_{q,r}\ell(K_{\bar{q}})}\ell(K)$$

$$= \frac{\zeta_{q,r}}{1 + (\zeta_{q,r} - 1)\ell(K_{\bar{q}})}\ell(K).$$

For $q \in Q$ and $r \in \{0, 1\}$, each of the functions $v_{q,r} : t \mapsto v_{q,r}(t)$ defined for $\zeta_{q,r} > 1$ and $t \in]0, 1[$ by

$$v_{q,r}(t) = \frac{\zeta_{q,r}}{1 + (\zeta_{q,r} - 1)t}$$

is decreasing and takes values > 1. Thus, $v = \min\{v_{q,r} \mid q \in Q, r \in \{0, 1\}\}$ satisfies (1)–(3) in Definition 10.19. □

As far as the questioning rule is concerned, it is intuitively clear that, from the standpoint of the observer, it would not be efficient to choose a question q with a likelihood $\mathbf{L}_n(K_q)$ of a correct response close to zero or one. Actually, it makes good sense to choose, as in the half-split questioning rule, a question q with $\mathbf{L}_n(K_q)$ as far as possible to zero or one. A much weaker form of this idea is captured in the next definition.

10.21. Definition. Let ν be a real valued function defined on the open interval $]0,1[$, with two numbers $\gamma, \delta > 0$ satisfying $\gamma + \delta < 1$, such that

(1) ν is strictly decreasing on $]0, \gamma[$;

(2) ν is strictly increasing on $]1 - \delta, 1[$;

(3) $\nu(t) > \nu(t')$ whenever $\gamma \leq t' \leq 1 - \delta$ and either $0 < t < \gamma$ or $1 - \delta < t < 1$.

An example of such a function ν is shown in Figure 10.5.

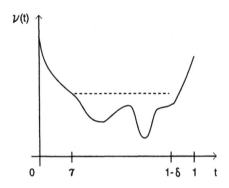

Fig. 10.5. An example of function ν as in Definition 10.21.

Define, for $\ell \in \Lambda$,

$$S(\nu, \ell) = \big\{ q \in Q \,|\, \nu(\ell_n(\mathcal{K}_q)) \leq \nu(\ell_n(\mathcal{K}_{q'})) \text{ for all } q' \in Q \big\}.$$

A questioning rule Ψ is said to be *inner* if such a function ν exists with

$$\Psi(q, \ell) = \frac{\iota_{S(\nu, \ell)}(q)}{|S(\nu, \ell)|}.$$

10.22. Theorem. *The half-split questioning rule is inner.*

This follows readily from the definitions, using $\nu(t) = |2t - 1|$.

10.23. Theorem. *If the updating rule is convex with a constant parameter θ, the informative questioning rule is inner. Moreover, the informative and the half-split questioning rules induce the same drawing of questions.*

The proofs of this theorem and of the next one are gathered in a later section of this chapter, which contains starred material. The next theorem states the main result of this chapter.

10.24. Theorem. *Let* $(\mathbf{R}_n, \mathbf{Q}_n, \mathbf{L}_n)$ *be a stochastic assessment procedure parametrized by* u, Ψ *and* K_0, *with* u *regular and* Ψ *inner. Then,* K_0 *is uncoverable in the sense that:*

$$\mathbf{L}_n(K_0) \overset{\text{a.s.}}{\to} 1.$$

10.25. Corollary. *A latent knowledge state is uncoverable by a stochastic assessment procedure with an updating rule which is either convex or multiplicative, and a questioning rule which is half-split.*

PROOF. This results from Theorems 10.20, 10.22, 10.24 and the definitions.

□

10.26. Corollary. *A latent knowledge state is uncoverable by a stochastic assessment procedure with a convex updating rule having a constant parameter* θ, *and an informative questioning rule.*

PROOF. Use Theorems 10.20, 10.23, 10.24 and the definitions.   □

Notice that the results of this Chapter are not complete: we do not have a proof of the almost sure convergence $\mathbf{L}_n(K_0) \to 1$ for the fourth special case, namely, a multiplicative updating rule with an informative questioning rule.

Refining the Assessment

10.27. Remark. In practice, the output of an assessment algorithm in the style of this chapter takes the form of a probability distribution $\mathbf{L}_n$ (for some final trial number n) on the collection of states. In principle, such a probability distribution will have a mass concentrated on one or a handful of closely related states. For example, the procedure may have selected three largely overlapping knowledge states K_1, K_2 and K_3. As a simple illustration, suppose that

$$K_1 = \{a, b, d, e\}, \quad K_2 = \{b, c, d\}, \quad K_3 = \{a, b, c, e\},$$

with

$$\mathcal{L}_n(K_1) = \mathcal{L}_n(K_2) = .25, \quad \mathcal{L}_n(K_3) = .40,$$

and the rest of the mass of $\mathbf{L}_n$ being scattered on the remaining states. On the basis of such information, the best bet for the state of the subject is K_3. However, one may wish to refine this assessment by reconsidering at that time the full sequence of questions asked and responses observed during the application of the procedure. The selection of one among these three states could be based on a Bayesian heuristic. Reviving the local independence assumption (Axiom [N] in Chapter 8), we can recompute the conditional likelihood of observing each of the selected states K_1, K_2 and K_3, given the actual sequence of responses to the questions asked. The states yielding the greatest likelihood can then be taken as the final result of the assessment. Obviously, such a computation makes sense only when good estimates are available for the probabilities β_q and η_q of careless errors and lucky guesses, respectively.

Suppose that the subject has been asked questions c, d, and e, and has provided an incorrect response to c, and a correct response to the two other questions. Let us denote these data by the letter 'D'. According to the local independence Axiom [N], the conditional probabilities of these data, given the three states, are

$$P(D\,|\,K_1) = (1-\eta_c)(1-\beta_d)(1-\beta_e), \tag{21}$$

$$P(D\,|\,K_2) = \beta_c(1-\beta_d)\eta_e, \tag{22}$$

$$P(D\,|\,K_3) = \beta_c\eta_d(1-\beta_e). \tag{23}$$

Using a Bayesian rule to recompute the probabilities of the states, we get for $i = 1, 2, 3$:

$$P(K_i\,|\,D) = \frac{P(D\,|\,K_i)\mathbf{L}_n(K_i)}{\sum_{j=1}^{3} P(D\,|\,K_j)\mathbf{L}_n(K_j)}. \tag{24}$$

Assume that from previous analyses, the following estimates have been obtained for the careless error and lucky guess parameters:

$$\widehat{\beta}_c = \widehat{\beta}_d = .05, \qquad \widehat{\beta}_e = .10,$$

$$\widehat{\eta}_c = \widehat{\eta}_d = .10, \qquad \widehat{\eta}_e = .05.$$

Replacing the $P(D\,|\,K_j)$ in (24) by their estimates in terms of the $\widehat{\beta}_q$'s and $\widehat{\eta}_q$'s, via Equations (21), (22) and (23), we obtain approximately

$$\widehat{P}(K_1\,|\,D) = .988,$$

$$\widehat{P}(K_2\,|\,D) = .003,$$

$$\widehat{P}(K_3\,|\,D) = .009.$$

The picture resulting from such a computation is quite different from that based on $\mathbf{L}_n$ alone: the overwhelmingly most plausible state is now K_1. (It should be noted that, even though this Bayesian computation is heuristically defensible, it is not strictly founded in theory; see Problem 7 in this connection.)

Proofs*

10.28. Proof of Theorem 10.23. (All logarithms are in base 2.) By definition of the informative questioning rule in 10.14 and Equation (12), we have to minimize, over all q in Q, the quantity

$$\tilde{H}(q,\ell) = \ell(\mathcal{K}_q)H\big(u(1,q,\ell)\big) + \ell(\mathcal{K}_{\bar{q}})H\big(u(0,q,\ell)\big), \tag{25}$$

where

$$\ell(\mathcal{K}_q) = \sum_{K \in \mathcal{K}_q} \ell(K), \quad \ell(\mathcal{K}_{\bar{q}}) = \sum_{K \in \mathcal{K}_{\bar{q}}} \ell(K).$$

This quantity depends upon the updating rule u. As the updating rule is assumed to be convex with a constant parameter θ, we obtain from Equation (5)

$$H\big(u(1,q,\ell)\big)$$
$$= -\sum_{K \in \mathcal{K}} u_K(1,q,\ell) \log u_K(1,q,\ell)$$
$$= -\sum_{K \in \mathcal{K}_q} \ell(K)\left(1 - \theta + \frac{\theta}{\ell(\mathcal{K}_q)}\right)\left(\log \ell(K) + \log\left(1 - \theta + \frac{\theta}{\ell(\mathcal{K}_q)}\right)\right)$$
$$\quad - \sum_{J \in \mathcal{K}_{\bar{q}}} \ell(J)(1 - \theta)\big(\log \ell(J) + \log(1 - \theta)\big) \tag{26}$$

and

$$H\big(u(0,q,\ell)\big)$$
$$= -\sum_{J \in \mathcal{K}_{\bar{q}}} \ell(J)\left(1 - \theta + \frac{\theta}{\ell(\mathcal{K}_{\bar{q}})}\right)\left(\log \ell(J) + \log\left(1 - \theta + \frac{\theta}{\ell(\mathcal{K}_{\bar{q}})}\right)\right)$$
$$\quad - \sum_{K \in \mathcal{K}_q} \ell(K)(1 - \theta)\big(\log \ell(K) + \log(1 - \theta)\big). \tag{27}$$

Using Equations (25), (26) and (27) and grouping appropriately leads to

$$\tilde{H}(q,\ell) = H(\ell) - 2\ell(K_q)\ell(K_{\bar{q}})(1-\theta)\log(1-\theta)$$
$$-\ell(K_q)\big((1-\theta)\ell(K_q)+\theta\big)\log\left(1-\theta+\frac{\theta}{\ell(K_q)}\right)$$
$$-\ell(K_{\bar{q}})\big((1-\theta)\ell(K_{\bar{q}})+\theta\big)\log\left(1-\theta+\frac{\theta}{\ell(K_{\bar{q}})}\right).$$

That is, with $\ell(K_q) = t$ and for $t \in\ ]0,1[$

$$g(t) = -2t(1-t)(1-\theta)\log(1-\theta) - t\big((1-\theta)t+\theta\big)\log\left(1-\theta+\frac{\theta}{t}\right)$$
$$- (1-t)\big((1-\theta)(1-t)+\theta\big)\log\left(1-\theta+\frac{\theta}{1-t}\right), \qquad (28)$$

we have

$$\tilde{H}(q,\ell) = H(\ell) + g(t). \qquad (29)$$

Notice that g is symmetric around $\frac{1}{2}$, that is, $g(t) = g(1-t)$ for $0 < t < 1$. To establish the two assertions of the Theorem, it suffices now to prove that the function g is convex on $]0,1[$ and has a strict extremum at $\frac{1}{2}$. (Thus, g will serve as the function ν in Definition 10.21.) Since g is symmetric around $\frac{1}{2}$, we only have to show that the second derivative is strictly positive on $]0,\frac{1}{2}[$. We shall derive this from the fact that $g'''(t) < 0$ for $0 < t < \frac{1}{2}$, together with $g''(\frac{1}{2}) > 0$. To compute the derivatives, we simplify the expression of g. Setting

$$a(t) = (1-\theta)t, \qquad b(t) = (1-\theta)t + \theta,$$
$$f(t) = -a(t)b(t)\log\frac{b(t)}{a(t)},$$

Equation (28) simplifies into

$$g(t) = -\log(1-\theta) + \frac{1}{1-\theta}\big(f(t) + f(1-t)\big). \qquad (30)$$

Using $a'(t) = b'(t) = 1 - \theta$, we obtain for the derivatives of f

$$f''(t) = (1-\theta)^2\left(\theta\frac{a(t)+b(t)}{a(t)\,b(t)} - 2\log\frac{b(t)}{a(t)}\right),$$
$$f'''(t) = -\frac{(1-\theta)^3\,\theta^3}{a(t)^2\,b(t)^2} < 0.$$

From Equation (30), it follows $g'''(t) < 0$ for $0 < t < \frac{1}{2}$. On the other hand, with

$$h(\theta) = \frac{2\theta}{1 - \theta^2} - \log\frac{1 + \theta}{1 - \theta},$$

we have

$$g''(\frac{1}{2}) = 4(1 - \theta)h(\theta).$$

Since $\lim_{\theta \to 0+} h(\theta) = 0$ and $h'(\theta) > 0$ for $0 < \theta < 1$, we have $g''(\frac{1}{2}) > 0$.

$\square$

10.29. Proof of Theorem 10.24. Let $\widetilde{\Omega}$ be the set of all realizations ω for which, for every trial n,

  (i) $\mathbf{Q}_n \in S(\nu, \mathbf{L}_n)$, with S as in Definition 10.21;

  (ii) $\mathbf{R}_n = \iota_{K_0}(\mathbf{Q}_n)$.

Notice that $\widetilde{\Omega}$ is a measurable set of the sample space Ω, and that $\mathbf{P}(\widetilde{\Omega}) = 1$. Writing $\mathbf{L}_n^{\omega}(K_0)$ for the value of the random variable $\mathbf{L}_n(K_0)$ at the point $\omega \in \Omega$, we only have to establish that for any point $\omega \in \widetilde{\Omega}$, we have

$$\lim_{n \to \infty} \mathbf{L}_n^{\omega}(K_0) = 1.$$

Take $\omega \in \widetilde{\Omega}$ arbitrarily. It follows readily from the assumptions that $\mathbf{L}_n^{\omega}(K_0)$ is nondecreasing, and thus converges. Therefore, it suffices to show that $\mathbf{L}_{n_i}^{\omega}(K_0) \to 1$ for at least one subsequence $s = (n_i)$ of the positive integers. Since $\mathcal{K}$ is finite and $\mathbf{L}_n(K) \in]0, 1[$, we can take $s = (n_i)$ such that $\mathbf{L}_{n_i}^{\omega}(K)$ converges for all $K \in \mathcal{K}$. In the rest of this proof, we consider a fixed subsequence s satisfying those conditions.

We define a function $f_{\omega,s} : Q \to [0, 1]$ by

$$f_{\omega,s}(q) = \begin{cases} \lim_{i \to \infty} \mathbf{L}_{n_i}^{\omega}(\mathcal{K}_q) & \text{if } q \in K_0; \\ \lim_{i \to \infty} \mathbf{L}_{n_i}^{\omega}(\mathcal{K}_{\bar{q}}) & \text{if } q \notin K_0. \end{cases}$$

We also set

$$\widetilde{Q}_{w,s} = \{q \in Q \mid f_{\omega,s}(q) < 1\}.$$

If $\widetilde{Q}_{w,s}$ is empty, the fact that $\mathbf{L}_{n_i}^{\omega}(K_0) \to 1$ follows readily from Lemma 3 (see below). The core of the proof of Theorem 10.24 consists in establishing that $\widetilde{Q}_{w,s} = \varnothing$, which is achieved by Lemma 2.

Lemma 1. *If* $f_{\omega,s}(q) < 1$, *then* $\{i \in \mathbf{N} \,|\, \mathbf{Q}_{n_i}(\omega) = q\}$ *is a finite set.*

PROOF. Assume that $q \in K_0$. Since $f_{\omega,s}(q) < 1$, there is $\epsilon > 0$ such that $f_{\omega,s}(q) + \epsilon < 1$. If $\mathbf{Q}_{n_i}(\omega) = q$ and i is large enough to ensure that $\mathbf{L}_{n_i}^{\omega}(\mathcal{K}_q) \leq f_{\omega,s}(q) + \epsilon$, we derive

$$
\begin{aligned}
\mathbf{L}_{n_{i+1}}^{\omega}(K_0) &\geq \mathbf{L}_{n_i+1}^{\omega}(K_0) \\
&\geq v\big(\mathbf{L}_{n_i}^{\omega}(\mathcal{K}_q)\big)\mathbf{L}_{n_i}^{\omega}(K_0) \\
&\geq v\big(f_{\omega,s}(q) + \epsilon\big)\mathbf{L}_{n_i}^{\omega}(K_0).
\end{aligned}
$$

Since $v\big(f_{\omega,s}(q) + \epsilon\big) > 1$ does not depend on i and $\mathbf{L}_{n_i}^{\omega}(K_0) \leq 1$, we may have $\mathbf{Q}_{n_i}(\omega) = q$ for at most a finite number of values of i. The proof is similar in the case $q \notin K_0$. $\qquad\square$

Lemma 2. $\widetilde{Q}_{w,s} = \{q \in Q \,|\, f_{\omega,s}(q) < 1\} = \varnothing$.

PROOF. We proceed by contradiction. If $f_{\omega,s}(q) < 1$ for some $q \in Q$, we can assert the existence of some positive integer j and some $\epsilon > 0$ such that whenever $i > j$, we have

either $\qquad 0 < \ell_1(K_0) \leq \mathbf{L}_{n_i}^{\omega}(\mathcal{K}_q) < f_{\omega,s}(q) + \epsilon < 1 \quad$ if $q \in K_0$,

or $\qquad 0 < \ell_1(K_0) \leq \mathbf{L}_{n_i}^{\omega}(\mathcal{K}_{\bar{q}}) < f_{\omega,s}(q) + \epsilon < 1 \quad$ if $q \notin K_0$.

This means that for $i > j$, both $\mathbf{L}_{n_i}^{\omega}(\mathcal{K}_q)$ and $\mathbf{L}_{n_i}^{\omega}(\mathcal{K}_{\bar{q}})$ remain in some interval $]\gamma_q', 1 - \delta_q'[$ with $0 < \gamma_q', \delta_q'$. The above argument applies to all $q \in \widetilde{Q}_{w,s}$. Because of the finiteness of $\widetilde{Q}_{w,s}$, the index q may be dropped in γ_q', δ_q'. Moreover, referring to Definition 10.21, we can assert the existence of $\bar{\gamma}$ and $\bar{\delta}$ such that $0 < \bar{\gamma} < \gamma$, $0 < \bar{\delta} < \delta$ and $\mathbf{L}_{n_i}^{\omega}(\mathcal{K}_q) \in]\bar{\gamma}, 1 - \bar{\delta}[$ for $i > j$ and $q \in \widetilde{Q}_{w,s}$.

Since $\widetilde{Q}_{w,s}$ is finite, Lemma 1 may be invoked to infer the existence of a positive integer k such that $\mathbf{Q}_{n_i}(\omega) \notin \widetilde{Q}_{w,s}$ whenever $i > k$. Note that, by definition of $\widetilde{Q}_{w,s}$, $f_{\omega,s}(q') = 1$ for all $q' \notin \widetilde{Q}_{w,s}$. Since $\mathbf{L}_{n_i}^{\omega}(\mathcal{K}_{q'})$ converges, there is $i^\star > j, k$ such that neither $\mathbf{L}_{n_i}^{\omega}(\mathcal{K}_{q'})$ nor $\mathbf{L}_{n_i}^{\omega}(\mathcal{K}_{\overline{q'}})$ are points of $]\bar{\gamma}, 1 - \bar{\delta}[$ for all $i > i^\star$ and $q' \notin \widetilde{Q}_{w,s}$. By definition of $\widetilde{\Omega}$, we must have $\mathbf{Q}_{n_i}(\omega) \in \widetilde{Q}_{w,s}$, contradicting $i > k$. $\qquad\square$

Define, for every $K \in \mathcal{K}$,

$$
A_{\omega,s}(K) = \{q \in K \,|\, f_{\omega,s}(q) = 1\}.
$$

Lemma 3. *Suppose that, for some $K \in \mathcal{K}$, $A_{\omega,s}(K) \neq A_{\omega,s}(K_0)$. Then*

$$\lim_{i \to \infty} \mathbf{L}_{n_i}^{\omega}(K) = 0.$$

PROOF. Assume that there is some $q \in K_0 \setminus K$ such that $f_{\omega,s}(q) = 1$; that is $\lim_{i \to \infty} \mathbf{L}_{n_i}^{\omega}(\mathcal{K}_q) = 1$. This implies $\lim_{i \to \infty} \mathbf{L}_{n_i}^{\omega}(\mathcal{K}_{\bar{q}}) = 0$ and the thesis since $K \in \mathcal{K}_{\bar{q}}$. The other case, $q \in K \setminus K_0$, is similar.   □

By Lemma 2, $K \neq K_0$ implies $A_{\omega,s}(K) \neq A_{\omega,s}(K_0)$. It follows from Lemma 3 that, for all $K \neq K_0$, $\lim_{i \to \infty} \mathbf{L}_{n_i}^{\omega}(K) = 0$, yielding $\mathbf{L}_{n_i}^{\omega}(K_0) \to 1$.
This concludes the proof of Theorem 10.24.   □

10.30. Remark. A careful study of the proof shows that the assumption of an inner questioning rule can be replaced by the following one. For any $\gamma, \delta \in]0, 1[$, denote by $E_{n,q}(\gamma, \delta)$ the event that $\gamma < \mathbf{L}_n(\mathcal{K}_q) < 1 - \delta$, and let $E_n(\gamma, \delta) = \cup_{q \in Q} E_{n,q}(\gamma, \delta)$. The condition states that there exists $\sigma > 0$ such that, for all $\gamma, \delta \in]0, \sigma[$,

$$\mathbf{P}(\mathbf{Q}_n = q' \,|\, \overline{E_{n,q'}(\gamma, \delta)} \cap E_n(\gamma, \delta)) = 0.$$

In other words, and somewhat loosely: no question q will be chosen with $\mathbf{L}_n(\mathcal{K}_q)$ in a neighborhood of one or zero when this can be avoided.

Original Sources and Related Works

This chapter follows closely the paper by Falmagne and Doignon (1988a). The first applications of the algorithms described in this chapter were made by Villano (1991), who has tested them extensively in his dissertation (Villano, 1991; see also Villano et al., 1987), and by Kambouri (1991). They form a key component of the knowledge assessment engine of the Aleks system briefly described in Chapter 0.

As mentioned earlier, the updating operators involved in the algorithms have been inspired by some operators of mathematical learning theory (Bush and Mosteller, 1955; Norman, 1972). Specifically, the convex updating rule is related to a Bush and Mosteller learning operator (Bush and Mosteller, 1955), and the multiplicative updating rule is close to the learning operator of the so-called beta learning model of Luce (1959). Bayesian updating rules in intelligent tutoring systems have been discussed by Kimbal (1982).

Problems

1. Show that the convex updating rule defined by Equation (5) is not permutable in the sense of Equation (6).

2. Check that the multiplicative updating rule of Equation (7) is permutable in the sense of Equation (6).

3. Complete the proof of Theorem 10.16 and show that the process $(\mathbf{R}_n, \mathbf{Q}_n, \mathbf{L}_n)$ is Markovian.

4. (Continuation.) Prove that the process $(\mathbf{Q}_n, \mathbf{L}_n)$ is also Markovian.

5. In the knowledge structure of Example 10.18, suppose that the half-split questioning rule and the multiplicative updating rule with a constant parameter $\zeta_{q,r}$ have been used. Verify Eqs. (15), (16) and (17) for $n = 1, 2$. You should assume that $\mathbf{L}_1(K) = .2$ for any state K.

6. Let $Q = \{a, b, c\}$ and $\mathcal{K} = \{\varnothing, \{a, b\}, \{b, c\}, \{a, c\}, Q\}$. Assume that the subject's knowledge state oscillates randomly between trials according to the probability distribution ϕ defined by $\phi(\{a, b\}) = \phi(\{a, c\}) = \phi(Q) = \frac{1}{3}$. Suppose that the half-split questioning rule and the convex updating rule with a constant parameter θ are used. Prove that $\lim_{n \to \infty} E(\mathbf{L}_n(\{b, c\})) > 0$ (or better, compute this limit). Argue that, in view of this result, even the domain of ϕ cannot be uncovered by the assessment procedure.

7. Discuss the Bayesian computation proposed in Remark 10.27 for refining the assessment from a theoretical viewpoint.

8. Give a detailed proof of Corollary 10.25.

9. Give a detailed proof of Corollary 10.26.

10. Suppose that the subject knowledge state changes once during the assessment. Discuss in detail the impact of such a change on the efficiency and the accuracy of the multiplicative assessment procedure.

Chapter 11

Uncovering the State of an Individual: A Markov Chain Procedure

This chapter presents an assessment procedure that is similar in spirit to those described in Chapter 10, but different in a key aspect: it is based on a finite Markov chain rather than on a Markov process with an uncountable set of Markov states. As a consequence, the procedure requires less storage and computation and can thus be implemented on a small machine.

Outline of a Markov Chain Procedure

A fixed knowledge structure $(Q, \mathcal{K})$ is used by the assessor. We suppose that, on any discrete time in the course of the assessment, some of the knowledge states in $\mathcal{K}$ are considered as plausible from the standpoint of the assessor. These 'marked states' are collected in a family which is the value of a random variable $\mathbf{M}_n$, where the index $n = 1, 2, \ldots$ indicates the trial number. During the first phase of the procedure, this family decreases in size until a single marked state remains. In the second phase, the single 'marked state' evolves in the structure. This last feature allows the assessor, through a statistical analysis of the observed sequence of problems and answers, to estimate the 'true' state. (A formal definition of 'true' is given in 11.4.) Note that, in some cases, a useful estimate can be obtained even if the 'true' state estimate is not part of the structure. Before stepping into technicalities, we will illustrate the basic ideas by tracing an exemplary realization of the Markov chain to be described.

11.1. Example. We take the same knowledge structure as in Example 10.1 (cf. Figure 10.1):

$$\mathcal{K} = \{ \varnothing, \{a\}, \{c\}, \{a,c\}, \{b,c\}, \{a,b\}, \{a,b,c\},$$
$$\{a,b,c,d\}, \{a,b,c,d,e\} \}. \quad (1)$$

We suppose that the assessor is initially unbiased: all the knowledge states are regarded as plausible. Thus, all nine states of $\mathcal{K}$ are marked and we set by convention $\mathbf{M}_1 = \mathcal{K}$. (In some situations, a smaller subset of states could be marked that would reflect some a priori information on the student population.) During the first phase of the procedure, each question asked is selected in such a manner that, no matter which response is given (correct or incorrect), the number of marked states is decreased as much as possible. This goes on until only a single marked state remains. For example, suppose that $\mathbf{M}_n = \mathcal{M}$ is the set of marked states on trial n. Assume that some question q is chosen on that trial and that a correct response is given. The set of states marked on trial $n + 1$ would contain all those states in $\mathcal{M}$ which also contain q, namely $\mathcal{M}_q$. If the response is incorrect, then the set of marked states on trial $n + 1$ would be $\mathcal{M}_{\bar{q}}$, containing all those states of $\mathcal{M}$ that avoid q. It makes sense to select q in order to minimize the maximum of the two possible numbers of states kept. This clearly amounts to selecting a question q that divides as equally as possible the presently marked states into those containing q and those not containing q. In other words, q should render

$$\big| |\mathcal{M}_q| - |\mathcal{M}_{\bar{q}}| \big|$$

as small as possible. In our example, we have for $q = a$ and $\mathcal{M} = \mathcal{K}$

$$\big| |\mathcal{K}_a| - |\mathcal{K}_{\bar{a}}| \big| = |6 - 3| = 3.$$

Similar calculations for the other questions give the counts:

$$\begin{aligned}
\text{for } b: &\qquad |5 - 4| = 1, \\
\text{for } c: &\qquad |6 - 3| = 3, \\
\text{for } d: &\qquad |2 - 7| = 5, \\
\text{for } e: &\qquad |1 - 8| = 7.
\end{aligned}$$

Thus, b should be the first question asked. Denoting the question asked on trial n by $\mathbf{Q}_n$, a random variable, we set here $\mathbf{Q}_1 = b$ with probability 1. Suppose that we observe an incorrect answer. We denote this fact by writing $\mathbf{R}_1 = 0$. In general, we define $\mathbf{R}_n$ as a random variable taking the value 1 if the question asked on trial n is correctly answered, and 0 otherwise. In our example, the family of marked states becomes on trial 2

$$\mathbf{M}_2 = \big\{ \varnothing, \{a\}, \{c\}, \{a, c\} \big\}.$$

The new counts for selecting the next question are:

$$
\begin{aligned}
\text{for } a: \quad & |2 - 2| = 0, \\
\text{for } b: \quad & |0 - 4| = 4, \\
\text{for } c: \quad & |2 - 2| = 0, \\
\text{for } d: \quad & |0 - 4| = 4, \\
\text{for } e: \quad & |0 - 4| = 4.
\end{aligned}
$$

Thus, either a or c should be asked. We choose randomly between them, with equal probabilities: $\mathbf{P}(\mathbf{Q}_2 = a) = \mathbf{P}(\mathbf{Q}_2 = b) = .5$. Suppose that we ask item a and get a correct answer, that is $\mathbf{Q}_2 = a$ and $\mathbf{R}_2 = 1$. The family of marked states on trial 3 is

$$
\mathbf{M}_3 = \big\{ \, \{a\}, \{a, c\} \, \big\}.
$$

With probability 1, we get $\mathbf{Q}_3 = c$. If $\mathbf{R}_3 = 1$, we are left with the single marked state $\{a, c\}$. That is, we have

$$
\mathbf{M}_4 = \big\{ \, \{a, c\} \, \big\}.
$$

In this example, the second phase of the procedure starts on trial 4. From here on, the set of marked state will always contain a single state which may vary from trial to trial according to the question asked and the response given. The choice of a question on any trial $n \geq 4$ is based on the set of all the states in the neighborhood of the current single marked state. Specifically, we use the neighborhood formed by all states in $\mathcal{K}$ situated at a distance at most 1 from that marked state (cf. 2.7). In the case of the single marked state $\{a, c\}$, this neighborhood is

$$
N(\{a, c\}, 1) = \big\{ \, \{a, c\}, \{a\}, \{c\}, \{a, b, c\} \, \big\}.
$$

As in phase 1, the next question q is selected in order to split as equally as possible this set of states into those which contain q and those which do not contain q. Here, a, b or c will be randomly selected (with equal probability). If the answer collected confirms the current single marked state, we keep it as the only marked state. Otherwise, we change this state into another one, according to the new information. For concreteness,

$\mathbf{M}_4$	$\mathbf{Q}_4$	$\mathbf{R}_4$	$\mathbf{M}_5$
$\{\{a,c\}\}$	a	0	$\{\{c\}\}$
$\{\{a,c\}\}$	a	1	$\{\{a,c\}\}$
$\{\{a,c\}\}$	b	0	$\{\{a,c\}\}$
$\{\{a,c\}\}$	b	1	$\{\{a,b,c\}\}$

Table 11.1. Four generic cases generating $\mathbf{M}_5$.

we consider the four generic cases given in Table 11.1. Suppose first that $\mathbf{Q}_4 = a$ and $\mathbf{R}_4 = 0$ (row 1 of Table 11.1). As the response to a is incorrect, we remove a from the single marked state $\{a,c\}$, which yields the single marked state $\{c\}$, with $\mathbf{M}_5 = \{\{c\}\}$.

In row 2 of Table 11.1 , $\mathbf{Q}_4 = a$ and a is correctly solved. Thus $\{a,c\}$ is confirmed and we keep it as the single marked state, that is $\mathbf{M}_5 = \{\{a,c\}\}$. In row 3, we also end up with the same single marked state $\{a,c\}$ as a result of a confirmation, but this time b is asked and the answer is incorrect. Finally, in row 4, b is also asked but yields a correct answer. As a result, we add b to the current single marked state. A third possible question on trial 4 is c; we leave this case to the reader. Table 11.2 summarizes an exemplary realization of the process in the early trials. By convention, we set $\mathbf{M}_1 = \mathcal{K}$. Rows 4, 5 in Table 11.2 correspond to row 1 in Table 11.1.

trial n	$\mathbf{M}_n$	$\mathbf{Q}_n$	$\mathbf{R}_n$
1	$\mathcal{K}$	b	0
2	$\{\varnothing, \{a\}, \{c\}, \{a,c\}\}$	a	1
3	$\{\{a\}, \{a,c\}\}$	c	1
4	$\{\{a,c\}\}$	a	0
5	$\{\{c\}\}$	$\ldots$	$\ldots$

Table 11.2. An exemplary realization of the process in the early trials.

11.2. Remarks. a) In this example, it is easy to check that the marking rule sketched here will always yield a single marked state after trial 4, whatever the question asked and the answer observed. Clearly, some assumptions about the knowledge structure are needed in order to establish these results in general. We will assume the structure to be well-graded (cf. Definition 2.7 and Theorem 2.9), and that the question is selected in the fringe of the single marked state. Under these conditions, only one state will remain marked (see Theorem 11.6).

b) Our rationale for adopting these transition rules is dictated by caution. Some misleading answers by the subject may occur in the course of the procedure (due to lucky guesses or to careless errors, for example), resulting in a failure to uncover the 'true' state among those in the structure. It is also possible that the knowledge structure itself is mistaken to some extent: some states might be missing. Ideally, the final outcome of our procedure should compensate for both kind of errors. This can be achieved by analyzing the data collected during the second phase of the procedure, after the single state has been reached.

For instance, suppose that some particular state S has been omitted from the structure $\mathcal{K}$ used by the assessor. This missing state may nevertheless closely resemble some of the states in $\mathcal{K}$, say K_1, K_2 and K_3. In such a case, the sequence of single marked states, in the second phase of the procedure, may very well consist in transitions between these three states. The intersection and the union of these three states may provide lower and upper bounds for the state S to be uncovered.

On the other hand, if we believe that the state to be uncovered is one of the states visited during the second phase, then the choice between them may be made by standard statistical methods: we may simply chose that state maximizing the likelihood of the sequence of responses observed, using the conditional response probabilities β_q and η_q introduced in Chapter 7 in the context of the *local independence* assumption (Definition 7.2). In general, the single marked states visited during the second phase of the procedure may be used to estimated the 'true' state whether or not this state is contained in the structure used by the assessor.

Using the neighborhoods to guide both the choice of the question and the determination of the marked states is a sound idea not only for the second phase, but in fact for the whole process. The axioms given in the next section formalize this concept. The procedure described here shares many features with those presented in Chapter 10. The diagram displayed in Figure 11.1 highlights the differences and the similarities.

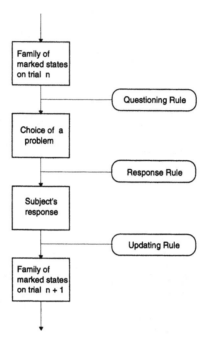

Fig. 11.1. Diagram of the transitions for the Markov chain procedure.

The Stochastic Assessment Process

The stochastic assessment procedure is described by four sequences of jointly distributed random variables $(\mathbf{R}_n, \mathbf{Q}_n, \mathbf{K}_n, \mathbf{M}_n)$, where $n = 1, 2, \ldots$ stands for the trial number.

The subject's unobservable state on trial n is denoted by $\mathbf{K}_n$. We assume that $\mathbf{K}_n$ takes its values in a fixed finite knowledge structure $\mathcal{K}$ with domain Q. (Examples will be given later in which this assumption is weakened; see 11.19.) The question asked and the response observed on trial n are denoted by $\mathbf{Q}_n$ and $\mathbf{R}_n$; these random variables take their values respectively in Q and $\{0, 1\}$, with 0 standing for incorrect and 1 for correct. Finally, the random variable $\mathbf{M}_n$ stands for the family of *marked states* on trial n. As indicated in our introductory section, these marked states are those which are still regarded as plausible candidates for the subject's unknown state. Thus, the values of $\mathbf{M}_n$ lie in $2^{\mathcal{K}}$. The stochastic process is defined by the sequence of quadruples $(\mathbf{R}_n, \mathbf{Q}_n, \mathbf{K}_n, \mathbf{M}_n)$, $n = 1, 2, \ldots$ The complete history of the process from trial 1 to trial n is abbreviated as

$$\mathbf{W}_n = ((\mathbf{R}_n, \mathbf{Q}_n, \mathbf{K}_n, \mathbf{M}_n), \ldots, (\mathbf{R}_1, \mathbf{Q}_1, \mathbf{K}_1, \mathbf{M}_1)),$$

with $\mathbf{W}_0$ denoting the empty history.

The axioms given below specify the probability measure recursively. We first state general requirements, involving several unspecified functions and parameters. In the next section, severe restrictions will be put on these functions and parameters. Note that, as in Chapter 7, we use the local independence assumption (cf. Definition 7.2) to specify the conditional response probabilities in Axiom [R].

11.3. Axioms.

[K]. **Knowledge State Rule.** The knowledge state varies from trial to trial according to a fixed probability distribution π on $\mathcal{K}$, independent of the trial: for all natural numbers n,

$$\mathbf{P}(\mathbf{K}_n = K \mid \mathbf{M}_n, \mathbf{W}_{n-1}) = \pi(K).$$

[T]. **Questioning Rule.** The question asked on trial n depends only on the marked states. That is, there is a function $\tau : Q \times 2^{\mathcal{K}} \to [0,1]$ such that, for all natural numbers n,

$$\mathbf{P}(\mathbf{Q}_n = q \mid \mathbf{K}_n, \mathbf{M}_n, \mathbf{W}_{n-1}) = \tau(q, \mathbf{M}_n).$$

[R]. **Response Rule.** The response on trial n only depends upon the knowledge state and the question asked on that trial. The dependence is specified, for each question q in Q, by a *(careless) error probability* β_q and a *(lucky) guessing probability* η_q, with $\beta_q < 1$ and $\eta_q < 1$. Specifically, we have for all nonnegative integers n,

$$\mathbf{P}(\mathbf{R}_n = 1 \mid \mathbf{Q}_n = q, \mathbf{K}_n = K, \mathbf{M}_n, \mathbf{W}_{n-1}) = \begin{cases} 1 - \beta_q & \text{if } q \in K; \\ \eta_q & \text{if } q \notin K. \end{cases}$$

[M]. **Marking rule.** The marked states on trial $n+1$ only depend upon the following events on trial n: the marked states, the question asked and the response collected. That is, there is a function $\mu : 2^{\mathcal{K}} \times \{0,1\} \times Q \times 2^{\mathcal{K}} \to [0,1]$ such that

$$\mathbf{P}(\mathbf{M}_{n+1} = \Psi \mid \mathbf{W}_n) = \mu(\Psi, \mathbf{R}_n, \mathbf{Q}_n, \mathbf{M}_n).$$

11.4. Definition. A process $(\mathbf{R}_n, \mathbf{Q}_n, \mathbf{K}_n, \mathbf{M}_n)$ satisfying Axioms [K], [T], [R], and [M] above is a *stochastic assessment process paramatrized by* π, τ, β, η, and μ. The case in which $\eta_q = 0$ for each question q will be called *fair*. This case deserves special attention in the theory since in many practical applications the questions may be designed in order to make lucky guesses impossible. If in addition $\beta_q = 0$ for each q, then this case is *straight*.

The knowledge states K for which $\pi(K) > 0$ will be called the *true* states; they form the *support* of K. Our tentative view is that in practice the support will not contain more than a handful of states, which are moreover close to each other in a sense made precise in the next section. If the support contains only one state, this state is the *unit support* of π.

It will be convenient to refer to the functions τ and μ appearing in Axioms [T] and [M] as the *testing rule* and *marking rule*, respectively. Special cases of the process defined by the four general axioms will arise from particularizing the testing rule and the marking rule. In general, as indicated by a cursory examination of these axioms, the process $(\mathbf{R}_n, \mathbf{Q}_n, \mathbf{K}_n, \mathbf{M}_n)$ is a Markov chain. The same remark holds for various other subprocesses, such as $(\mathbf{M}_n)$ and $(\mathbf{Q}_n, \mathbf{M}_n)$. Note in passing the implicit assumption that the subject's state distribution is not affected by the testing procedure. The Markov chain $(\mathbf{M}_n)$ will be referred to as the *marking process* and is of central interest. Its behavior is affected by several sources of errors (or randomness), in particular the error probabilities β_q, the guessing probabilities η_q, and the subject's distribution π.

Combinatorial Assumptions on the Structure

To implement the concepts introduced in Example 11.1, some combinatorial machinery is required which extends tools introduced in Chapter 2. Let $(Q, \mathcal{K})$ be a knowledge structure. Recall from Definition 2.4 that the essential distance between two states K and L is given by

$$e(K, L) = |K^* \bigtriangleup L^*|,$$

where K^* and L^* are as in 1.4, that is, $K^* = \{q^* \mid q \in K\}$ is the set of all notions q^* included in K, with $q^* = \{r \in Q \mid \mathcal{K}_q = \mathcal{K}_r\}$. Thus, $e(K, L)$ counts the notions by which K and L differ. If the structure were discriminative (cf. Definition 1.4), then $e(K, L)$ would coincide with the symmetric-difference distance (see 0.23)

$$d(K, L) = |K \triangle L|.$$

For $\Psi \subseteq \mathcal{K}$ and $\varepsilon \geq 0$, we define the ε-neighborhood of Ψ as

$$N(\Psi, \varepsilon) = \{K' \in \mathcal{K} \,|\, e(K, K') \leq \varepsilon, \text{ for some } K \text{ in } \Psi\}.$$

The elements of $N(\Psi, \varepsilon)$ are called ε-neighbors of Ψ. Those that contain item q are (q, ε)-neighbors of Ψ, and similarly those that do not contain q are $(\bar{q}, \varepsilon)$-neighbors; we set

$$N_q(\Psi, \varepsilon) = N(\Psi, \varepsilon) \cap \mathcal{K}_q \quad \text{and} \quad N_{\bar{q}}(\Psi, \varepsilon) = N(\Psi, \varepsilon) \cap \mathcal{K}_{\bar{q}},$$

calling these sets respectively the (q, ε)-neighborhood and $(\bar{q}, \varepsilon)$-neighborhood of Ψ. When $\Psi = \{K\}$, we abbreviate $N(\{K\}, \varepsilon)$ into $N(K, \varepsilon)$, and similarly write in shorthand $N_q(K, \varepsilon)$ and $N_{\bar{q}}(K, \varepsilon)$.

To exercise these concepts, here are a few elementary facts whose proofs are left to the reader (see Problem 4). For $y = q$ or $y = \bar{q}$, we always have $N_y(\Psi, 0) = \Psi_y \subseteq \Psi$; the last inclusion is strict except in two cases: (i) $y = q \in \cap \Psi$; or (ii) $y = \bar{q}$ and $q \notin \cup \Psi$. Also, $N_y(\Psi, 0)$ is empty if $y = q \notin \cup \Psi$, or $y = \bar{q}$ and $q \in \cap \Psi$.

These neighborhood concepts will be used to specify the testing rule and the marking rule along the lines introduced in Example 11.1. Loosely speaking, a question q is chosen on trial n by considering an ε-neighborhood $N(\mathbf{M}_n, \varepsilon)$ of the set $\mathbf{M}_n$ of marked states (where ε may depend on the size of the set of marked states), and attempting to split this ε-neighborhood into two subsets $N_q(\mathbf{M}_n, \varepsilon)$ and $N_{\bar{q}}(\mathbf{M}_n, \varepsilon)$ as equal in size as feasible. We base the choice of the question on $N(\mathbf{M}_n, \varepsilon)$ rather than on the potentially smaller set $\mathbf{M}_n$ in order to account for the possibility of errors committed by the assessment procedure in earlier steps.

11.5. Definition. Let $\varepsilon : \mathbf{N} \cup \{0\} \to \mathbf{R}^+ \cup \{0\}$ be a given function. Denote by $\mathbf{T}_n$ the set of all questions q in Q minimizing, for $s = \varepsilon(|\mathbf{M}_n|)$, the quantity

$$\nu_q(\mathbf{M}_n, s) = \Big| |N_q(\mathbf{M}_n, s)| - |N_{\bar{q}}(\mathbf{M}_n, s)| \Big|.$$

We say that the testing rule τ in Axiom [T] is ε-*half-split* iff

$$\tau(q, \mathbf{M}_n) = \frac{\iota_{\mathbf{T}_n}(q)}{|\mathbf{T}_n|},$$

where ι_A is the indicator function of the set A, that is

$$\iota_A(q) = \begin{cases} 1 & \text{if } q \in A; \\ 0 & \text{if } q \notin A. \end{cases}$$

Thus, $\varepsilon(|\mathbf{M}_n|)$ is a parameter, the value of which may depend on the size of $\mathbf{M}_n$. Note in passing that if $|\mathbf{M}_n| = 1$ and $\varepsilon(1) = 0$, or if $\mathbf{M}_n = \varnothing$, then $\mathbf{T}_n = Q$; in these cases, all the questions in Q have the same probability of being chosen.

The assumption $\varepsilon(1) = 1$ will be central in the sequel, together with the closely related concept of fringe $K^{\mathcal{F}}$ of a knowledge state K defined in 2.7 by

$$K^{\mathcal{F}} = K^{\mathcal{I}} \cup K^{\mathcal{O}},$$

where $K^{\mathcal{I}}$ and $K^{\mathcal{O}}$ are respectively the inner and outer fringes of K:

$$\begin{aligned} K^{\mathcal{I}} &= \{q \in K \mid K \setminus q^* \in \mathcal{K}\}, \\ K^{\mathcal{O}} &= \{q \in Q \setminus K \mid K \cup q^* \in \mathcal{K}\}. \end{aligned}$$

11.6. Theorem. *For any state K, any $q \in K^{\mathcal{F}}$ and any $r \in Q \setminus K^{\mathcal{F}}$, we have:*

$$\nu_q(K, 1) = |N(K, 1)| - 2 < \nu_r(K, 1) = |N(K, 1)|. \qquad (2)$$

Moreover, if the testing rule is half-split with $\varepsilon(1) = 1$, and $K^{\mathcal{F}} \neq \varnothing$, then for any positive integer n

$$\mathbf{P}(\mathbf{Q}_n = q \mid \mathbf{K}_n, \mathbf{M}_n = \{K\}, \mathbf{W}_{n-1}) = \begin{cases} 1/|K^{\mathcal{F}}| & \text{if } q \in K^{\mathcal{F}}; \\ 0 & \text{otherwise}, \end{cases}$$

and thus also

$$\mathbf{P}(\mathbf{Q}_n \in K^{\mathcal{F}} \mid \mathbf{K}_n, \mathbf{M}_n = \{K\}, \mathbf{W}_{n-1}) = 1.$$

PROOF. Since the second assertion follows from the first one, we establish only Equation (2). Suppose $q \in K^{\mathcal{F}}$. We treat successively two cases. If $q \in K$, then there is a state L such that $K = L \cup q^*$ and $N_{\bar{q}}(K,1) = \{L\}$. This yields

$$\nu_q(K,1) = |N(K,1)| - 2. \tag{3}$$

If $q \notin K$, then the same equality obtains with this time $L = K \cup q^*$ and $N_q(K,1) = \{L\}$.

On the other hand, suppose now $r \in Q \setminus K^{\mathcal{F}}$. Then there are only two cases: $r \notin \cup N(K,1)$ or $r \in \cap N(K,1)$. In the first case, we get

$$N_{\bar{r}}(K,1) = N(K,1), \qquad N_r(K,1) = \varnothing$$

and

$$\nu_r(K,1) = |N(K,1)|. \tag{4}$$

In the second case, we get

$$N_r(K,1) = N(K,1), \qquad N_{\bar{r}}(K,1) = \varnothing,$$

yielding again Equation (4). The first assertion of the statement follows from Equations (3) and (4). $\square$

We now turn to the marking rule. The idea here is that suggested by Example 11.1, namely, we will retain as marked states on trial $n + 1$ only those states in a δ-neighborhood of $\mathbf{M}_n$ which are consistent with the question asked and the response observed. Thus, with probability one,

$$\mathbf{M}_{n+1} = N_y(\mathbf{M}_n, \delta), \tag{5}$$

where y is q if the answer is correct, and $\bar{q}$ if the answer is incorrect. The value of the parameter δ in Equation (5) may vary with the size of $\mathbf{M}_n$. It may also depend upon whether the question $\mathbf{Q}_n$ belongs to at least one of the marked states in $\mathbf{M}_n$ and whether the response was correct. This generates the four cases in the next definition. Notice that enlarging the family of marked states makes sense: the true state could have escaped the actual collection of marked states, and we would like the procedure to correct this omission.

11.7. Definition. Let δ_1, δ_2, $\bar{\delta}_1$, and $\bar{\delta}_2$ be four functions defined on the nonnegative integers, with nonnegative real values. The marking rule μ of Axiom [M] is *selective with parameter* $\delta = (\delta_1, \delta_2, \bar{\delta}_1, \bar{\delta}_2)$ if it satisfies:

$$\mu(\mathbf{M}_{n+1}, \mathbf{R}_n, \mathbf{Q}_n, \mathbf{M}_n) = \begin{cases} 1 & \text{in the four cases (i)–(iv) below;} \\ 0 & \text{in all other cases.} \end{cases}$$

(i) $\mathbf{R}_n = 1$, $\mathbf{Q}_n = q \in \bigcup \mathbf{M}_n$, and $\mathbf{M}_{n+1} = N_q(\mathbf{M}_n, \delta_1(|\mathbf{M}_n|))$;

(ii) $\mathbf{R}_n = 1$, $\mathbf{Q}_n = q \notin \bigcup \mathbf{M}_n$, and $\mathbf{M}_{n+1} = N_q(\mathbf{M}_n, \delta_2(|\mathbf{M}_n|))$;

(iii) $\mathbf{R}_n = 0$, $\mathbf{Q}_n = q \in \bigcup \mathbf{M}_n$, and $\mathbf{M}_{n+1} = N_{\bar{q}}(\mathbf{M}_n, \bar{\delta}_1(|\mathbf{M}_n|))$;

(iv) $\mathbf{R}_n = 0$, $\mathbf{Q}_n = q \notin \bigcup \mathbf{M}_n$, and $\mathbf{M}_{n+1} = N_{\bar{q}}(\mathbf{M}_n, \bar{\delta}_2(|\mathbf{M}_n|))$.

Note that this requirement could be generalized by letting the functions δ_1, δ_2, $\bar{\delta}_1$, and $\bar{\delta}_2$ depend on the question q in Q.

Let us apply Definition 11.7 in the case K_0 is the unit support and $\mathbf{M}_n = \{K\}$, for some K, $K_0 \in \mathcal{K}$.

11.8. Theorem. *Suppose that the marking rule is selective with parameter* $\delta = (\delta_1, \delta_2, \bar{\delta}_1, \bar{\delta}_2)$ *satisfying*

$$\delta_1(1) = \bar{\delta}_2(1) = 0, \qquad \delta_2(1) = \bar{\delta}_1(1) = 1. \tag{6}$$

Writing $A_n(K, K_0)$ *for the joint event* $(\mathbf{M}_n = \{K\}, \mathbf{K}_n = K_0)$, *we have*

$$\mathbf{P}(\mathbf{M}_{n+1} = \{K\} \,|\, \mathbf{Q}_n \in K^{\mathcal{I}} \cap K_0, A_n(K, K_0)) = 1 - \beta_{\mathbf{Q}_n};$$

$$\mathbf{P}(\mathbf{M}_{n+1} = \{K \setminus \mathbf{Q}_n^*\} \,|\, \mathbf{Q}_n \in K^{\mathcal{I}} \cap K_0, A_n(K, K_0)) = \beta_{\mathbf{Q}_n};$$

$$\mathbf{P}(\mathbf{M}_{n+1} = \{K\} \,|\, \mathbf{Q}_n \in K^{\mathcal{I}} \setminus K_0, A_n(K, K_0)) = \eta_{\mathbf{Q}_n};$$

$$\mathbf{P}(\mathbf{M}_{n+1} = \{K \setminus \mathbf{Q}_n^*\} \,|\, \mathbf{Q}_n \in K^{\mathcal{I}} \setminus K_0, A_n(K, K_0)) = 1 - \eta_{\mathbf{Q}_n};$$

$$\mathbf{P}(\mathbf{M}_{n+1} = \{K \cup \mathbf{Q}_n^*\} \,|\, \mathbf{Q}_n \in K^{\mathcal{O}} \cap K_0, A_n(K, K_0)) = 1 - \beta_{\mathbf{Q}_n};$$

$$\mathbf{P}(\mathbf{M}_{n+1} = \{K\} \,|\, \mathbf{Q}_n \in K^{\mathcal{O}} \cap K_0, A_n(K, K_0)) = \beta_{\mathbf{Q}_n};$$

$$\mathbf{P}(\mathbf{M}_{n+1} = \{K \cup \mathbf{Q}_n^*\} \,|\, \mathbf{Q}_n \in K^{\mathcal{O}} \setminus K_0, A_n(K, K_0)) = \eta_{\mathbf{Q}_n};$$

$$\mathbf{P}(\mathbf{M}_{n+1} = \{K\} \,|\, \mathbf{Q}_n \in K^{\mathcal{O}} \setminus K_0, A_n(K, K_0)) = 1 - \eta_{\mathbf{Q}_n}.$$

The proof is left as Problem 5. Note that, using Theorems 11.6 and 11.8, we can calculate all possible transition probabilities from $\mathbf{M}_n = \{K\}$ to $\mathbf{M}_{n+1} = \{K'\}$ if the questioning rule is ε-halfsplit with $\varepsilon(1) = 1$ and also $K^{\mathcal{F}} \neq \varnothing$.

Markov Chains Terminology

Our results concern Markov chains. To avoid ambiguities, we will use the term m-*state* to denote the Markov states of these chains, reserving the expression (knowledge) states for elements of $\mathcal{K}$. For Markov chain concepts, we refer the reader to e.g. Feller (1968), Kemeny and Snell (1960), or Parzen (1962). Except when otherwise indicated, we follow the terminology of Kemeny and Snell (for example, we say "ergodic" rather than "recurrent" or "persistent"). Here is a brief glossary of the terminology, recalling some concepts encountered in Chapter 7.

11.9. Definition. Let $(\mathbf{X}_n)_{n\in\mathbf{N}}$ be a Markov chain on a finite set E of m-states, with *transition probability matrix* $M = (M_{ij})_{i,j\in E}$ and *initial probability distribution* $v = (v_i)_{i\in E}$; thus

$$v_i = \mathbf{P}(\mathbf{X}_1 = i), \qquad\qquad i \in E;$$
$$M_{ij} = \mathbf{P}(\mathbf{X}_{n+1} = j \,|\, \mathbf{X}_n = i), \qquad n = 1, 2, \ldots$$

An m-state j is *reachable* from an m-state i when there is a natural number n such that $(M^n)_{ij} > 0$ (note that j is not necessarily reachable from itself). A subset C of E is a *closed set* of m-states if any m-state outside C cannot be reached from any m-state in C. An m-state is *absorbing* when it is the single element of a closed set. A *class* is a subset C of E which is maximal for the property that, for any two m-states i, j in C, j is reachable from i; in particular, j is reachable from itself. Because of the finiteness of E, we may define an *ergodic m-state* as an m-state that belongs to some closed class. A closed class is sometimes called an *ergodic set*. An m-state which is not ergodic is *transient*. The chain $(\mathbf{X}_n)$ is *regular* when all its m-states form a single class and moreover $(\mathbf{X}_n)$ is not *periodic*, the latter requirement meaning that $(M^n)_{ij} > 0$ for some $n \in \mathbf{N}$ and all $i, j \in E$. Finally, a probability distribution p on E is *stationary* (or *invariant*) when $\sum_{i\in E} p_i M_{ij} = p_j$.

If the chain $(\mathbf{X}_n)$ is regular, then it has a unique stationary distribution p which is called the *limit* or *asymptotic distribution*, with

$$p_j = \lim_{n\to\infty} \mathbf{P}(\mathbf{X}_n = j) = \lim_{n\to\infty} \sum_{i\in E} v_i (M^n)_{ij}.$$

The limit distribution p does not depend on the initial probability distribution v on E.

Results for the Fair Case

As before, let $\mathcal{K}$ be a knowledge structure with finite domain Q. Let $(\mathbf{R}_n, \mathbf{Q}_n, \mathbf{K}_n, \mathbf{M}_n)$ be a stochastic assessment procedure parametrized by π, τ, β, η and μ. We begin the investigation of its behavior with a simple general result for the straight case (i.e. the error probability β_q and the guessing probability η_q are zero for all questions q in Q).

11.10. Theorem. *In the straight case, assume that there is only one true state K_0 and that the marking rule is selective. Then, for all natural numbers n:*

$$\mathbf{P}(K_0 \in \mathbf{M}_{n+1} \mid K_0 \in \mathbf{M}_n) = 1. \tag{7}$$

In other words, the set of all m-states containing K_0 is a closed set of the Markov chain $(\mathbf{M}_n)$.

It suffices to prove Equation (7), which is left as Problem 6. We consider next a situation in which the parameters ε and δ are chosen so as to narrow down quickly (during phase 1, as we considered in Example 11.1) the set of marked states.

11.11. Theorem. *Suppose that the questioning rule is ε-half-split, that the marking rule is selective with parameter δ, and that $\varepsilon(l) = \delta_k(l) = \bar{\delta}_k(l) = 0$ for $k = 1, 2$, and all integers $l > 1$. Then*

$$\mathbf{P}\big(|\mathbf{M}_{n+1}| < |\mathbf{M}_n| \,\big|\, |\mathbf{M}_n| > 1\big) = 1. \tag{8}$$

If, moreover, $\delta_1(1) = 0$, $\delta_2(1) \leq 1$, $\bar{\delta}_1(1) \leq 1$, $\bar{\delta}_2(1) = 0$, it follows that for some natural number r, we have for all $n \geq r$

$$\mathbf{P}(|\mathbf{M}_n| \leq 1) = 1. \tag{9}$$

In particular, in the straight case, if K_0 is the unit support and $\delta_2(1) = \bar{\delta}_1(1) = 0$, then there is a positive integer r such that whenever $n \geq r$,

$$\mathbf{P}(\mathbf{M}_n = \{K_0\} \mid K_0 \in \mathbf{M}_1) = 1, \tag{10}$$

and

$$\lim_{n \to \infty} \mathbf{P}(\mathbf{M}_n = \varnothing) = 1 \qquad \text{iff} \qquad K_0 \notin \mathbf{M}_1. \tag{11}$$

In fact, the Markov chain $(\mathbf{M}_n)$ has exactly two absorbing m-states which are $\{K_0\}$ and $\varnothing$.

PROOF. Equation (8) results immediately from the axioms and the hypotheses. Since $\mathbf{M}_n$ is finite for every positive integer n, Equation (9) follows. Applying Theorem 11.10 gives (10), and then (11) is easily derived.   □

The assumption $\delta_k(1) = \bar{\delta}_k(1) = 0$ practically locks the set of marked states as soon as not more than one such state remains. Thus, the unit support is either found quickly if it belongs to $\mathbf{M}_1$, or missed otherwise. We now study a more flexible approach that allows in phase 2 a single marked state K to evolve in the structure, thus making possible a gradual construction (in the straight case) or approximation (in the fair case) of the unit support by K. For convenience, we give below a label to this set of conditions; notice the requirement $\varepsilon(1) = 1$ which was motivated in Example 11.1 and allows only small changes of the single marked state K.

11.12. Definition. Let $(\mathbf{R}_n, \mathbf{Q}_n, \mathbf{K}_n, \mathbf{M}_n)$ be a stochastic assessment procedure parametrized by π, τ, β, η and μ, with an ε-halsplit testing rule and a marking rule which is selective with parameter δ. Suppose that the following conditions are satisfied:

(1) the knowledge structure $(Q, \mathcal{K})$ is a well-graded space;
(2) $\varepsilon(n) = 0$ for $n > 1$; also, $\varepsilon(1) = 1$;
(3) $\delta = 0$ except in two cases: $\delta_2(1) = 1$, and $\bar{\delta}_1(1) = 1$.

Then $(\mathbf{R}_n, \mathbf{Q}_n, \mathbf{K}_n, \mathbf{M}_n)$ will be called unitary.

11.13. Convention. In the rest of this section, we consider a fair, unitary, stochastic assessment procedure $(\mathbf{R}_n, \mathbf{Q}_n, \mathbf{K}_n, \mathbf{M}_n)$ in the sense (and notation) of Definition 11.12. We assume that there is a unique support K_0; accordingly, the certain event $\mathbf{K}_n = K_0$ will not be mentioned in the statement of results.

Terms such as 'm-state' and 'ergodic set' refer to the Markov chain $\mathbf{M}_n$, which is our principal object of investigation. This chain thus satisfies Equations (8) and (9). As soon as some state K remains the single marked state (which is bound to happen because of Theorem 11.11), question $\mathbf{Q}_n$ will be drawn in the fringe of K. This fringe $K^{\mathcal{F}}$ is nonempty because $\mathcal{K}$ is well-graded (cf. Theorem 2.9(iv)). We first give the possible transitions from $\mathbf{M}_n = \{K\}$, leaving the proof to the reader (see Problem 8).

11.14. Theorem. *There is a natural number n_0 such that for $n \geq n_0$,*

$$\mathbf{P}(|\mathbf{M}_n| = 1) = 1. \qquad (12)$$

Moreover, for any natural number n,

$$\mathbf{P}(|\mathbf{M}_{n+1} = 1| \, | \, |\mathbf{M}_n| = 1) = 1.$$

More precisely, with

$$A = |K^{\mathcal{O}} \setminus K_0| + \sum_{q \in K^{\mathcal{O}} \cap K_0} \beta_q + \sum_{q \in K^{\mathcal{I}} \cap K_0} (1 - \beta_q),$$

we have

$$\mathbf{P}(\mathbf{M}_{n+1} = \{K'\} \, | \, \mathbf{M}_n = \{K\}) =$$

$$\begin{cases} (1/|K^{\mathcal{F}}|)(1 - \beta_q) & \text{if } K' = K \cup q^*, \text{ with } q \in K^{\mathcal{O}} \cap K_0, \\ (1/|K^{\mathcal{F}}|) & \text{if } K' = K \setminus q^*, \text{ with } q \in K^{\mathcal{I}} \setminus K_0, \\ (1/|K^{\mathcal{F}}|)\beta_q & \text{if } K' = K \setminus q^*, \text{ with } q \in K^{\mathcal{I}} \cap K_0, \\ (1/|K^{\mathcal{F}}|)A & \text{if } K = K', \\ 0 & \text{otherwise.} \end{cases}$$

Thus, in particular, for any $K \not\subseteq K_0$,

$$\mathbf{P}(\mathbf{M}_{n+1} = \{K\} \, | \, \mathbf{M}_n = \{K_0\}) = 0.$$

In the straight case, we have $\beta_{\mathbf{Q}_n} = 0$, which implies that with probability one $e(\mathbf{M}_{n+1}, K_0) \leq e(\mathbf{M}_n, K_0)$ when $|\mathbf{M}_n| = 1$. In general, if Ψ is a nonempty family of subsets of Q and K a subset of Q, we set

$$e(\Psi, K) = \min\{e(K', K) \, | \, K' \in \Psi\}.$$

11.15. Theorem. *In the straight case, for any choice of $\mathbf{M}_1$ nonempty, we have*

$$\lim_{n \to \infty} \mathbf{P}(\mathbf{M}_n = \{K_0\}) = 1. \qquad (13)$$

Moreover, for any state K with $e(K, K_0) = j > 0$, and any ℓ, n in $\mathbf{N}$

$$\mathbf{P}(\mathbf{M}_{\ell+n} = \{K_0\} \, | \, \mathbf{M}_n = \{K\}) \geq \sum_{k=0}^{\ell-j} \binom{j+k-1}{k} \lambda^j (1-\lambda)^k \qquad (14)$$

where λ is defined by

$$\lambda = \min \frac{|(K \,\triangle\, K') \cap K^{\mathcal{F}}|}{|K^{\mathcal{F}}|},$$

for $K, K' \in \mathcal{K}$ with $K \neq K'$.

PROOF. To prove Equation (14) first, we consider a sequence of Bernoulli trials with "success" meaning a step of size one towards K_0, that is at trial n we have $e(\mathbf{M}_{n+1}, K_0) = e(\mathbf{M}_n, K_0) - 1$. If $\mathbf{M}_n = \{K\}$ is realized, and $\mathbf{K}_n = K_0$ with $e(K_0, K) = j$, then at least j successes are necessary in the following ℓ trials to achieve $\mathbf{M}_{\ell+n} = \{K_0\}$, with the probability of each success being at least λ. The wellgradedness implies $\lambda > 0$. Now, in a sequence of Bernoulli trials with probability of success equal to λ, the number $j + k$ of trials required to achieve exactly j successes has a probability specified by

$$\binom{j + k - 1}{k} \lambda^j (1 - \lambda)^k.$$

Thus, k is a value of a random variable, which is distributed as a negative binomial with parameters j and λ. Consequently, Equation (14) follows from Equation (12). Since the right member of (14) tends to 1 as ℓ tends to ∞, we conclude that Equation (13) holds by using Theorem 11.14. $\square$

We also have:

11.16. Theorem. *There exists a positive integer n_0 such that whenever $n \geq n_0$, then $\mathbf{P}(|\mathbf{M}_n| = 1) = 1$. Moreover, the Markov chain $(\mathbf{M}_n)$ has a unique ergodic set E_0 which contains $\{K_0\}$ and possibly some m-states $\{K\}$ such that $K \subseteq K_0$, but no other m-states. If, in addition $\beta_q > 0$ for all $q \in K_0$, then E_0 is in fact the family of all those m-states $\{K\}$ such that $K \subseteq K_0$.*

PROOF. By Theorem 11.14 (Equation 12), an ergodic m-state contains exactly one knowledge state. Using the transition probabilities described in Theorem 11.14 and the wellgradedness of $\mathcal{K}$, we see that it is possible to reach the m-state $\{K_0\}$ from any m-state $\{K\}$. This clearly implies the uniqueness of the ergodic set E_0, with moreover $\{K_0\} \in E_0$. The remaining assertions also follow readily from Theorem 11.14. $\square$

Uncovering a Stochastic State: Examples

Most of the results presented in the previous section suppose a unit support. It is also interesting to investigate the case in which the probability distribution π on the family $\mathcal{K}$ is not concentrated on a single state.

In an ideal situation, the Markov chain $\mathbf{M}_n$ admits a unique ergodic set ξ that contains the support $\sigma(\pi)$ of the probability distribution π; then, a sensitive strategy is to analyze the statistics of occupation times of the m-states in ξ in order to assess the probability $\pi(K)$ of each true state K. (In practice, because we cannot ask many questions, we aim only at ballpark estimates of these probabilities.) This strategy will be illustrated in two examples. The first one is based on a discriminative chain $\mathcal{K}$ of states. Its items are thus linearly ordered[1].

In all the examples, we consider a fair, unitary stochastic assessment procedure (cf. Definitions 11.4 and 11.12).

11.17. Example. Suppose that $\mathcal{K}$ is a chain of states $L_0 = \varnothing$, $L_1 = \{q_1\}$, ..., $L_m = \{q_1, q_2, \ldots, q_m\}$. By Theorem 11.11, Equation (9), all the m-states of the Markov chain $(\mathbf{M}_n)$ containing more than one state are transient. Any m-state containing a single true state is ergodic. (All assertions left unproved in this example are dealt with by Problem 9). The same holds for any m-state $\{K\}$ such that $K' \subseteq K \subseteq K''$ for two true states K' and K''. When $\beta_q > 0$ for each q in $\cap\sigma(\pi)$, we see that $\{K\}$ is an ergodic m-state for any state K included in a true state. On the other hand, if $\beta_q = 0$ for $q \in \cap\sigma(\pi)$, then the only ergodic m-states are of the form $\{K\}$ with K a state between two true states. Thus, in the straight case, if the true states form a subchain L_j, L_{j+1}, ..., L_k of $\mathcal{K}$ for some j, k with $0 \le j \le k \le m$, the ergodic m-states are essentially the true states. As will be indicated in an example, the statistics of the occupation times of the states in the recurrent class can be used to estimate the probabilities $\pi(K)$. Suppose that we have exactly three true states L_{i-1}, L_i, and L_{i+1}, with $1 < i < m$. Thus, setting

$$\pi_{i-1} = \pi(L_{i-1}) \qquad \text{and} \qquad \pi_{i+1} = \pi(L_{i+1}),$$

we assume $\pi(L_i) = 1 - \pi_{i-1} - \pi_{i+1} > 0$, and $\pi_{i-1} > 0$, $\pi_{i+1} > 0$. Suppose

---

[1] Needless to say, this case could have been treated by other methods (such as "tailored testing"; see Lord, 1974; Weiss, 1983).

also $\beta_q = 0$ for $q \in Q$ (straight case). From these assumptions, it follows that there are three ergodic m-states, namely $\{L_{i-1}\}$, $\{L_i\}$, and $\{L_{i+1}\}$. The probabilities for the observable variables on trial n are given by the tree-diagram of Figure 11.2 for the transient m-states $\{L_j\}$, for $0 < j < i-1$, and $\{L_k\}$, for $i+1 < k < m$ (we leave the cases of $\{\varnothing\}$ and $\{Q\}$ to the reader). A similar tree-diagram for the ergodic m-states is provided in Figure 11.3. The possible transitions between all m-states of the form $\{K\}$, with their probabilities, are shown in Figure 11.4.

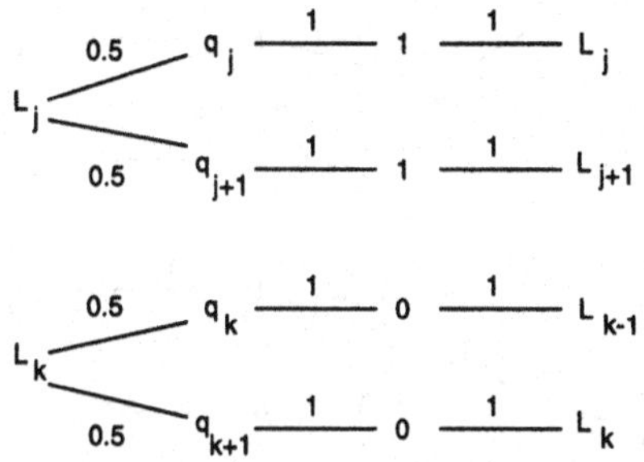

Fig. 11.2. Probabilities for the observable variables at trial n in Example 11.17, when leaving a transient state $\{L_j\}$ or $\{L_k\}$, for $0 < j < i-1$ or $i+1 < k < m$.

Asymptotically, we have a Markov chain on the three m-states L_{i-1}, L_i, and L_{i+1}, which is regular. Setting

$$p_j = \lim_{n \to \infty} \mathbf{P}(\mathbf{M}_n = \{L_j\}),$$

we thus have $p_j \neq 0$ iff $i - 1 \leq j \leq i + 1$. The stationary distribution of this chain on three states is the unique solution to the following system of linear equations:

$$(p_{i-1}, p_i, p_{i+1}) \cdot \begin{pmatrix} \dfrac{1 + \pi_{i-1}}{2} & \dfrac{1 - \pi_{i-1}}{2} & 0 & 1 \\[2ex] \dfrac{\pi_{i-1}}{2} & \dfrac{2 - \pi_{i-1} - \pi_{i+1}}{2} & \dfrac{\pi_{i+1}}{2} & 1 \\[2ex] 0 & \dfrac{1 - \pi_{i+1}}{2} & \dfrac{1 + \pi_{i+1}}{2} & 1 \end{pmatrix}$$

$$= (p_{i-1}, p_i, p_{i+1}, 1).$$

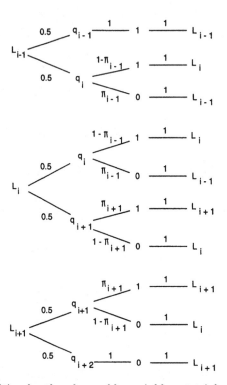

Fig. 11.3. Probabilities for the observable variables at trial n in Example 11.17, when leaving an ergodic state $\{L_{i-1}\}$, $\{L_i\}$, or $\{L_{i+1}\}$.

As the first equation gives

$$\frac{1+\pi_{i-1}}{2}p_{i-1} + \frac{\pi_{i-1}}{2}p_i = p_{i-1},$$

we infer

$$\pi_{i-1} = \frac{p_{i-1}}{p_{i-1} + p_i}.$$

A similar computation with the third equation gives

$$\pi_{i+1} = \frac{p_{i+1}}{p_i + p_{i+1}}.$$

To be complete, we also give

$$\pi(L_i) = \frac{p_i^2 - p_{i-1}p_{i+1}}{p_i + p_{i-1}p_{i+1}}.$$

Since in this case the asymptotic probabilities of the knowledge states can be estimated from the proportions of visits to the m-states in any realization,

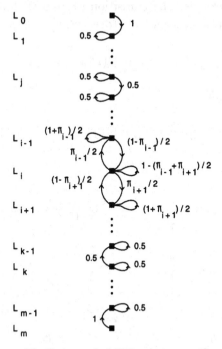

Fig. 11.4. Transitions with their probabilities between the m-states of the form $\{L_j\}$, see Example 11.17.

we have the possibility of obtaining at least a rough estimate of the unknown probabilities $\pi(L_j)$ from the data of any particular subject.

Our next example shows that these techniques are not restricted to the case of a chain of states.

11.18. Example. Let $Q = \{a, b, c, d, e\}$ be a set of five questions, with a knowledge space $\mathcal{K}$ derived from the partial order in Figure 11.5 (a). We have thus

$$\mathcal{K} = \{ \varnothing, \{c\}, \{e\}, \{b,c\}, \{c,e\}, \{a,b,c\}, \{b,c,e\},$$
$$\{c,d,e\}, \{a,b,c,e\}, \{b,c,d,e\}, Q \}.$$

Suppose that the straight case holds, and that $\{c\}$ and $\{e\}$ are the only true states. Setting $\alpha = \pi(e)$, we have thus $\pi(c) = 1 - \alpha$. As the assessment procedure is assumed to be straight, we conclude that there is only one ergodic set, namely $\{\{\varnothing\}, \{\{c\}\}, \{\{e\}\}, \{\{c,e\}\}\}$.

Figure 11.5 (b) provides the transition probabilities among the four ergodic m-states. Setting, for $\{K\} \in \{\{\emptyset\}, \{\{c\}\}, \{\{e\}\}, \{\{c,e\}\}\}$,

$$p(K) = \lim_{n\to\infty} \mathbf{P}(\mathbf{M}_n = \{K\}),$$

we obtain for example

$$p(\emptyset) = \frac{1}{2}p(\emptyset) + \frac{1}{3}\alpha p(\{c\}) + \frac{1}{2}(1-\alpha)p(\{e\}).$$

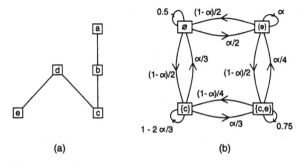

Fig. 11.5. (a) The Hasse diagram of the partial order in Example 11.18.
(b) Transition probabilities among the four ergodic m-states in that Example.
(One layer of brackets is omitted, i.e. $\{a\}$ means $\{\{a\}\}$.)

Solving for α, we get

$$\alpha = \frac{p(\emptyset) - p(\{e\})}{(2/3)p(\{c\}) - p(\{e\})}.$$

Here again, the unknown quantity α can be estimated from the asymptotic probabilities of the ergodic m-states of $(\mathbf{M}_n)$, via the proportions of visits of the m-states in any realization of the procedure. We will consider in the next section some circumstances in which such an estimation is theoretically feasible.

Let us now turn to an example in which the knowledge structure is not described accurately at the start. For instance, the student's state could have been omitted.

11.19. Example. We use the same structure as in Example 11.18. Suppose that the student has mastered questions c and d only; thus, the knowledge

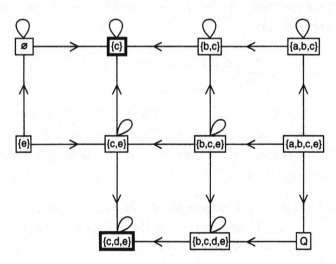

Fig. 11.6. Possible transitions among the m-states from Example 11.19. (One layer of brackets is omitted.)

state of that student is $\{c,d\}$ which is not an element of $\mathcal{K}$. When running a fair, unitary procedure to assess the knowledge of this subject, we assume that the response is correct if questions c or d were asked, and incorrect otherwise. From an analysis of this case, it turns out that $(\mathbf{M}_n)$ is a Markov chain; Figure 11.6 indicates its reachability relation, together with the transition probabilities. There are two absorbing m-states, namely $\{\{c\}\}$ and $\{\{c,d,e\}\}$, which are singled out in the figure. Thus, depending on the starting point, the chain will end remaining in one of these two m-states with probability one. As a conclusion, we see that observing one realization of the process may lead the observer to diagnose an incorrect state (unavoidably), but one not too far from the student's actual state.

Intractable Cases

In general, the goal of the assessment procedure that we have been studying is to estimate the probability distribution on the knowledge structure associated with a particular subject whose knowledge state is, for some reason or other, randomly varying in the course of the procedure. Assume that the error probabilities β_q and guessing probabilities η_q are zero for all $q \in Q$ (straight case). The observable probabilities $\rho(q)$ of correct answers to questions q are then completely determined by the subject's distribution π through the formula:

$$\sum_{K \in \mathcal{K}_q} \pi(K) = \rho(q), \qquad \text{for } q \in Q \tag{15}$$

(where as before $\mathcal{K}_q = \{K \in \mathcal{K} \mid q \in K\}$). This is a system of linear equations in the unknown quantities $\pi(\mathcal{K})$, and with the constant terms $\rho(q)$ in the right hand members. Notice that the coefficients of the unknowns take only the value 0 or 1.

11.20. Definition. The *incidence matrix* M of the collection $\mathcal{K}$ of subsets of the finite domain Q of questions is the matrix, with rows indexed by questions q in Q and columns indexed by states K in $\mathcal{K}$, that satisfies: the entry $M_{q,K}$ of M takes the value 1 if $q \in K$, and 0 otherwise.

Given a vector $\rho = (\rho(q))$ of correct response probabilities, where $q \in Q$, we ask when there exists at most one vector solution $\pi = (\pi(K))$ (with $K \in \mathcal{K}$) to Equation (15) that moreover satisfies $\pi(K) \geq 0$ and $\sum_{K \in \mathcal{K}} \pi(K) = 1$. Assume first that some vector solution $\hat{\pi}$ to Equation (15) exists, with $0 < \hat{\pi}(K) < 1$ for all $K \in \mathcal{K}$. The next theorem shows when such a solution is unique.

11.21. Theorem. *In the straight case, a strictly positive probability distribution $\hat{\pi}$ on the knowledge structure $\mathcal{K}$ can be recovered from the vector of correct response probabilities through Equation (15) iff the rank of the incidence matrix of the collection $\mathcal{K}^* = \mathcal{K} \setminus \{\varnothing\}$ is equal to $|\mathcal{K}^*|$.*

PROOF. Denote by Λ the simplex formed by all the probability distributions π on $\mathcal{K}$; thus Λ lies in the vector space $\mathbf{R}^{\mathcal{K}}$, where a vector has one real coordinate for each element in $\mathcal{K}$. The strictly positive probability distribution $\hat{\pi}$ mentioned in the statement is a relative interior point of Λ. Notice that the vector subspace S_0 consisting of all solutions π (without a positive condition) to the homogeneous system

$$\sum_{K \in \mathcal{K}_q} \pi(K) = 0, \qquad \text{for } q \in Q, \tag{16}$$

always contains the vertex of Λ corresponding to the distribution having all its mass concentrated on the empty set. Hence this subspace S_0, which contains also the origin of $\mathbf{R}^{\mathcal{K}}$, is never parallel to the hyperplane defined by $\sum_{K \in \mathcal{K}} \pi(K) = 1$.

The affine subspace S of all solutions to Equation (15) is the translate of S_0 passing through the point $\hat{\pi}$. As $\hat{\pi}$ is a relative interior point of Λ, we see that $\hat{\pi}$ is uniquely determined iff the dimension of S_0 is 1. This happens exactly if the incidence matrix of $\mathcal{K}$ has rank $|\mathcal{K}| - 1$, or equivalently the incidence matrix of $\mathcal{K}^*$ has rank $|\mathcal{K}^*|$.   □

Theorem 11.21 covers the case in which the support of $\sigma(\pi)$ equals $\mathcal{K}$. Dropping this condition but assuming a fixed support $\sigma(\pi)$ containing $\varnothing$, the same argument shows that we cannot recover any latent distribution π from its induced response probabilities ρ when the rank of the incidence matrix of $\sigma(\pi)$ is not equal to $|\sigma(\pi)| - 1$.

These considerations lead to the following problem: characterize those collections $\mathcal{K}$ of nonempty subsets of a finite domain Q for which the incidence matrix has rank over the reals equal to $|\mathcal{K}|$. An obvious necessary condition reads $|Q| \geq |\mathcal{K}|$, that is: for any probability distribution on $\mathcal{K}$ to be recoverable by a stochastic assessment procedure, the number of nonempty states cannot exceed the number of questions.

Many families of subsets satisfy the rank condition; here is a large class of examples.

11.22. Theorem. *If $\mathcal{P}$ is a partial order on the finite domain Q, then the family of its principal ideals*

$$\mathcal{I} = \{\{q \in Q \mid q \mathcal{P} r\} \mid r \in Q\}$$

has an incidence matrix of rank $|Q|$.

PROOF. There exists a linear extension T of $\mathcal{P}$ (Szpilrajn, 1930; see also Trotter, 1992). List the columns of the incidence matrix in the order of T. Since the principal ideal generated by an item r in Q is the first element of $\mathcal{I}$ that contains r, the column indexed by r is linearly independent from the preceding columns.   □

Original Sources and Related Works

This chapter closely follows the paper by Falmagne and Doignon (1988b), except for one detail: we did not assume here that the knowledge structures were discriminative. Notice that the combinatorial part of this paper was presented in Chapter 2.

Problems

1. Work out several other realizations of the assessment procedure in Example 11.1, by making various choices of question when possible.

2. By carefully selecting an exemplary knowledge structure different from those encountered in the chapter, explain why the wellgradedness assumption is crucial. Give examples in which the Markov chain $(\mathbf{M}_n)$ can reach an m-state of the form $\{K\}$, with K not a true state, and remain there with probability 1. Use your examples to find out which results in this chapter remain true for knowledge structures that are not well-graded.

3. Show that the various processes mentioned at the end of Definition 11.4, such as $(\mathbf{M}_n)$ and $(\mathbf{Q}_n, \mathbf{M}_n)$, are indeed Markov chains.

4. Prove the following assertions from the beginning of section entitled "Combinatorial assumptions on the structure." For $y = q$ or $y = \bar{q}$, we always have $N_y(\Psi, 0) = \Psi_y \subseteq \Psi$; the last inclusion is strict except in two cases: (i) $y = q \in \bigcap \Psi$; or (ii) $y = \bar{q}$ and $q \notin \bigcup \Psi$. Also, $N_y(\Psi, 0)$ is empty if $y = q \notin \bigcup \Psi$, or $y = \bar{q}$ and $q \in \bigcap \Psi$.

5. Provide proofs for the various cases of Theorem 11.8.

6. Prove Equation (7).

7. Assume a unitary assessment procedure is ran on a well-graded knowledge structure. If there are two true states incomparable for inclusion, then the Markov chain $(\mathbf{M}_n)$ has at least three ergodic m-states; true or false? Prove your answer.

8. Give a proof of Theorem 11.14.

9. Establish all the assertions left unproved in Example 11.17.

10. Analyze Example 11.18 in case the support is another family of states, say of three states.

11. Analyze Example 11.19 in case the student's knowledge is another subset of Q, also not belonging to $\mathcal{K}$.

12. Evaluate the computer storage needed to implement the stochastic assessment procedure from this chapter (assuming student's answers are collected at the keyboard). Compare the memory required with that needed by the procedure from Chapter 10.

13. Determine all the collections $\mathcal{K}$ of nonempty subsets on a two-element set whose incidence matrix has rank over the reals equal to 2. Try to find (up to isomorphism) all the similar collections on a three-element domain with rank 3.

Chapter 12

Building the Knowledge Structure
in Practice

In Chapter 5, we established the equivalence of two seemingly quite different concepts: on the one hand the knowledge spaces, and on the other hand the entailments for Q. Recall that the latter are the relations $\mathcal{P} \subseteq (2^Q \setminus \{\varnothing\}) \times Q$ that satisfy the following two conditions: for all $q \in Q$ and $A, B \in 2^Q \setminus \{\varnothing\}$,

(1) if $q \in A$, then $A\mathcal{P}q$;
(2) if $A\mathcal{P}b$ and $B\mathcal{P}q$ holds whenever $b \in B$, then $A\mathcal{P}q$

(see Theorem 5.5). The unique entailment $\mathcal{P}$ derived from some particular space $\mathcal{K}$ is defined by the formula

$$A\mathcal{P}q \quad \Longleftrightarrow \quad \left\{ \begin{array}{l} \text{any state of } \mathcal{K} \text{ containing } q \text{ also} \\ \text{contains at least one element of } A, \end{array} \right. \tag{1}$$

where $A \in 2^Q \setminus \{\varnothing\}$ and $q \in Q$. An empirical interpretation of an entailment is suggested by this formula, in terms of the responses to a class of questions or queries that an expert may be asked. In the field of education, these queries may take the form:

[Q1] *Suppose that a student under examination has just provided wrong responses to all the questions in some set A. Is it practically certain that this student will also fail item q? Assume that the conditions are ideal in the sense that errors and lucky guesses are excluded.*

In this chapter, the set A appearing in the formula $A\mathcal{P}q$ or in [Q1] will often be referred to as an antecedent set. This type of query may be abbreviated here as: *Does failing all the questions in some antecedent set A entail failing also question q?*

It is clear from (1) that, for a given knowledge structure $\mathcal{K}$, the responses to all the queries of the form [Q1] are determined. By hypothesis, the student under consideration must be in some state not intersecting the antecedent set A. If none of those states having an empty intersection with A contains q, respond 'yes'; otherwise, respond 'no.'

Conversely, if the responses to all such queries are given, then in view of Theorem 5.5, a unique knowledge space is determined. (In the terminology introduced in Definition 5.6, we say then that this particular knowledge space is "derived from" the given entail relation.) These considerations suggest a practical technique for constructing a knowledge space. We suppose that, when asked a query of the form [Q1], an expert relies (explicitly or implicitly) on a personal knowledge space to manufacture a response. It suffices thus to ask the expert all the queries of the form [Q1]. An obvious difficulty is that the number of possible queries of the form [Q1] is considerable: if Q contains m items, then there are $(2^m - 1) \cdot m$ queries of the form [Q1]. Fortunately, in practice, only a minute fraction of these queries need to be asked, because many responses to new queries are either trivial or can be inferred from previous responses. This chapter begins with a description of the QUERY routine, an interactive algorithm capable of querying an expert and based on the idea just mentioned. We then discuss the results of an exemplary application of the QUERY routine due to Kambouri, Koppen, Villano, and Falmagne (1993; see also Kambouri, 1991). The experts queried by the routine are four experienced teachers and the experimenter, Kambouri. The items were taken from the standard high school curriculum in mathematics. Our presentation follows closely Kambouri et al. (1993). This application of the QUERY routine was only partly successful. This led Cosyn and Thiéry (1999) to elaborate the QUERY routine into a comprehensive technique using both judgments from education experts and student data. Their results are also summarized here.

Koppen's Algorithm for Constructing the Knowledge Space

Our discussion of this algorithm will cover the main ideas. Full details can be found in Koppen (1993).

12.1. Outline. In principle, the construction of the knowledge space could proceed in two steps.

(a) Draw up the list of all the subsets of Q.

(b) Examine all the responses to the queries of the form [Q1]. Whenever a response $A \mathcal{P} q$ is observed, remove from the list of subsets all the sets containing q, and having an empty intersection with A.

The unique knowledge space consistent with the expert's responses is ultimately generated by this procedure. An example of elimination of subsets from the domain as potential states is given in Table 12.1. The domain is $Q = \{a, b, c, d, e\}$, and we only consider the result of the three responses: $\{a\}\mathcal{P}e, \{b\}\mathcal{P}e$ and $\{c\}\mathcal{P}e$. The sets eliminated by each of the three responses are marked in the last three columns of the table by the symbol '$\sqrt{}$.' The sets remaining after elimination are marked by the symbol '$\star$' appearing in some of the lines of the table.

12.2. Remarks. 1) Not only is the final output of the procedure a knowledge space, but the collection of sets remaining at each intermediate step is also closed under union. Indeed, suppose that some union $S \cup T \subseteq Q$ is eliminated because of some particular positive response $A\mathcal{P}q$. Thus, we must have $q \in S \cup T$ and $A \cap (S \cup T) = \varnothing$. This means that q must belong to at least one of the subsets S and T, but both of these subsets have an empty intersection with A. Thus, the observation $q \in A$ must remove at least one of S and T (if these sets are still part of the list at that point).

2) The procedure outlined in 12.1 contains the essential ideas of the QUERY routine. In practice, critical refinements are required to render this idea workable. To begin with, we cannot literally keep a running list of the sets which are still feasible states at each step of the procedure, because, in the early stages of the procedure and at least for non trivial applications, that list is much too large. (For 50 items, the initial list would have 2^{50} entries.) Rather than keeping a running list of the remaining sets, the QUERY routine only stores the positive and negative responses to the queries of the form [Q1]. A list of the feasible states remaining at any point can be generated whenever such a list is actually needed and its length is not prohibitive. A second refinement stems from the observation that only a very small subset of the queries must be asked, for two reasons. One is that the responses to some queries are known a priori. For example, if $q \in A$, then failing all the items in the set A implies failing q. Thus, all the positive responses of the form $A\mathcal{P}q$ with $q \in A$ are trivially true. (These trivial positive responses do not eliminate any subsets, though.) This cuts the number of possible queries in half (see Problem 1.) More importantly, many queries need not be asked because the responses can be inferred from the responses to other queries. The mechanism of these inferences is not obvious, and it will be useful to examine a few examples.

	$\{a\}\mathcal{P}e$	$\{b\}\mathcal{P}e$	$\{c\}\mathcal{P}e$	
$\emptyset$				*
$\{a\}$				*
$\{b\}$				*
$\{c\}$				*
$\{d\}$				*
$\{e\}$	✓	✓	✓	
$\{a,b\}$				*
$\{a,c\}$				*
$\{a,d\}$				*
$\{a,e\}$		✓	✓	
$\{b,c\}$				*
$\{b,d\}$				*
$\{b,e\}$	✓		✓	
$\{c,d\}$				*
$\{c,e\}$	✓	✓		
$\{d,e\}$	✓	✓	✓	
$\{a,b,c\}$				*
$\{a,b,d\}$				*
$\{a,b,e\}$			✓	
$\{a,c,d\}$				*
$\{a,c,e\}$		✓		
$\{a,d,e\}$		✓	✓	
$\{b,c,d\}$				*
$\{b,c,e\}$	✓			
$\{b,d,e\}$	✓		✓	
$\{c,d,e\}$	✓	✓		
$\{a,b,c,d\}$				*
$\{a,b,c,e\}$				*
$\{a,b,d,e\}$			✓	
$\{a,c,d,e\}$		✓		
$\{b,c,d,e\}$	✓			
$\{a,b,c,d,e\}$				*

Table 12.1. Inferences from the responses $\{a\}\mathcal{P}e$, $\{b\}\mathcal{P}e$ and $\{c\}\mathcal{P}e$ (see 12.1). The states eliminated by a response are indicated by the symbol '✓'. The symbol '*' indicates the sets remaining after elimination.

We denote by $\overline{\mathcal{P}}$ the 'negation' of $\mathcal{P}$:

$$A\overline{\mathcal{P}}q \quad \Longleftrightarrow \quad \begin{cases} \text{there is a state containing } q \\ \text{and none of the elements of } A. \end{cases} \tag{2}$$

Formally, for a given knowledge space $\mathcal{K}$

$$A\mathcal{P}q \quad \Longleftrightarrow \quad (\forall K \in \mathcal{K} : q \in K \Rightarrow A \cap K \neq \varnothing), \tag{3}$$
$$A\overline{\mathcal{P}}q \quad \Longleftrightarrow \quad (\exists K \in \mathcal{K} : q \in K \text{ and } A \cap K = \varnothing). \tag{4}$$

We also introduce the convention to write $p\mathcal{P}q$ to mean $\{p\}\mathcal{P}q$.

12.3. Examples of inferences. a) If we observe $A\mathcal{P}q$, we can infer $A'\mathcal{P}q$ for any set $A' \supseteq A$. This is easily derived from Equation (3) (see Problem 2). In fact, all the sets that could be eliminated from a positive response $A'\mathcal{P}q$ have already vanished in step [b] of the Outline 12.1 on the basis of the positive response $A\mathcal{P}q$. Indeed, this elimination concerns all the sets S containing q and having a nonempty intersection with A. Since $A \subseteq A'$ and $S \cap A \neq \varnothing$ implies $S \cap A' \neq \varnothing$, a positive response $A'\mathcal{P}q$ would not eliminate any new set, and can thus be omitted.

b) If the positive responses $p\mathcal{P}q$ and $q\mathcal{P}r$ have been observed, the positive response $p\mathcal{P}r$ can be inferred: the restriction of the relation $\mathcal{P}$ to pairs of items is transitive (see Problem 3). The corresponding query should not be asked, since the expected positive response would not lead to any new elimination of sets.

c) The transitivity of the relation $\mathcal{P}$ (restricted to pairs), also permits inferences to be drawn from negative responses. As in the previous example, we start with the observation $p\mathcal{P}q$, but then observe $p\overline{\mathcal{P}}r$. Should the routine ask whether failing q entails failing r? This query should not be asked since, by the argument in Example (b) above, a positive response would lead to $p\mathcal{P}r$, contradicting $p\overline{\mathcal{P}}r$.

d) Examples (b) and (c) describes cases concerning pairs of item, but can be generalized. Suppose that we have observed the positive responses $A\mathcal{P}p_1$, $A\mathcal{P}p_2$, ..., $A\mathcal{P}p_k$. Thus, failing all the items in the set A entails failing also $p_1, p_2, \ldots, p_k$. Suppose moreover that the expert has provided the positive response $\{p_1, p_2, \ldots, p_k\}\mathcal{P}q$. As $\mathcal{P}$ is an entailment, we may then infer $A\mathcal{P}q$ and omit the corresponding query.

The argument of the last example can be more easily stated in terms of entail relations.

12.4. Entail relations. We recall from Chapter 5 (cf. Theorem 5.7 and Definition 5.8) that to any entailment $\mathcal{P} \subseteq (2^Q \setminus \{\varnothing\}) \times Q$ corresponds a unique entail relation $\hat{\mathcal{P}}$ on $2^Q \setminus \{\varnothing\}$ with, for $A, B \in 2^Q \setminus \{\varnothing\}$:

$$A\hat{\mathcal{P}}B \iff (\forall b \in B : A\mathcal{P}b). \tag{5}$$

12.5. Warning. In the sequel, we shall drop the 'hat' specifying the entail relation $\hat{\mathcal{P}}$ associated with an entailment $\mathcal{P}$, and use the same notation for both relations. As the entail relation extends the entailment in an obvious way, this abuse of notation will never be a source of ambiguity. Note that in practice we shall always write $A\mathcal{P}q$ instead of $A\mathcal{P}\{q\}$.

The inference described in Example 12.3, written as

$$\text{whenever } A\mathcal{P}B \text{ and } B\mathcal{P}q, \text{ then } A\mathcal{P}q \tag{6}$$

directly follows from the transitivity of the entail relation $\mathcal{P}$. The special case: $p\mathcal{P}q$ and $q\mathcal{P}r$ imply $p\mathcal{P}r$, was the justification of the elimination carried out in Example 12.3(b).

We give one last example.

12.6. Example. Condition (6) is logically equivalent to

$$\text{whenever } A\mathcal{P}B \text{ and } A\overline{\mathcal{P}}q, \text{ then } B\overline{\mathcal{P}}q. \tag{7}$$

Consequently, whenever the expert has provided the positive responses coded as $A\mathcal{P}B$ and the negative response $A\overline{\mathcal{P}}q$, the negative response $B\overline{\mathcal{P}}q$ may be inferred, and the corresponding query should not be asked.

These five examples illustrate some of the inferences used in the refined QUERY routine. All the inferences are subsumed by the four rules given in Table 12.2, which can be derived from the transitivity of $\mathcal{P}$ and the implication

$$A\mathcal{P}b \text{ and } A \cup \{b\}\mathcal{P}C \text{ imply } A\mathcal{P}C.$$

The justification of these four rules are left to the reader (Problems 4 to 7).

	From	we can infer	when it has been established that
[IR1]	$A\mathcal{P}p$	$B\mathcal{P}q$	$(A \cup \{p\})\mathcal{P}q$ and $B\mathcal{P}A$
[IR2]	$A\mathcal{P}p$	$B\overline{\mathcal{P}}q$	$A\overline{\mathcal{P}}q$ and $(A \cup \{p\})\mathcal{P}B$
[IR3]	$A\mathcal{P}p$	$B\overline{\mathcal{P}}q$	$B\overline{\mathcal{P}}p$ and $(B \cup \{q\})\mathcal{P}A$
[IR4]	$A\overline{\mathcal{P}}p$	$B\overline{\mathcal{P}}q$	$(B \cup \{q\})\mathcal{P}p$ and $A\mathcal{P}B$

Table 12.2. The four rules of inference [IR1]-[IR4] permitting the deletion of redundant queries to the expert (see 12.6).

12.7. Generating the space from the entailment table. The procedure keeps track of the queries asked and responses given in the form of a subset-by-item table: for each set A and item q, it records whether $A\mathcal{P}q$ or $A\overline{\mathcal{P}}q$, or whether the answer is still unknown. The reason for maintaining such a table is that it contains all the information required for constructing the knowledge space. (Or more precisely, at any time, the current table can be used to generate the knowledge space containing all the states which are still feasible at that time.) In principle, this table is inconveniently large. As shown in the next paragraph, however, the full table is redundant and can be summarized very efficiently by a much smaller subtable. We first explain how the knowledge space can be generated from the table. For the time being, we suppose that the full, completed table is available.

Each row of the table is indexed by a subset $A \subseteq Q$. The columns correspond to the items. For each row q, the entry in the cell (A, q) contains either $A\mathcal{P}q$ or $A\overline{\mathcal{P}}q$. Let

$$A^+ = \{q \in Q \mid A\mathcal{P}q\} \tag{8}$$
$$A^- = \{q \in Q \mid A\overline{\mathcal{P}}q\}. \tag{9}$$

Thus, A^+ contains all the items that the subject would fail who is known to have failed all the items in A. We have necessarily

$$A \subseteq A^+, \qquad A^+ \cap A^- = \varnothing, \quad \text{and} \quad A^+ \cup A^- = Q.$$

It is easy to see that not only is the set A^- a feasible knowledge state, but **any** state K must be equal to some set A^- defined as in Equation (9) (Problem 8). Generating the knowledge space from the complete table is thus a simple matter. The full table contains $(2^m - 1) \times m$ entries and is obviously much too large for convenient storage. Fortunately, the information in the full table can be recovered from a much smaller subtable, by application of the inferences discussed above.

12.8. Constructing the minimal subtable. The subtable of inferences is organized into *blocks* generated successively. Block 1 contains all the responses of the expert to questions of the form "Does failing p entails failing q?" The information in Block 2 concerns the generic question "Does failing both p_1 and p_2 entails failing q?" In general, Block n is defined here by the number n of items in the antecedent set A involved in queries of type [Q1]: "Does failing all the items in the set A entails failing also item q?" This numbering of the blocks reflects the order in which the queries are asked to the expert by the QUERY routine. In other words, the queries in Block 1 are asked first, then come the queries in Block 2, and so on. This ordering of the queries seems sensible. The querying routine starts with the queries which are certainly the easiest ones for the expert to resolve: Block 1 only involves two items and a possible relationship between the two. Block 2 concerns somewhat more difficult queries, with three items and the relationship (or lack of) between two of these items, forming the antecedent set, and the third item. Things get gradually worse as the block number increases. However, the data collected in the early blocks yield inferences affecting later blocks, removing thus from the list of open queries some that are among the most difficult for the expert to answer. The impact of these inferences may be dramatic. For example, in the application described in this chapter, which involves 50 items, the final knowledge space of each of the five expert was obtained in less than 6 blocks. Moreover, most of the construction was already accomplished after 3 blocks (see Table 12.3).

Not only are the queries asked by blocks, but a new block of queries is generated only after the previous blocks have been completed. The information collected in these previous blocks is then used to decide which queries must be included in the new block. Typically, Block n will consist of only a fraction of the total number of queries with antecedent set of size n. This is of course not the case for Block 1; in the absence of any information we must consider here all queries containing just one item. Then we ask the queries pertaining to this block, drawing inferences along the way. Note, however, that the inference rules can only be applied to the queries appearing in this block, for the simple reason that these are the only queries that have been generated thus far. After the first block is finished, the second block of queries is constructed.

Block 2 consists of 2-item antecedent queries. As indicated earlier, which 2-item antecedent queries are included in this block depends upon the responses collected in Block 1. In other words, before asking queries from Block 2, the routine implements the inferences having two antecedents that can be drawn from the information collected in Block 1. It is only when this preliminary pruning of Block 2 is terminated that QUERY starts asking the remaining open queries from Block 2, again drawing inferences with respect to other subsets in this second block. The cycle repeats: Block 3 is constructed on the basis of Blocks 1 and 2, that is, inferences obtainable from the previous blocks are 'transported' to the new block, after which the remaining open queries are asked, and so on. The process terminates when the newly constructed block does not contain any subsets at all. This indicates that all information has been collected in the subtable constructed so far. In particular, the knowledge space that we wanted to uncover is given by the collection of all sets $A^- = \{p \in Q \mid A\overline{\mathcal{P}}p\}$, where A runs through the rows of this subtable. We have not discussed here how the information in the previous blocks determines which subsets will appear in the new block, or how the postponed inferences for the new block are found. All these specifics are provided in Koppen (1993). Figure 12.1 displays a flowchart representing the overall design of the algorithm outlined above.

12.9. Choosing the next query. As we have seen in the preceding subsection, the order in which the queries are asked depends to a considerable degree on the block structure of the procedure. Within a block, however, any of the remaining open queries can be chosen, and we can take advantage of this freedom. In particular, it makes sense to select the queries so as to minimize the number of further queries to the expert. Conceivably, this may be achieved by maximizing (in some sense) the total number of inferences that can be made from the response to the queries. Pursuing this objective is far from straightforward, however.

First, since it is difficult to look more than one step ahead, the choice of a question by the QUERY routine is only guided by the number of inferences

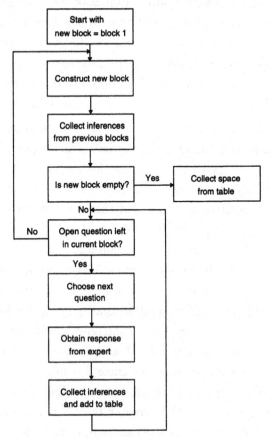

Fig. 12.1. Overall design of the QUERY algorithm.

that would be obtained if the query was asked at that moment. Second, because the inference rules are only applied in the current block (see 12.7), "number of inferences" in the previous sentence is to be interpreted with this qualification. Third, even in this restricted sense, we do not know how many inferences a query will yield, because we do not know what response the expert will give. The inferences following a positive response will differ from those after a negative response. In the sequel, we restrict consideration to the cases in which the choice of a query will depend upon the two classes of inferences associated with the positive or negative responses to the query, and more specifically, the numbers of inferences in each of the two classes.

There are various ways in which these two numbers may be used to determine which query is best. For example, one may try to maximize the expected gain. In the absence of any information we may assume that the expert is equally likely to give a positive or a negative response. Accordingly, the expected number of inferences associated with a query is given by the mean of the numbers of inferences in the two classes. This means (equivalently) that the sum of the two numbers becomes the criterion: we choose a query for which this sum is maximal over all remaining open queries.

In the application presented in the next section, another selection rule was adopted, based on a "maximin" criterion. The purpose is not so much to maximize the direct gain as to minimize the possible cost of the "wrong" (in terms of efficiency) response to the query. For each query, the number of inferences for a positive response and that for a negative response are computed. First, we only consider the minimum of these two numbers: query q_1 is chosen over query q_2 if this minimum number for q_1 is higher than that for q_2. Only when the minimum for the two queries is the same do we look at the other number and we choose the query for which this other number is higher. For instance, suppose that query q_1 yields 3 inferences in the case of a positive response and 2 inferences for a negative, while for queries q_2 and q_3 these numbers are 1 and 7, and 2 and 4, respectively. Then query q_1 is chosen over query q_2 since $2 = \min\{3,2\}$ is greater than $1 = \min\{1,7\}$, but query q_3 is chosen over q_1 since they have the same minimum 2, but $4 = \max\{2,4\}$ is greater than $3 = \max\{3,2\}$. In short, we select a query with the best worse case and, from among these, one with the best better case.

It is not always feasible to apply this selection process to all open queries. If there are too many open queries, we just pick a pseudorandom sample and choose the best query from this sample. We suspect that we do not lose much this way, since the range of the numbers of inferences from one query is rather restricted. Therefore, if there are many open queries, there will be many that are "best" or approximately so and the important thing is to avoid a particularly poor choice.

Kambouri's Experiment

The QUERY routine was used to construct the knowledge spaces of five subjects: four expert-teachers of high school mathematics, and the experimenter, Maria Kambouri (whose name is abbreviated M.K. in the sequel).

In this section we describe the specific domain chosen to apply this theory, the subjects employed for this task, the procedure used, and the results obtained from this application of the algorithm. (For a more detailed presentation, see Kambouri, 1991.)

12.10. Domain. Our working domain is within high school mathematics. As items, we have used standard problems in arithmetic, algebra and geometry from the 9th, 10th and 11th grade curriculum of New York State high schools. Specifically, the list of items was built around the Regents Competency Test in Mathematics (RCT). Passing this examination is a minimum requirement for graduation from New York State high schools and it is usually given at the end of the 9th grade. Statistics from the New York City Board of Education show that almost 70% of the students pass the test the first time around; those who fail are allowed to retake it a few times (sometimes up to the 12th grade).

At the time of the study, the full test contained a total of sixty items and was divided into two parts. The first one consisted of twenty completion (open-ended) items, while the second was made of forty multiple choice items for which the student had to select one answer from four alternatives. To pass, students were required to provide correct solutions to at least 39 problems. There was no time limit for this test. Most students returned their copy after less than three hours. For our purposes, only the first part of the test, containing the open-ended items, was of interest. The twenty open-ended problems of the June 1987 Math RCT provided the core of the material on which experts were to be questioned by the QUERY procedure. In order to span a broad range of difficulty, this set of problems was extended to 50 by adding 10 simple problems of arithmetic and 20 more complex problems of first-year algebra. The 30 additional items were reviewed by one of the experts, who had extensive experience tutoring students for this type of material. A sample of the 50 items is contained in the next paragraph. The topics covered by the final set of 50 problems included: addition, subtraction, multiplication and division of integers, decimals and ratios; percentage word problems; evaluating expressions; operations with signed numbers; elementary geometry and simple graphs; radicals; absolute values; monomials; systems of linear equations; operations with exponents and quadratic equations (by factorization).

12.11. Some sample items. Two examples among the ten easy items:

(1) *Add:* $34 + 21 =$?

(2) *What is the product when 3 is multiplied by 5?*

Three of the 20 open items from the June 1987 RCT:

(3) *Write the numeral for twelve thousand thirty seven.*

(4) *In the triangle ABC, the measure of angle A is 30 degrees and the measure of angle B is 50 degrees. What is the number of degrees in the measure of angle C?*

(5) *Add:* $546 + 1248 + 26 =$?

Three of the 20 more difficult items:

(7) *Write an equation of the line which passes through* $(1,0); (3,6)$.

(8) *Solve for* x:

$$\frac{2x+1}{5} + \frac{3x-7}{2} = 7.$$

(9) *What is the product of* $6x^3$ *and* $12x^4$?

12.12. The experts. Three highly qualified and experienced teachers and one graduate student who had had extensive practice in tutoring mathematics were selected as experts. The experimenter was included as a fifth expert. All five experts were accustomed to interacting with students in one-to-one settings. Moreover, each expert had taught students coming from widely different populations, ranging from gifted children to students with learning disabilities. At the time of this study, the three teachers were working in the New York City school system. A more detailed description of each of the five subjects follows. The initials of all subjects but the experimenter have been changed to protect their identity.

(A.A.) This subject had an M.Sc. in Biology and an M.A. in Education (Teachers College), both from Columbia University. He had taught in a public high school in Harlem for a total of 9 years: 2 in mathematics and the rest in science. He had taught a remedial RCT class (students of grades 9-12) and a Fundamental Mathematics class (9th grade), as a preparation for the June 1987 RCT exam. A.A. also taught summer schools and had been on the RCT grading committee for the three summers preceding this study.

(**B.B**) This subject had a B.A. in Psychology and Education from Brooklyn College and an M.A. in Special Education for the emotionally handicapped from Teachers College of Columbia University. She also held a Post-Graduate Certificate on Education, Administration and Supervision. She had over ten years of experience in teaching both mathematics and reading to students with different kinds of learning problems including the learning disabled and those with autistic tendencies. At the time of this study, she was also working as a teacher-trainer in special education and a consultant (Staff Development) for the N.Y.C. Board of Education. Her work involved tutoring students (up to 18 years old) in mathematics. B.B. helped teachers set up diagnostic tests and plan the curriculum at the beginning of the school year.

(**C.C.**) This expert held a Bachelor's degree in Political Science (Russian) from Barnard School at Columbia University and an M.A. in School Psychology from N.Y.U. She had 15-years of experience tutoring students (mainly high school) in mathematics: algebra, geometry, trigonometry, precalculus, as well as other topics. In addition, C.C. had been a consultant in various pediatric psychological centers, N.Y.C schools and maintained a private practice, working with learning disabled individuals of all ages on general organizational skills and study techniques.

The other two subjects who served as experts, a graduate student and the experimenter, originated from different educational systems. However, they had studied (and had tutored) all the topics covered by the 50 items chosen for this project.

(**D.D.**) This subject held a B.Sc. in Pure Mathematics from Odessa University (former USSR). At the time of the experiment, she was a student in the master's program at the Courant Institute of Mathematical Sciences, New York University. D.D. has taught a class of 25 gifted children aged 13 to 15 for three years, in such topics as calculus, number theory and logic. She also had experience tutoring high-school students who needed additional help in their Geometry and Algebra classes.

(**M.K.**) The experimenter held a B.Sc. and a M.Sc. in Statistics from the University of London. She had some experience in tutoring high school Mathematics and had been a teaching assistant for undergraduate and graduate statistics and probability courses.

The three teachers received payment at an hourly rate.

12.13. Method. The experts were asked to respond to a series of questions generated by a computerized version of the QUERY routine. The QUERY routine was programmed in `Pascal` by Koppen, who is one of the authors of the source paper (Kambouri et al. 1993). The user-interface was written in `C` using the *curses* screen optimization library (Arnold, 1986) by another author (Villano).

Each step began with the presentation at a computer terminal of a query of the form [Q1]. The displayed question was accompanied by a request for the expert's response on the same screen. After the response had been entered at the keyboard, the program computed the query for the next step. Each step began with the display at a computer terminal of a query of type [Q1]. The displayed query was accompanied by a request for the expert's response on the same screen. After the response had been entered at the keyboard, the program computed the query for the next step. An example of a typical screen is shown in Figure 12.2.

The top part of the screen displays the problem(s) that a hypothetical student has failed to solve (in this case, the four problems a to d). Four problems can be displayed simultaneously in this section of the display. If the number of these problems exceeds four[1], the expert can toggle the display of the remaining items by pressing a key. In the middle of the screen, the text of the query regarding the antecedent items is displayed. This is followed by the new problem (e). At the bottom of the screen, the remaining text of the query appears, along with the prompt for the expert's response, labeled "Rating:". The question for the expert in Figure 12.2 is whether the information about failing the four items a to d is compelling the conclusion that the student would also fail the new item e. Based on the expert's answer, the routine computes the next query to ask and, after a short delay (typically on the order of a few seconds), presents the expert with the next screen. Especially after the third block, the delay between blocks was much longer due to the large amount of computation necessary to prepare the open queries for the next block.

---

[1] We shall see that no expert required more than 5 blocks to complete the task.

---

a.   378   b. $58.7 \times 0.97 = ?$

            $\times$ 605

             ?

---

c. $\frac{1}{2} \times \frac{5}{6} = ?$ d.   What is 30% of 34?

Suppose that a student under examination has just provided
a wrong response to problem(s) a, b, c, d

---

e.   Gwendolyn is $\frac{3}{4}$ as old as Rebecca
Rebecca is $\frac{2}{5}$ as old as Edwin.
Edwin is 20 years old.
How old is Gwendolyn?

Is is practically certain that this student will also fail
Problem e?   Rating:

Fig. 12.2. A typical screen for the QUERY routine.

Although the QUERY routine consists in asking the experts the queries
of the type [Q1], which involve a dichotomous (yes-no) decision, a rating
scale was used to ensure that the 'yes' responses to be used by the procedure
corresponded to a practical certainty on the part of the experts. Therefore,
an unequally spaced 3-point scale was adopted with the numbers 1, 4 and 5.
A '5' represented a firm 'yes' answer to an instance of [Q1], a '4' indicated
less certainty, and a '1' represented a resounding 'no.' Only the '5' response
was interpreted by the program as a 'yes' (positive) answer to a query of
type [Q1]. Both '4' and '1' were converted to a 'no' (negative). The points
2 and 3 were not accepted by the program as valid responses.

Prior to the start of the experiment, the experts were given the text of the
50 items to review at home. Before they began the task, they received a set
of written instructions which included a short explanation of the purpose of
this experiment. Next, the list of the 50 items was displayed to acquaint the
experts with the appearance of the items on the screen. The procedure was

then explained and exemplified. A short, on-line training session followed, consisting of 10 sample steps intended to familiarize the subjects with the task, and in particular with the use of the 3 point scale. After each example, the expert received feedback concerning the interpretation of the rating scale. This introduction to the task lasted approximately thirty minutes. Before starting the main phase of the experiment, the experts were given time to ask queries about the procedure and discuss any difficulties they might have in answering the queries with a rating.

The experts determined the number of steps for a given session. They were advised to interrupt the experiment whenever they felt that their concentration was decreasing. At the beginning of the next session, the program would return them to the same query from which they had exited. At any step, the experts were given the option to go back one query and reconsider their response. They also had the choice to skip a query that they found particularly hard to answer. A skipped query could either come back at a later step, or be eliminated by inference from some other query. Depending on the expert, the task took from 11 to 22 hours (not including breaks).

Results

12.14. Number of queries asked. Table 12.3 displays the number of queries effectively asked by the procedure. The first column shows the block number, followed by the theoretical maximum number of queries in each block, which would been asked if indeed all the 2^{50} subsets are states. The remaining columns contain, for each expert, the actual number of queries they were asked. Compared to the theoretical maximum, the reduction is spectacular.

One of the experts (B.B.) did not complete the procedure; as shown in Table 12.3, the number of queries asked in the third block was still increasing (the other experts' results show a gradual decrease after the second block). Furthermore, we observed that the fourth block of queries to be answered by expert B.B. was even larger than the third. It was therefore decided to interrupt the routine at that point. Of the remaining experts, one (D.D.) finished after 4 blocks, and the other three (A.A., C.C. and M.K.) after 5 blocks.

Block	Number of queries asked				
	A.A.	B.B.	C.C.	D.D.	M.K.
1. $\binom{50}{1} \times 49 = 2{,}450$	932	675	726	664	655
2. $\binom{50}{2} \times 48 = 58{,}800$	992	1,189	826	405	386
3. $\binom{50}{3} \times 47 = 921{,}200$	260	1,315	666	162	236
4. $\binom{50}{4} \times 46 \approx 10^7$	24	-	165	19	38
5. $\binom{50}{5} \times 45 \approx 10^8$	5	-	29	0	2

Table 12.3. Theoretical maxima of the number of queries to be asked through the QUERY procedure for 50 items (these maxima obtain when all subsets are states), and the actual number of queries asked, for each block and each of the 5 experts. The symbol "-" in a cell indicates that the procedure was interrupted.

We noted in Remark 12.2(1) that, at each intermediate step, the current list of states constitutes a space. The algorithm described in the previous section can be adapted to construct such intermediate spaces. Conceptually, this is achieved by replacing, at the point of interest, the real expert with a fictitious one who only provides negative responses. That is, from this point on we let the procedure run with automatic negative responses to all open queries. This way, we can construct the knowledge spaces after each block (including the one for the interrupted expert B.B. after Block 3, representing the final data in her case).

12.15. Number of states. Table 12.4 presents the gradual reduction of the number of states in each block. At the outset, all 2^{50} ($\approx 10^{15}$) subsets of the 50 items set are considered as potential states. For experts A.A. and B.B. the number of states after the first block was over 100,000. At the end, the number of remaining states ranges from 881 to 3,093 (7,932 for the unfinished space of B.B.). This amounts to less than one billionth of one percent of the 2^{50} potential states considered initially. It is remarkable that the reduction after Block 3 is minimal for the four experts having completed the task.

12.16. Comparisons of the data across experts. Despite a generally good overall agreement between the experts, there are significant discrep-

Block	Number of states per expert				
	A.A	B.B.	C.C.	D.D.	M.K.
1	> 100,000	> 100,000	93,275	7,828	2,445
2	3,298	15,316	9,645	1,434	1,103
3	1,800	7,932	3,392	1,067	905
4	1,788	-	3,132	1,058	881
5	1,788	-	3,093	1,058	881

Table 12.4. Number of knowledge states at the end of each block. The initial number is 2^{50} for each expert. (The symbol "-" in a cell indicates that the procedure was interrupted.)

ancies concerning the details of their performances. In particular, we shall see that the five final knowledge spaces differ substantially.

We first examine the correlation between the ratings. The queries asked by the QUERY procedure depend on the responses previously given by the expert. In Block 1, however, the number of common queries asked for any two experts was sufficient to provide reliable estimates of the correlation between the ratings. For two particular experts, such a correlation is computed from a 3×3 contingency table containing the observed responses to the queries that the experts had in common. The correlation was estimated, for each pair of experts, using the polychoric coefficient. (See Tallis, 1962; Drasgow, 1986. This correlation coefficient is a measure of bivariate association which is appropriate when ordinal scales are used for both variables.) The values obtained for this coefficient ranged between .53 and .63, which are disappointingly low values hinting at a possible lack of reliability or validity of the experts.

These results only concerned the common queries that any two experts were asked. Obviously, the same kind of correlation can also be computed on the basis of the inferred responses, provided that we only consider the two categories 'Yes-No' used internally by the QUERY routine. In fact, the tetrachoric coefficient was computed by Kambouri for all the responses— manifest or inferred—to all $2450 (= 50 \cdot 49)$ theoretically possible queries in Block 1. In this case, the data for each of the 10 pairs of experts form a 2×2 contingency table in which the total of all 4 entries is equal to 2450. The entries correspond to the 4 cases: both experts responded positively (YY);

one expert responded positively, and the other negatively (2 cases, YN and NY); both experts responded negatively (NN). As indicated in Table 12.5, the values of the tetrachoric coefficient are then higher, revealing a better agreement between the experts. We note in passing that the number of NN pairs far exceeds the number of YY pairs. A typical example is offered by the contingency matrix of Experts A.A. and B.B. We find 523 YY pairs, 70 YN pairs, 364 NY pairs, and 1493 NN pairs. (There were thus 1493 queries—out of 2450—to which both Experts A.A. and B.B. responded negatively.)

	A	B	C	D	K
A	-	.62	.61	.67	.67
B	-	-	.67	.74	.73
C	-	-	-	.72	.73
D	-	-	-	-	.79

Table 12.5. Values of the tetrachoric coefficient between the experts' ratings for all the responses (manifest and inferred) from the full Block 1 data. Each correlation is based on a 2×2 table, with a total cell count of 2450.

12.17. Descriptive Analysis by Item. Experienced teachers should be expected to be good judges of the relative difficulty of the items. The individual knowledge spaces contain an implicit evaluation of item difficulty. Consider some item q and the knowledge space of a particular expert. This item is contained in a number of states. Suppose that the smallest of these states contains k items. This means that at least $k - 1$ items must be mastered before mastering q. The number $k - 1$ constitutes a reasonable index of the difficulty of the item q, as reflected by the knowledge space of the expert. In general, we shall call the *height* of an item q the number $h(q) = k - 1$, where k is the number of items in a minimal state containing the item q. Thus, a height of zero for an item means that there is a state containing just that item.

Kambouri checked whether the experts generally agreed in their assessment of the difficulty of the items, evaluated by their heights. The height of each of the 50 items was obtained from the knowledge spaces of the five experts. The correlation between these heights was then computed, for each pair of experts. The results are contained in Table 12.6.

	A	B	C	D	K
A	-	.81	.79	.86	.81
B	-	-	.80	.85	.87
C	-	-	-	.83	.77
D	-	-	-	-	.86

Table 12.6. Correlations (Pearson) between the heights of the 50 items computed for each of the 5 experts.

The correlations between the indices of items in Table 12.6 are remarkably high. It may then come as a surprise that a comparison between the knowledge spaces exposes sharp disparities.

12.18. Comparison of the knowledge spaces. As indicated in Table 12.4, the number of states differs substantially across the 5 constructed knowledge spaces. In itself, however, this is not necessarily a sign of strong disagreement. For instance, it might happen that all the states recognized by one expert are also states in another expert's space. Unfortunately, the picture is not that simple.

The source paper contains the distribution of the size of the states in the different knowledge spaces. For each space, the number of knowledge states containing k items was computed, for $k = 0, \ldots, 50$. Two exemplary histograms of these distributions, concerning the two experts C.C. and A.A. are displayed in Figure 12.3. Notice that the histogram of A.A. is bimodal. This is the case for 4 of the 5 experts. No explanation was given for this fact which, in any event, points out a noticeable difference between one expert and the others. The source paper also contains a comparison of the knowledge spaces of the five experts based on a computation of a 'discrepancy index.' The idea is to use the distance $d(A, B) = |A \triangle B|$ between sets A and B introduced in 0.23.

Consider two arbitrary knowledge spaces $\mathcal{K}$ and $\mathcal{K}'$. If these two knowledge spaces resemble each other, then, for any knowledge state K in $\mathcal{K}$, there should be some state K' in $\mathcal{K}'$ which is either identical to K or does not differ much from it; that is, a state K' such that $d(K, K')$ is small.

This suggests, for any state K in $\mathcal{K}$, to compute for all states K' in $\mathcal{K}'$, the distance $d(K, K')$, and then to take the minimum of all such distances.

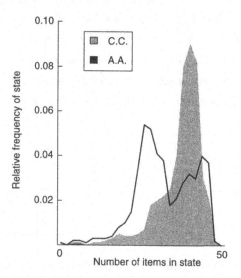

Fig. 12.3. Histograms of the relative frequencies of states containing a given number of elements, for the two experts C.C. and A.A.

In set theory, the resulting minimum distance is sometimes referred to as the *distance between* K *and* K' and is then denoted by $d(K, K')$. As an illustration, take the two knowledge spaces

$$\mathcal{K} = \{\,\varnothing, \{c\}, \{d\}, \{c,d\}, \{b,c,d\}, \{a,b,c,d\}\,\},$$
$$\mathcal{K}' = \{\,\varnothing, \{a\}, \{a,b\}, \{a,b,c\}, \{a,b,c,d\}\,\}. \tag{9}$$

on the same domain $\{a,b,c,d\}$. For the state $\{c\}$ of $\mathcal{K}$, we have

$$d(\{c\}, \mathcal{K}') = \min\{d(\{c\}, \varnothing), \; d(\{c\}, \{a\}), \; d(\{c\}, \{a,b\}),$$
$$d(\{c\}, \{a,b,c\}), \; d(\{c\}, \{a,b,c,d\})\}$$
$$= \min\{1, 2, 3\} = 1.$$

Performing this computation for all states of $\mathcal{K}$, we obtain a frequency distribution of these minimum distances. (We have one such distance for each state of $\mathcal{K}$.) Thus, this frequency distribution concerns the number $f_{\mathcal{K},\mathcal{K}'}(n)$ of states of $\mathcal{K}$ lying at a minimum distance n, with $n = 0, 1, \ldots,$ to any state of $\mathcal{K}'$. For two identical knowledge spaces, this frequency distribution is concentrated at the point 0; that is, $f_{\mathcal{K},\mathcal{K}}(0) = |\mathcal{K}|$. In general, for two knowledge structures $\mathcal{K}$, $\mathcal{K}'$ on the same domain Q, the distance between a state of $\mathcal{K}$ and a state of $\mathcal{K}'$ is at most one half the number of items. This means that we have $f_{\mathcal{K},\mathcal{K}'}(n) > 0$ only if $0 \leq n \leq h(Q) = \lfloor \frac{1}{2}|Q| \rfloor$ (where $\lfloor r \rfloor$ is the largest integer smaller or equal to the number r). Two examples of such frequency distributions are given in Figure 12.4 for the two knowledge spaces of Equation (9). We see that three of the states of $\mathcal{K}$ lie at a minimum distance of 1 to any state of $\mathcal{K}'$, namely $\{d\}$, $\{c\}$ and $\{d,c,b\}$.

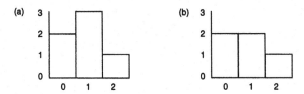

Fig. 12.4. Two frequency distributions of the distances: (a) from $\mathcal{K}$ to $\mathcal{K}'$; and (b) from $\mathcal{K}'$ to $\mathcal{K}$ (cf. Equation (9)).

Note that the frequency distribution $f_{\mathcal{K},\mathcal{K}'}$ of the minimum distances from $\mathcal{K}$ to $\mathcal{K}'$ in Figure 12.4(a) is distinct from that from $\mathcal{K}'$ to $\mathcal{K}$ in Figure 12.4(b), which is denoted by $f_{\mathcal{K}',\mathcal{K}}$: the latter concerns the minimum distances from states of $\mathcal{K}'$ to states of $\mathcal{K}$.

Such frequency distributions were computed for all 20 pairs of spaces. For comparison purposes, it is convenient to carry out a normalization, and to convert all the frequencies into relative frequencies. In the case of $f_{\mathcal{K},\mathcal{K}'}$ this involves dividing all the frequencies by the number of states of $\mathcal{K}$. This type of distribution of relative frequencies will be referred to as the *discrepancy distribution* from the knowledge space $\mathcal{K}$ to the knowledge space $\mathcal{K}'$.

A *discrepancy index* from $\mathcal{K}$ to $\mathcal{K}'$ is obtained by computing the mean

$$di(\mathcal{K},\mathcal{K}') = \frac{1}{|\mathcal{K}|} \sum_{n=0}^{h(Q)} n f_{\mathcal{K},\mathcal{K}'}(n)$$

of the discrepancy distribution from $\mathcal{K}$ to $\mathcal{K}'$, where Q is the common domain of $\mathcal{K}$ and $\mathcal{K}'$.

The standard deviations of such discrepancy distributions are also informative. The first five columns of Table 12.7 contain the computed means and standard deviations (in parentheses) of all 20 discrepancy distributions.

	A	B	C	D	K	[K]
A	-	3.0 (1.3)	4.8 (2.0)	3.4 (1.4)	4.3 (1.6)	13.4 (5.5)
B	4.3 (1.6)	-	4.6 (1.4)	4.3 (1.6)	4.3 (1.6)	15.9 (3.8)
C	4.1 (1.8)	4.0 (1.4)	-	5.3 (1.5)	5.6 (1.7)	10.8 (4.4)
D	3.2 (1.3)	2.6 (1.2)	4.7 (1.6)	-	4.0 (1.7)	13.2 (5.6)
K	3.5 (1.5)	2.4 (1.1)	4.7 (1.8)	3.6 (1.7)	-	13.2 (6.2)

Table 12.7. Means (and standard deviations) of the discrepancy distributions for all pairs of knowledge spaces. The entry 4.3 (1.6) in the second row of the first column of the table refers to the mean and the standard deviation of the discrepancy distribution from the space of expert B.B. to the space of expert A.A. The last column contains the means and standard deviations of the discrepancy distributions from the knowledge spaces of each expert to the 'random' knowledge space [K].

(We recall that we have two discrepancy distributions for each pair of experts.)

These means appear to be rather high. As a baseline for evaluating these results, Kambouri (1991) also computed the discrepancy distributions from the knowledge space of each expert to a 'random' knowledge space, the construction of which is explained below. It was sensible to take, for such comparison purposes, a knowledge space with the same structure as that of one of the experts. The knowledge space of M.K. was selected, but all 50 items were arbitrarily relabeled. To minimize the risk of choosing some atypical relabeling, 100 permutations on the domain were randomly selected. The discrepancy distribution from the knowledge spaces of each of expert to each of these 100 random knowledge spaces was computed. Then, the average relative frequencies were computed. In other words, a mixture of the resulting 100 discrepancy distribution was formed. The numbers in the last column of Table 12.7 are the means and the standard deviations of these mixture distributions. These means are considerably higher that those appearing in the first five columns. (Note that, since a knowledge structure contains at least the empty set and the domain, the distance between any state of one knowledge space, to some other knowledge space is at most 25, half the number of items.)

Discussion of Kambouri's Results

On the basis of the data collected from the five experts, it must be concluded that the QUERY routine has proved to be applicable in a realistic setting. This was far from obvious a priori, in view of the enormous number of queries that had to be responded to. A close examination of the data reveals mixed results. On the positive side, there is a good agreement between experts concerning gross aspects of the results. In particular:

(i) the sizes of knowledge spaces have the same order of magnitude (a few thousand states—from around 900 to around 8,000—for the 50 items considered, see Table 12.4);

(ii) there is a good consistency across experts concerning the rating responses given for the same queries asked by the routine (Table 12.5);

(iii) high correlations between experts are also obtained for the difficulty of the items, as evaluated by their heights in the various spaces (Table 12.6).

Nevertheless, the discrepancy distributions reveal considerable differences between the knowledge spaces constructed by the QUERY routine for the 5 experts. For example, we see from Table 12.7 that most of the means of the discrepancy distributions exceed 4. (In other words, a state in one knowledge space differs, on average, by at least 4 items from the closest state in some other knowledge space.)

A sensible interpretation of these results is that there are considerable individual differences between experts concerning either their personal knowledge spaces or at least their ability to perform the task proposed to them by the QUERY routine, or both of these factors. It must be realized that this task is intellectually quite demanding. In the context of the QUERY routine, the notion of "expertise" has, in fact, two components. First, the expert has to be very familiar with the domain and the chosen population of students, so as to be (at least implicitly) aware of which knowledge states may appear in practice. Second, the expert must also be able to transmit this knowledge structure faithfully through her answers to the queries of the form [Q1]. Experts may differ on either of these components.

Assuming that some carefully selected expert is subjected to the QUERY routine, the questioning would not result in the correct knowledge space if the responses to the questions asked by QUERY do not, for some reason or other, faithfully stem from that subject's awareness of the feasible states. For example, it is conceivable that, when a query is experienced as too cognitively demanding, a tired subject may resort to some kind of shortcut strategy. An example given by Kambouri et al. (1993) involves the query displayed in Figure 12.2. An expert confronted with that query must examine the items a, b, c, d and e, and decide whether failing a to d would imply a failure on e. Rather than relying on the exact content of each of a to d the expert could just scan these items and arrive, somehow, at an estimate of some overall "difficulty level" for the set of items failed. Similarly, instead of looking at the precise content of the other item e, the expert might collapse the information into some "difficulty level" and then respond on the basis of a comparison of the two difficulty levels.

A variant of this strategy mentioned by Kambouri et al. (1993) may also play a role in which an expert would simply rely on the number of items failed. Referring again to the situation of Figure 12.2, an expert may be led to decide that a fifth item would also be failed, irrespective of its content. This tendency may be reinforced by the fact that in the QUERY routine this question is only asked when the expert has, before, given negative responses (directly or through inferences) to all queries involving e with a strict subset of the items a to d. So, the expert might read this query as a repetition, with an implied request to "finally say yes." (Notice, in this respect, that, with the one exception, all experts finished within 5 blocks, so the situation in Figure 12.2 is about as bad as it gets.)

These examples illustrate how experts equally well informed about the domain and the chosen population of students can nevertheless produce different knowledge structures by giving invalid responses to some of the questions posed by the QUERY routine. This phenomenon may explain some of the differences between experts that were observed in the experiment described in this chapter. This raises the question whether, in further use of this routine, anything can be done to reduce these effects. An answer to this question can go into one of two directions: (i) try to detect and correct such invalid responses, and (ii) try to avoid them as much as possible. Kambouri et al. (1993) discuss both of these possibilities in detail. We only outline the main ideas here.

A practical way of detecting at least some invalid responses is to postpone the implementation of the inferences associated with a response until either a confirmation or a contradiction of that response arises from a new

response. Only the confirmed responses would be regarded as valid, and
their inferences implemented. Although this may not necessarily detect all
invalid responses (a confirmation of an erroneous response may still occur),
it should certainly decrease the frequency of those erroneous responses due
to the unreliability of the subject. As for avoiding invalid responses, we
noticed earlier that the questions asked were not all of equal difficulty. The
questions from Block 1 are certainly easier to answer than those from Block
5, say. We may be able to avoid some invalid responses by limiting the
application of the QUERY routine to Block 1. We would end up with a
knowledge space that might be much larger than the real one, but that
could contain more valid states, possibly most of the states of the target
structure. This large knowledge structure could then be reduced from student data, using a technique from Villano (1991) based on the elimination
of states occurring infrequently in practice.

The rest of this chapter is devoted to the work of Cosyn and Thiéry (1999)
who investigate these ideas, and show that they lead to a feasible overall
procedure.

Cosyn and Thiéry's Work

Cosyn and Thiéry begin where Kambouri et al. (1993) left off. They design
a procedure based on the improvements of QUERY discussed in the last
section.

12.19. The PS-QUERY, or Pending-Status-QUERY. This routine is
a modification of QUERY in which a 'pending status' is conferred to any
response provided by the expert to a new question. The key mechanisms
involves two buffers in which the positive and negative inferences arising
from any new response are temporarily stored, pending confirmation or
disconfirmation. We only give an outline of the algorithm (for details, see
Cosyn and Thiéry, 1999). Some notation are needed.

We write $(S_1, q_1), \dots, (S_k, q_k), \dots$ for the sequence of questions proposed
to the expert. Thus, on step k of the procedure, $S_k \subseteq Q$ is the antecedent
set, $q_k \in Q$ is an item, and the expert is asked whether failing all the items
in S_k entails also failing q_k. We denote by $\mathcal{P}_{k-1}$ and $\overline{\mathcal{P}}_{k-1}$ the entailment
relation and its negation (in the sense of Equation (2)), respectively, before

asking question (S_k, q_k) to the expert. We have thus by definition

$$\mathcal{P}_0 = \varnothing, \quad \text{and} \quad (S_k, q_k) \notin \mathcal{P}_{k-1} \cup \overline{\mathcal{P}}_{k-1} \quad \text{for } k = 1, 2, \dots$$

We also denote by P_k and $\overline{P}_k$ the sets of fresh positive and negative infer-ences, respectively, drawn from $\mathcal{P}_{k-1}$, $\overline{\mathcal{P}}_{k-1}$ and the expert's response to question (S_k, q_k)[2]. Finally, P^B_{k-1} and $\overline{P}^B_{k-1}$ stand for the states of the two buffers containing the pending positive and negative inferences; that is, the inferences from all steps up to step $k - 1$ inclusively that have neither been confirmed nor disconfirmed. A contradiction is detected by PS-QUERY ei-ther when a new positive inference from P_k is found to be already in the buffer $\overline{P}^B_{k-1}$ of pending negative inferences, or if a negative inference from $\overline{P}_k$ is found to be in the buffer P^B_{k-1} of pending positive inferences. The contradictory piece of evidence is removed from the relevant buffer, the ex-pert's response is discarded and we have $\mathcal{P}_k = \mathcal{P}_{k-1}$, $\overline{\mathcal{P}}_k = \overline{\mathcal{P}}_{k-1}$. If no contradiction is detected, PS-QUERY looks for possible confirmations, that is, inferences which are either in $P_k \cap P^B_{k-1}$ or in $\overline{P}_{k-1} \cap \overline{P}^B_{k-1}$. Those con-firmed inferences, if any, are added to $\mathcal{P}_{k-1}$ or $\overline{\mathcal{P}}_{k-1}$ accordingly. They are also removed from the corresponding buffers of pending inferences. More-over, for each confirmed pair in $P_k \cap P^B_{k-1}$ or in $\overline{P}_{k-1} \cap \overline{P}^B_{k-1}$, PS-QUERY computes all the inferences from that pair and those in $\mathcal{P}_{k-1}$ and $\overline{\mathcal{P}}_{k-1}$. The two buffers are then updated, and $\mathcal{P}_k$ and $\overline{\mathcal{P}}_k$ are computed.

12.20. Simulation of PS-QUERY. In their paper, Cosyn and Thiéry compare the performance of QUERY and PS-QUERY by simulation. They use as the target knowledge structure one that had been constructed, using QUERY, by a real human expert. The domain is a set of 50 items covering the arithmetic curriculum from grade 4 to grade 8. We denote by $\mathcal{K}^r$ this *reference structure*, and by $\mathcal{K}^{r,1}$ the superset of $\mathcal{K}^r$ obtained from the data of Block 1 of the real expert. Thus, $\mathcal{K}^{r,1}$ is a quasi ordinal knowledge space (cf. 1.47), which in the sequel is called the *reference order*. Note that $\mathcal{K}^r$ and $\mathcal{K}^{r,1}$ contain 3043 and 14346 states, respectively. The simulated expert was assumed to use $\mathcal{K}^r$ to provide the response to the questions asked by PS-QUERY.

---

[2] Notice that P_k contains the positive response to question (S_k, q_k), if any. Similarly, $\overline{P}_k$ contain the negative response to that question, if any.

A real or simulated expert may commit two kinds or error in responding to the questions asked by QUERY or PS-QUERY: (1) the *false positive* responses, consisting in responding 'Yes' when the correct response according to the reference structure is 'No'; (2) the *misses*, which are erroneous 'No' responses. In Cosyn and Thiéry's simulation, the probabilities of two kinds of errors were set equal to 0.05. The first block of both QUERY and PS-QUERY was simulated for nineteen fictitious experts. On the average, the number of queries required for terminating the first block of PS-QUERY was roughly the double of that for terminating the first block of QUERY: 1,480 against 662. Table 12.8 displays these numbers, together with the mean discrepancy indices computed over the nineteen simulations and comparing the target structure $\mathcal{K}^r$ to the knowledge spaces obtained at the end of Block 1. The notation $\mathcal{K}^{E,1}$ in the heading of Table 12.8 refers to the set of all Block 1 data obtained from the nineteen simulated experts.

	No of trials	$\overline{di(\mathcal{K}^r, \mathcal{K}^{E,1})}$	$\overline{di(\mathcal{K}^{E,1}, \mathcal{K}^r)}$
QUERY	662 (39)	1.51 (0.65)	3.00 (1.21)
PS-QUERY	1,480 (65)	0.16 (0.11)	1.43 (0.17)

Table 12.8. Means of the number of trials required to terminate the first block of QUERY and PS-QUERY with the two error probabilities set equal to 0.05, and of the two mean discrepancy indices. All the means are computed over the nineteen simulations. The standard deviations are in parentheses.

The numbers in columns 2 and 3 refer to the average of the distributions of discrepancy indices. (Each average is computed over nineteen discrepancy indices.) These averages are denoted by $\overline{di(\mathcal{K}^r, \mathcal{K}^{E,1})}$ (column 2) and $\overline{di(\mathcal{K}^{E,1}, \mathcal{K}^r)}$ (column 3). The table also contains (in parentheses) the standard deviations of these distributions. The critical column of the table is the second one. The number 1.51 in the second row indicates that the states in $\mathcal{K}^r$ differ on the average by 1.51 items from those contained in the quasi ordinal knowledge structures in $\mathcal{K}^{E,1}$ generated from the first block of QUERY. By contrast, the number 0.16 in the last line shows that, when $\mathcal{K}^{E,1}$ is generated by PS-QUERY, the states in $\mathcal{K}^r$ only differ on the average by 0.16 from those of $\mathcal{K}^{E,1}$. In other words, $\mathcal{K}^r$ is almost included in $\mathcal{K}^{E,1}$. Thus, in principle, almost all the states of the target structure $\mathcal{K}^r$ can be recovered by suitably selecting from the states of $\mathcal{K}^{E,1}$. How such a selection might proceed is discussed in the next section. A more precise picture of the situation is provided by Figure 12.5 which is adapted from Cosyn and Thiéry's paper.

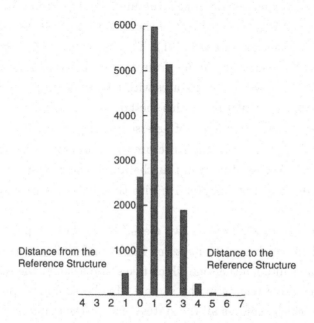

Fig. 12.5. Average discrepancy distributions $f_{\mathcal{K}^r,\mathcal{K}^{E,1}}$ (left) and $f_{\mathcal{K}^{E,1},\mathcal{K}^r}$ (right) obtained from the nineteen simulations of PS-QUERY, with the two error probabilities set equal to 0.05. (Adapted from Cosyn and Thiéry, 1999.)

Figure 12.5 displays two histograms. To the left of the zero point on the abscissa we have the histogram of the average discrepancy from $\mathcal{K}^r$ to the nineteen quasi ordinal spaces in $\mathcal{K}^{E,1}$. We see from the graph that, on the average (computed over the 19 simulations), of the 3043 states of $\mathcal{K}^r$, about 2600 are also in the expert's space, and about 600 are at a distance of 1 to that space. These number are consistent with the mean 0.16 in the last line of the third column of Table 12.8. To the right of the zero point in the abscissa of Figure 12.5, we have a similar histogram for the average discrepancy from the spaces in $\mathcal{K}^{E,1}$ to $\mathcal{K}^r$.

In the next section, we examine how Cosyn and Thiéry go about refining the ordinal knowledge space obtained from Block 1 of one expert so as to obtain a knowledge structure closely approximating the target structure $\mathcal{K}^r$.

Refining a Knowledge Structure

Cosyn and Thiéry start from the assumption that, using PS-QUERY or some other technique, a knowledge structure $\mathcal{K}^{e,1}$ has been obtained (from a single expert), that is a superset[3] of the target structure $\mathcal{K}^r$. They apply then a procedure due to Villano (1991). The idea is to use $\mathcal{K}^{e,1}$ for assessing students in a large enough sample from the population, and to use their data to prune $\mathcal{K}^{e,1}$ by removing states with low probabilities of occurrence. Cosyn and Thiéry simulation of this method show, surprisingly, that this can be achieved with a number of assessed students considerably smaller than the number of states in the structure to be pruned. (Plausible reasons for this fact are discussed later in this chapter.) The assessment procedure used is that described in Chapter 10, with the multiplicative updating rule (cf. 10.10).

The refinement is achieved in two steps. In the first step, a 'smoothing rule' is applied which transforms some initial probability distribution[4] on the set of states into another one which takes into account the results of the assessment of the students in the sample. In the second step, a 'pruning rule' is used which removes all the states—except the empty states and the domain—having a probability lower than some critical cut-off value. The probabilities of the remaining states are then normalized so as to obtained a probability distribution on the subset of remaining states. (Essentially, this amounts to computing the conditional probabilities of the remaining states conditional to the event that one of them occurs.) The next two sections contain the details.

12.21. The Smoothing Rule. Suppose that a large number of students $s_1, \ldots, s_k$ from a representative sample have been assessed by the procedure of Chapter 10, and that the assessment has provided a corresponding

---

[3] This would happen with an expert who would make no error in responding to the questions asked by PS-QUERY, an unrealistic assumption. Cosyn and Thiéry nevertheless make that assumption so as to clearly separate the investigation of the refinement procedure from the analysis of PS-QUERY.

[4] This may be the uniform distribution on $\mathcal{K}^r$.

sequence of probability distributions $\ell_1, \ldots, \ell_k$. Each of these probability distributions has most of its mass concentrated on one or a few states, and summarizes the assessment for one student. These probability distributions are used to transform an initial probability distribution $\mathcal{O}_1$. For concreteness, we may take $\mathcal{O}_1$ to be the uniform distribution $\mathcal{U}$ on some initial knowledge structure. For example, if this initial structure is the knowledge structure $\mathcal{K}^{e,1}$ deduced from the responses of an expert to the PS-QUERY procedure and containing m states, then $\mathcal{O}_1(K) = \frac{1}{m}$ for any K in $\mathcal{K}^{e,1}$. We recall that $\mathcal{K}^{e,1}$ is assumed to be a superset of the target structure $\mathcal{K}^r$.

Keeping track of all the probability distributions ℓ_j, $j = 1, \ldots, k$ is cumbersome if k is large. Accordingly, the effects of ℓ_1, ℓ_2, etc. on $\mathcal{O}_1$ are computed successively. The transformations used by Cosyn and Thiéry (1999) are as follows:

$$\mathcal{O}_1 = \mathcal{U} \tag{10}$$

$$\mathcal{O}_{j+1} = \frac{j\mathcal{O}_j + \ell_j}{j+1} \qquad \text{for } j = 1, 2, \ldots \tag{11}$$

Thus, each of the probability distributions $\mathcal{O}_{j+1}$ is a mixture of $\mathcal{O}_j$ and ℓ_j with probabilities $\frac{j}{j+1}$ and $\frac{1}{j+1}$. Note that for an initial knowledge structure $\mathcal{K}^{e,1}$ and any $j = 1, 2, \ldots$: if $\sum_{K \in \mathcal{K}^{e,1}} \mathcal{O}_j(K) = 1$, then

$$\sum_{K \in \mathcal{K}^{e,1}} \mathcal{O}_{j+1}(K) = \frac{j \sum_{K \in \mathcal{K}^{e,1}} \mathcal{O}_j(K) + \sum_{K \in \mathcal{K}^{e,1}} \ell_j(K)}{j+1} = 1.$$

Presumably, if a sufficiently large number k of students are assessed, then $\mathcal{O}_k$ will differ from $\mathcal{O}_1$ in that most states in $\mathcal{K}^{e,1} \setminus \mathcal{K}^r$ will have a lower probability than most states in $\mathcal{K}^r$. Thus, by removing all those states K of $\mathcal{K}^{e,1}$ with $\mathcal{O}_j(K)$ smaller than some appropriately chosen threshold τ (except the empty state and the domain), one can hope to uncover the structure $\mathcal{K}^r$ to a satisfactory approximation.

12.22. The Pruning Rule. The pruning rule

$$v : (\mathcal{K}, \mathcal{O}, \tau) \mapsto v_\tau(\mathcal{K}, \mathcal{O}) \in 2^{\mathcal{K}}$$

is defined for any knowledge structure $\mathcal{K}$ on some domain Q, any probability distribution $\mathcal{O}$ on $\mathcal{K}$ and any real number $\tau \in [0, 1]$ by the equation

$$v_\tau(\mathcal{K}, \mathcal{O}) = \{K \in \mathcal{K} \mid \mathcal{O}(K) \geq \tau\} \cup \{\varnothing, Q\}. \tag{12}$$

Thus, $v_\tau(\mathcal{K}, \mathcal{O})$ is a knowledge structure, and a probability distribution $\mathcal{O}'$ on $v_\tau(\mathcal{K}, \mathcal{O})$ can be defined by the normalization

$$\mathcal{O}'(K) = \frac{\mathcal{O}(K)}{\sum_{L \in v_\tau(\mathcal{K}, \mathcal{O})} \mathcal{O}(L)}.$$

Note that the order in which the successive transformations of $\mathcal{O}_j$ are carried out are immaterial because the operator combining the successive distributions ℓ_j—which is implicitly defined by Equation (11)—is commutative (cf. Problem 9). The last student assessed has thus the same effect on building the resulting knowledge structure as the first one.

It remains to show that this scheme can work in practice. Again, Cosyn and Thiéry answer this question by a simulation. In passing they determine a suitable value for the threshold τ. Some of their analyses are based on computing the *average discrepancy index* between a structure $\mathcal{K}$ and a structure $\mathcal{K}'$, that is, the quadratic mean

$$\overline{di}(\mathcal{K}, \mathcal{K}') = \sqrt{di^2(\mathcal{K}, \mathcal{K}') + di^2(\mathcal{K}', \mathcal{K}')} \tag{13}$$

between the two discrepancy indices for these structures.

Simulations of Various Refinements

Cosyn and Thiéry (1999) consider a knowledge structure $\mathcal{K}^r$ representing the full set of knowledge states in some fictitious population of reference. They suppose that an expert has been tested by PS-QUERY, and that the first block of responses provided a knowledge structure $\mathcal{K}^{e,1}$ (which is thus quasi ordinal). Specifically, they take $\mathcal{K}^r$ to be the reference knowledge structure used in the simulation of PS-QUERY and described in an earlier section (see 12.20), and they suppose that $\mathcal{K}^{e,1} = \mathcal{K}^r_1$. Thus, $\mathcal{K}^r$ and $\mathcal{K}^{e,1}$ have 3043 and 14346 states, respectively.

Various samples of fictitious subjects were generated by random sampling in the set of 3043 states of $\mathcal{K}^r$. The probability distribution used for this sampling was the uniform distribution $\mathcal{O}_1 = \mathcal{U}$ on $\mathcal{K}^r$. The sample sizes ranged from 1,000 to 10,000. Each of these fictitious subjects was tested by the assessment procedure of Chapter 10, using the multiplicative rule (Definition 10.10). It was assumed that these subjects were never able to guess the response of an item that was not in their state. On the other hand,

the careless errors probability for any item q was assumed to be either 0
or 0.10. (In the notation of 8.12, we have thus $\eta_q = 0$ and either $\beta_q = 0$
or $\beta_q = 0.10$.) For each of the samples, the transformed distribution $\mathcal{O}_k$
(where k is the size of the sample) was computed by the Smoothing Rule
defined by Equations (10)-(11).

12.23. The value of the threshold. The first task was to determine
an adequate value for the threshold τ of the Pruning Rule, according to
which a state K of $\mathcal{K}^{e,1}$ is retained if and only if $\mathcal{O}_k(K) \geq \tau$ or K is the
domain or is empty. Various values for the threshold τ were compared using
a sample of 1,000 fictitious subject. The knowledge structure generated
by pruning $\mathcal{K}^{e,1}$ with a threshold τ is thus $\mathcal{V}_\tau(1001) = v_\tau(\mathcal{K}^{e,1}, \mathcal{O}_{1001})$.
The criterion used for the comparison was the average discrepancy index
$\overline{di}(\mathcal{K}^r, \mathcal{V}_\tau(1001))$ between $\mathcal{K}^r$ and $\mathcal{V}_\tau(1001)$, that is, the quadratic mean
between the two discrepancy indices for these structures (cf. Equation (13)).
This index was computed for τ varying over a wide range. It was found
that $\overline{di}(\mathcal{K}^r, \mathcal{V}_\tau(1001))$ was very nearly minimum for $\tau = 1/|\mathcal{K}^{e,1}|$, the value
of the uniform distribution on $\mathcal{K}^{e,1}$. This estimate of τ was adopted for all
further simulations in Cosyn and Thiéry's paper. Accordingly, we drop the
notation for the threshold and write $\mathcal{V}(k) = \mathcal{V}_{\frac{1}{|\mathcal{Q}|}}(k)$ in the sequel.

12.24. The number of subjects. The effect of the number of subjects
assessed on the accuracy of the recovery of the target structure was also
investigated by simulation. The number k of subjects was varied from 0
to 10,000, for two careless error probabilities: $\beta = 0$ and $\beta = 0.10$. The
two discrepancy indices $di(\mathcal{K}^r, \mathcal{V}(k))$ and $di(\mathcal{V}(k), \mathcal{K}^r)$ were computed in
all cases and form the basis of the evaluation.

Conclusions. The two discrepancy indices decrease as the number k of
subjects increase (for $k > 100$). Specifically, with $\beta = 0$, the discrepancy
index $\overline{di}(\mathcal{K}^r, \mathcal{V}(k))$ decreases from 0.62 initially, to 0.01 after about 3000
subjects. With $\beta = 0.10$, as many as 8000 are required for this discrepancy
index to reach the same value of 0.01.

The behavior of the other discrepancy index $\overline{di}(\mathcal{V}(k), \mathcal{K}^r)$ is different. Its
value decreases mostly for k between 0 and 1000 and appears to reach an
asymptote at about $k = 1,500$. The asymptotic value are around 0.2 and
0.4 for $\beta = 0$ and $\beta = 0.10$, respectively. The results indicate that, in both

cases, most of the superfluous states were discarded at about $k = 1000$.

Overall, the refinement procedure studied by Cosyn and Thiéry (1999) manage to recover 92% of the reference states. The role of the careless error rate is worth noticing. On the one hand, the asymptotic value of $\overline{di}(\mathcal{V}(k), \mathcal{K}^r)$ increases with the error probability β. On the other hand, $\overline{di}(\mathcal{K}^r, \mathcal{V}(k))$ appears to tend to zero regardless of the error rate. It appears thus that, while a large error rate of the subjects may produce a large refined structure, most of the referent states will nevertheless be recovered if enough subjects are tested.

Original Sources and Related works

The bulk of this chapter is derived from two papers: Kambouri et al. (1993) and Cosyn and Thiéry (1999). Dowling has a paper in which she develops similar ideas (Müller, 1989; see also Dowling, 1994). In another paper (Dowling, 1993a), she combines these ideas into an algorithm that moreover exploits the base to store a space economically. We also stress the seminal rôle played by Villano's dissertation (Villano, 1991) in the work of Cosyn and Thiéry (1999).

Problems

1. Certain queries need not be asked by the QUERY routine because the responses are known a priori. For instance, and positive response $A\mathcal{P}q$ with $q \in A$ must be taken for granted. Why is that reducing the number of possible queries by one-half (cf. Remark 12.2(2))?

2. Prove using Equation (3) that if $A \subseteq A' \subseteq Q$ and $A\mathcal{P}x$, then $A'\mathcal{P}x$ (see Example 12.3(a)).

3. Show that the relation $\mathcal{P}$ defined by (3) restricted to pairs of items is transitive.

In the four following problems, we ask the reader to provide a formal justification for each of the four rules of inference contained in Table 12.2.

4. Prove [IR1].

5. Prove [IR2].

6. Prove [IR3].

7. Prove [IR4].

8. Let $\mathcal{P}$ be the unique entail relation corresponding to a knowledge space $(Q, \mathcal{K})$, and let A^+, A^- and $\overline{\mathcal{P}}$ be defined as in Equations (8), (9) and (2), respectively. Prove the following facts: (i) $A \subseteq A^+$; (ii) $A^+ \cap A^- = \varnothing$; (iii) $A^+ \cup A^- = Q$; (iv) A^- is a knowledge state; (v) any knowledge state in $\mathcal{K}$ must be equal to some set A^-. (You may find Theorem 5.5 useful for answering the last two questions.)

9. Show that the operator implicitly defined by Equation (11) and combining any two distribution ℓ_j and ℓ_{j+1} is commutative.

References

ACZÉL, J. (1966) *Lectures on Functional Equations and their Applications.* Mathematics in Science and Engineering, vol. 19. New York: Academic Press. [↦ 188, 230]

ADKE, S.R., & MANJUNATH, S.M. (1984) *An Introduction to Finite Markov Processes.* New York: Wiley. [↦ 191]

ALBERT, D. (Ed.) (1994) *Knowledge Structures.* Berlin–Heidelberg: Springer–Verlag. [↦ viii]

ALBERT, D., & HOCKEMEYER, C. (1997) Dynamic and adaptive hypertext tutoring systems based on knowledge space theory. In du Boulay, B., & Mizoguchi, R. (Eds.), *Artificial Intelligence in Education: Knowledge and Media in Learning Systems.* Frontiers in Artificial Intelligence and Applications, vol. 39 (pp. 553–555). Amsterdam: IOS Press. [↦ 10]

ALBERT, D., & LUKAS, J. (Eds.) (1998) *Knowledge Spaces: Theories, Empirical Research, Applications.* Berlin–Heidelberg: Springer–Verlag. [↦ viii]

ALBERT, D., SCHREPP, M., & HELD, TH. (1992) Construction of knowledge spaces for problem solving in chess. In Fischer, G.H., & Laming, D. (Eds.), *Contributions to Mathematical Psychology, Psychometrics, and Methodology.* (pp. 123–135). New York: Springer–Verlag. [↦ 86, 102]

ARMSTRONG, W.W. (1974) Dependency structures of data base relationships. *Information Processing,* **74**, 580–583. [↦ 117]

ARNOLD, K.C. (1986) Screen updating and cursor movement optimization: a library package. *UNIX Programmer's Supplementary Documents,* vol. 1. Berkeley: University of California, Computer Science Division. [↦ 288]

BARBUT, M., & MONJARDET, B. (1970) *Ordre et Classification: Algèbre et Combinatoire.* Paris: Hachette. [↦ 129, 139]

BARR, A., & FEIGENBAUM, E.A. (1981) *The Handbook of Artificial Intelligence.* London: Pitman. [↦ 83]

BIRKHOFF, G. (1937) Rings of sets. *Duke Mathematical Journal,* **3**, 443–454. [↦ 5, 39, 41]

BIRKHOFF, G. (1967) *Lattice Theory*. Providence, R.I.: American Mathematical Society. [↦ 28, 41, 123, 129, 139, 141]

BLYTH, T.S., & JANOWITZ, M.F. (1972) *Residuation Theory*. London: Pergamon. [↦ 139]

BRENT, R.P. (1973) *Algorithms for Minimization Without Derivatives*. Englewood Cliffs, N.J.: Prentice Hall. [↦ 149, 150]

BRUNK, H.D. (1965) *An Introduction to Mathematical Statistics*. (2nd. ed.) Waltham, Ma.: Blaisdell. [↦ 150]

BUEKENHOUT, F. (1967) Espaces à fermeture. *Bulletin de la Société Mathématique de Belgique*, **19**, 147–178. [↦ 41]

BUSH, R.R., & MOSTELLER, F. (1955) *Stochastic Models for Learning*. New York: Wiley. [↦ 245]

COGIS, O. (1982) Ferrers digraphs and threshold graphs. *Discrete Mathematics*, **38**, 33–46. [↦ 60]

COHN, P.M. (1965) *Universal Algebra*. New York: Harper and Row. [↦ 43]

COSYN, S., & THIÉRY, N. (1999) A practical procedure to build a knowledge structure. *Journal of Mathematical Psychology*, to appear. [↦ 105, 110, 172, 200, 275, 300, 303, 305, 306, 308]

CRAMÉR, H. (1963) *Mathematical Methods of Statistics*. Princeton, N.J.: Princeton University Press. [↦ 150]

DAVEY, B.A., & PRIESTLEY, H.A. (1990) *Introduction to Lattice and Orders*. Cambridge: Cambridge University Press. [↦ 28, 83]

DEGREEF, E., DOIGNON, J.-P., DUCAMP, A., & FALMAGNE, J.-CL. (1986) Languages for the assessment of knowledge. *Journal of Mathematical Psychology*, **30**, 243–256. [↦ 218, 220]

DOIGNON, J.-P. (1994a) Probabilistic assessment of knowledge. In Albert, D. (Ed.), *Knowledge Structures*. (pp. 1–57). Berlin–Heidelberg: Springer–Verlag. [↦ 16]

DOIGNON, J.-P. (1994b) Knowledge spaces and skill assignments. In Fischer, G.H., & Laming, D. (Eds.), *Contributions to Mathematical Psychology, Psychometrics, and Methodology*. (pp. 111–121). New York: Springer–Verlag. [↦ 102]

DOIGNON J.-P., DUCAMP, A., & FALMAGNE, J.-CL. (1984) On realizable biorders and the biorder dimension of a relation. *Journal of Mathematical Psychology*, **28**, 73–109. [↦ 52, 60]

DOIGNON J.-P., & FALMAGNE, J.-CL. (1985) Spaces for the assessment of knowledge. *International Journal of Man-Machine Studies*, **23**, 175–196. [↦ 16, 40, 82, 83, 139]

DOIGNON, J.-P., & FALMAGNE, J.-CL. (1987) Knowledge assessment: a set theoretical framework. In Ganter, B., Wille, R., & Wolfe, K.E. (Eds.), *Beiträge zur Begriffsanalyse, Vorträge der Arbeitstagung Begriffsanalyse, Darmstadt, 1986.* (pp. 129–140). Mannheim: B.I. Wissenschaftsverlag. [↦ 16]

DOIGNON, J.-P., & FALMAGNE J.-CL. (1988) Parametrization of knowledge structures. *Discrete Applied Mathematics*, **21**, 87–100. [↦ 16]

DOIGNON, J.-P., & FALMAGNE, J.-CL. (Eds.) (1991) *Mathematical Psychology: Current Developments.* New York: Springer–Verlag. [↦ 312]

DOIGNON, J.-P., & FALMAGNE J.-CL. (1997) Well-graded families of relations. *Discrete Mathematics*, **173**, 35–44. [↦ 53, 55, 60, 61]

DOIGNON J.-P., MONJARDET, B., ROUBENS, M., & VINCKE, PH. (1986) Biorder families, valued relations and preference modelling. *Journal of Mathematical Psychology*, **30**, 435–480. [↦ 60]

DOWLING, C.E. (1991) Constructing knowledge spaces from judgements with differing degrees of certainty. In Doignon, J.-P., & Falmagne, J.-Cl. (Eds.), *Mathematical Psychology: Current Developments.* (pp. 221–231). New York: Springer–Verlag. [↦ 199]

DOWLING, C.E. (1993a) Applying the basis of a knowledge space for controlling the questioning of an expert. *Journal of Mathematical Psychology*, **37**, 21–48. [↦ 199, 308]

DOWLING, C.E. (1993b) On the irredundant construction of knowledge spaces. *Journal of Mathematical Psychology*, **37**, 49–62. [↦ 29, 30, 41, 139]

DOWLING, C.E. (1994) Integrating different knowledge spaces. In Fischer, G.H., & Laming, D. (Ed.), *Contributions to Mathematical Psychology, Psychometrics, and Methodology.* (pp. 151–158). New York: Springer–Verlag. [↦ 118]

DOWLING, C.E., HOCKEMEYER, C., & LUDWIG A.H. (1996) Adaptive assessment and training using the neighbourhood of knowledge spaces. In Frasson, C., Gauthier, G., & Lesgold, A. (Eds.), *Intelligent Tutoring Systems.* (pp. 578–586). Berlin: Springer–Verlag. [↦ 10]

DRASGOW, F. (1986) Polychoric and polyserial correlations. In Kotz, S., Johnson, N.L., & Read, C.B. (Eds.), *Encyclopedia of Statistical Sciences, vol. 7.* (pp. 68–74). New York: Wiley. [↦ 292]

DUCAMP, A., & FALMAGNE, J.-CL. (1969) Composite measurement. *Journal of Mathematical Psychology*, **6**, 359–390. [↦ 52, 60]

DUDA, R.O., & HART, P.E. (1973) *Pattern Classification and Scene Analysis.* New York: Wiley. [↦ 11, 16]

DUGUNDJI, J. (1966) *Topology.* Boston: Allyn and Bacon. [↦ 15, 93]

DÜNTSCH, I., & GEDIGA, G. (1995) Skills and knowledge structures. *British Journal of Mathematical and Statistical Psychology*, **48**, 9–27. [↦ 102]

DÜNTSCH, I., & GEDIGA, G. (1996) On query procedures to build knowledge structures. *Journal of Mathematical Psychology*, **40**, 160–168. [↦ 118]

DURNIN, J., & SCANDURA, J.M. (1973) An algorithmic approach to assessing behavioral potential: comparison with item forms and hierarchical technologies. *Journal of Educational Psychology*, **65**, 262–272. [↦ 17]

EDELMAN, P.H., & JAMISON, R.E. (1985) The theory of convex geometries. *Geometriae Dedicata*, **19**, 247–270. [↦ 59, 60, 83]

FALMAGNE, J.-CL. (1989a) A latent trait theory via a stochastic learning theory for a knowledge space. *Psychometrika*, **54**, 283–303. [↦ 172, 203]

FALMAGNE, J.-CL. (1989b) Probabilistic knowledge spaces: a review. In Roberts, F.S. (Ed.), *Applications of Combinatorics and Graph Theory to the Biological and Social Sciences, IMA, vol. 17.* (pp. 283–303). New York: Springer–Verlag. [↦ 16, 172]

FALMAGNE, J.-CL. (1993) Stochastic learning paths in a knowledge structure. *Journal of Mathematical Psychology*, **37**, 489–512. [↦ 175, 187, 203]

FALMAGNE, J.-CL. (1994) Finite Markov learning models for knowledge structures. In Fischer, G.H., & Laming, D. (Eds.), *Contributions to Mathematical Psychology, Psychometrics, and Methodology.* (pp. 75–89). New York: Springer–Verlag. [↦ 159, 172]

FALMAGNE, J.-CL. (1996) Errata to SLP. *Journal of Mathematical Psychology*, **40**, 169–174. [↦ 175, 194, 204]

FALMAGNE, J.-CL., & DOIGNON, J.-P. (1988a) A class of stochastic procedures for the assessment of knowledge. *British Journal of Mathematical and Statistical Psychology*, **41**, 1–23. [↦ 172, 222, 245]

FALMAGNE, J.-CL., & DOIGNON, J.-P. (1988b) A Markovian procedure for assessing the states of a system. *Journal of Mathematical Psychology*, **32**, 232–258. [↦ 59, 172, 271]

FALMAGNE, J.-CL., & DOIGNON, J.-P. (1993) A stochastic theory for system failure assessment. In Bouchon-Meunier, B., Valverde, L., & Yager, R.R. (Eds.), *Uncertainty in Intelligent Systems*. (pp. 431–440). Amsterdam: North-Holland. [↦ 314]

FALMAGNE, J.-CL., & DOIGNON, J.-P. (1997) Stochastic evolution of rationality. *Theory and Decision*, **43**, 107–138. [↦ 60]

FALMAGNE, J.-CL., & DOIGNON, J.-P. (1998) Meshing knowledge structures. In Dowling, C., Roberts, F.S., & Theuns, P. (Eds.), *Recent Progress in Mathematical Psychology*. (pp. 143–154). Hillsdale, N.J.: Erlbaum. [↦ 118]

FALMAGNE J.-CL., KOPPEN, M., VILLANO, M., DOIGNON, J.-P., & JOHANESSEN, L. (1990) Introduction to knowledge spaces: how to build, test and search them. *Psychological Review*, **97**, 201–224. [↦ 16, 86, 102, 117, 146, 172, 199, 200]

FALMAGNE, J.-CL., & LAKSHMINARAYAN, K. (1994) Stochastic learning paths—Estimation and simulation. In Fischer, G.H., & Laming, D. (Eds.), *Contributions to Mathematical Psychology, Psychometrics, and Methodology*. (pp. 91–110). New York: Springer–Verlag. [↦ 204]

FEINBERG, S.E. (1981) *The Analysis of Cross-Classified Categorical Data*. (2nd. ed.) Cambridge, Ma.: MIT Press. [↦ 149]

FELLER, W. (1968) *An Introduction to Probability and its Applications*. (vol. 1, vol. 2 in 1966) New York: Academic Press. [↦ 160, 172, 259]

FISHBURN, P.C. (1970) Intransitive indifference with unequal indifference intervals. *Journal of Mathematical Psychology*, **7**, 144–149. [↦ 53, 60]

FISHBURN, P.C. (1985) *Interval Orders and Interval Graphs*. New York: Wiley. [↦ 60]

FLAMENT, CL. (1976) *L'Analyse Booléenne de Questionnaires*. Paris–The Hague: Mouton. [↦ 83]

FRASER, D.A.S. (1958) *Statistics, An Introduction*. New York: Wiley. [↦ 150]

FRIES, S. (1997) Empirical validation of a Markovian learning model for knowledge structures. *Journal of Mathematical Psychology*, **41**, 65–70. [↦ 172]

FU, K.S. (1974) *Syntactic Methods in Pattern Recognition.* New York: Academic Pres. [↦ 11, 16]

GANTER, B. (1984) *Two basic algorithms in concept analysis.* (FB4–Preprint, number 831.) TH Darmstadt. [↦ 41, 118, 139]

GANTER, B. (1987) Algorithmen zur Formalen Begriffsanalyse. In Ganter, B., Wille, R., & Wolfe, K.E. (Eds.), *Beiträge zur Begriffsanalyse, Vorträge der Arbeitstagung Begriffsanalyse, Darmstadt, 1986.* (pp. 241–254). Mannheim: B.I. Wissenschaftsverlag. [↦ 41, 139]

GANTER, B., & REUTER, K. (1991) Finding all closed sets: a general approach. *Order*, **8**, 283–290. [↦ 139]

GANTER, B., & WILLE, R. (1996) *Formale Begriffsanalyse: Mathematische Grundlagen.* Berlin–Heidelberg: Springer–Verlag (English translation by C. Franzke: *Formal Concept Analysis: Mathematical Foundations*, Springer–Verlag, 1998) . [↦ 41, 129, 139]

GAREY, M.R., & JOHNSON, D.S. (1979) *Computers and Intractability: A guide to the Theory of NP-completeness.* New York: W.H. Freeman. [↦ 33, 209]

GEGENFURTNER, K. (1992) Praxis: Brent's algorithm for function minimization. *Behavior Research Methods Instruments and Computers*, **24**, 560–564. [↦ 150, 199]

GUIGUES, J.-L., & DUQUENNE, V. (1986) Familles minimales d'implications informatives résultant d'un tableau de données binaires. *Mathématiques et Sciences Humaines*, **97**, 5–18. [↦ 118]

GUTTMAN, L. (1944) A basis for scaling qualitative data. *American Sociological Review*, **9**, 139–150. [↦ 52, 60]

HOCKEMEYER, C. (1997) RATH - a relational adaptive tutoring hypertext WWW–environment. Institut für Psychologie, Karl-Franzens-Universität Graz, Austria, 1997. Technical Report No. 1997/3. [↦ 10]

HOCKEMEYER, C., HELD, T., & ALBERT, D. (1998) RATH - a relational adaptive tutoring hypertext WWW–environment based on knowledge space theory. *Proceedings of the CALISCE '98 meeting (Computer Aided Learning in Science and Engineering).* Göteborg, Sweden, June 15-17, 1998. [↦ 10]

HYAFILL, L., & RIVEST, R.L. (1976) Constructing optimal binary decision trees is NP-complete. *Information Processing Letters*, **5**, 15–17. [↦ 209]

JAMESON, K. (1992) Empirical methods for generative semiotics models: an application to the roman majuscules. In Watt, W.C. (Ed.), *Writing Systems and Cognition*. (Neuropsychology and Cognition Series.) Dordrecht: Kluwer Academic Publishers. [↦ 16, 43]

JAMISON-WALDNER, R.E. (1982) A perspective on abstract convexity: classifying alignments by varieties. *Convexity and related combinatorial geometry (Norman, Okla., 1980)*. Lecture Notes in Pure and Applied Mathematics, vol. 76. (pp. 113–150). New York: Dekker. [↦ 41]

KAMBOURI, M. (1991) *Knowledge Assessment: a Comparison of Human Experts and Computerized Procedures*. Ph. D. Dissertation, New York University. [↦ 245, 275, 285, 297]

KAMBOURI, M., KOPPEN, M., VILLANO, M., & FALMAGNE, J.-CL. (1993) Knowledge assessment: tapping human expertise by the QUERY routine. *International Journal of Human-Computer Studies*, **40**, 119–151. [↦ 199, 200, 275, 288, 299, 300, 308]

KELVIN, W.T. (1889) *Popular Lectures and Addresses (in 3 volumes)*. (Vol. 1: *Constitution of Matter*, Chapter *Electrical Units of Measurement*.) London: MacMillan. [↦ vii]

KEMENY, J.G., & SNELL, J.L. (1960) *Finite Markov Chains*. Princeton, N.J.: Van Nostrand. [↦ 160, 172, 259]

KIMBAL, R. (1982) A self-improving tutor for symbolic integration. In Sleeman, D., & Brown, J.S. (Eds.), *Intelligent tutoring systems*. Computer and People Series. London: Academic Press. [↦ 245]

KOPPEN, M. (1989) *Ordinal Data Analysis: Biorder Representation and Knowledge Spaces*. Doctoral Dissertation, Nijmegen University, The Netherlands. [↦ 83, 199]

KOPPEN, M. (1993) Extracting human expertise for constructing knowledge spaces: an algorithm. *Journal of Mathematical Psychology*, **37**, 1–20. [↦ 3, 275, 282]

KOPPEN, M. (1994) The construction of knowledge spaces by querying experts. In Fischer, G.H., & Laming, D. (Eds.), *Contributions to Mathematical Psychology, Psychometrics, and Methodology*. (pp. 137–147). New York: Springer–Verlag. [↦ 316]

KOPPEN, M. (1998) On alternative representations for knowledge spaces. *Mathematical Social Sciences*, to appear . [↦ 83]

KOPPEN, M., & DOIGNON, J.-P. (1988) How to build a knowledge space by querying an expert. *Journal of Mathematical Psychology*, **34**, 311–331. [↦ 3, 117, 139, 199]

LAKSHMINARAYAN, K. (1995) *Theoretical and Empirical Aspects of Some Stochastic Learning Models*. Ph. D. Dissertation, University of California, Irvine. [↦ 5, 146, 172, 175, 191, 193, 194, 196, 199, 200, 202, 204]

LAKSHMINARAYAN, K., & GILSON, F. (1993) *An Application of a Stochastic Knowledge Structure Model*. (Technical Report MBS.) University of California, Irvine: Institute for Mathematical Behavioral Sciences. [↦ 146, 172, 203]

LANDY, M.S., & HUMMEL, R.A. (1986) A brief survey of knowledge aggregation methods.*Proceedings, 1986 International Conference on Pattern Recognition* . [↦ 230]

LEHMAN, E.L. (1959) *Testing Statistical Hypotheses*. New York: Wiley. [↦ 150]

LINDGREN, B.W. (1968) *Statistical Theory*. (2nd. ed.) New York: Macmillan. [↦ 150]

LORD, F.M. (1974) Individualized testing and item characteristic curve theory. In Krantz, D.H., Atkinson, R.C., Luce, R.D., & Suppes, P. (Eds.), *Contemporary Developments in Mathematical Psychology, vol. 2, Measurement, Psychophysics, and Neural Information Processing*. (pp. 106-126). San Francisco, CA: W.H. Freeman. [↦ 17, 264]

LORD, F.M., & NOVICK, M.R. (1974) *Statistical Theories of Mental Tests*. (2nd. ed.) Reading, Ma.: Addison-Wesley. [↦ 17, 86, 186]

LUCE, R.D. (1956) Semiorders and a theory of utility discrimination. *Econometrica*, **24**, 178–191. [↦ 53, 60]

LUCE, R.D. (1959) *Individual Choice Behavior*. New York: Wiley. [↦ 245]

LUCE, R.D. (1964) Some one-parameter families of commutative learning operators. In Atkinson, A.C. (Ed.), *Studies in Mathematical Psychology*. (pp. 380–398). Stanford University Press. [↦ 230]

LUKAS, J., & ALBERT, D. (1993) Knowledge assessment based on skill assignment and psychological task analysis. In Strube, G., & Wender, K.F. (Eds.), *The Cognitive Psychology of Knowledge.* (pp. 139–159). Amsterdam: Elsevier. [↦ 86, 102]

MARLEY, A.A.J. (1967) Abstract one-parameter families of commutative learning operators. *Journal of Mathematical Psychology*, **4**, 414–429. [↦ 230]

MARSHALL, S.P. (1981) Sequential item selection: optimal and heuristic policies. *Journal of Mathematical Psychology*, **23**, 134–152. [↦ 86]

MATALON, B. (1965) *L'Analyse Hiérarchique.* Paris: Mouton. [↦ 129, 139]

MONJARDET, M. (1970) Tresses, fuseaux, préordres et topologies. *Mathématiques et Sciences Humaines*, **30**, 11–22. [↦ 121, 129, 139]

MÜLLER, C.E. (1989) A procedure for facilitating an expert's judgement on a set of rules. In Roskam, E.E. (Ed.), *Mathematical Psychology in Progress.* (pp. 157–170). Berlin–Heidelberg–New York: Springer–Verlag. [↦ 117, 199, 308]

NORMAN, M.F. (1972) *Markov Processes and Learning Models.* New York: Academic Press. [↦ 245]

OVCHINNIKOV, S. (1983) Convex geometry and group choice. *Mathematical Social Sciences*, **5**, 1–16. [↦ 60]

PARZEN, E. (1960) *Modern Probability Theory and its Applications.* New York: Wiley. [↦ 180]

PARZEN, E. (1962) *Stochastic Processes.* San Francisco, CA: Holden-Day. [↦ 160, 172, 259]

PEARSON, K. (1924) *The Life, Letters and Labours of Francis Galton.* (Vol. 2: *Researches of Middle Life.*) London: Cambridge University Press. [↦ v, vi]

PIRLOT, M., & VINCKE, PH. (1997) *Semiorders: Properties, Representations, Applications.* Amsterdam: Kluwer. [↦ 60]

POWELL, M.J.D. (1964) An efficient method for finding the minimum of a function in several variables without calculating derivatives. *Computer Journal*, **7**, 155–162. [↦ 149, 199]

RICH, E. (1983) *Artificial Intelligence.* Singapore: McGraw-Hill. [↦ 83]

RIGUET, J. (1951) Les relations de Ferrers. *Comptes Rendus des Séances de l'Académie des Sciences (Paris)*, **232**, 1729–1730. [↦ 60]

ROBERTS, F.S. (1979) *Measurement Theory with Applications to Decision Making, Utility and the Social Sciences*. Encyclopedia of Mathematics and its Applications, ed. Rota, G.-C., vol. 7: Mathematics and the Social Sciences. Reading, Ma.: Addison-Wesley. [↦ 13, 60]

ROUBENS, M., & VINCKE, PH. (1985) *Preference Modelling*. Lecture Notes in Economics and Mathematical Systems, vol. 250. Berlin–Heidelberg: Springer–Verlag. [↦ 60]

RUSCH, A., & WILLE, R. (1996) Knowledge spaces and formal concept analysis. In Bock, H.-H., & Polasek, W. (Eds.), *Data Analysis and Information Systems*. (pp. 427–436). Berlin–Heidelberg: Springer–Verlag. [↦ 139]

SCOTT, D., & SUPPES, P. (1958) Foundational aspects of theories of measurement. *Journal of Symbolic Logic*, **23**, 113–128. [↦ 53, 60]

SHORTLIFFE, E.H. (1976) *Computer-Based Medical Consultation: Mycin*. New York: American Elsevier. [↦ 10, 16]

SHORTLIFFE, E.H., & BUCHANAN, D.G. (1975) A model of inexact reasoning in medicine. *Mathematical Bioscience*, **23**, 351–379. [↦ 10, 16]

SHYRYAYEV, A.N. (1996) *Probability*. (2nd edition.) New York: Springer–Verlag. [↦ 160]

SIERKSMA, G. (1981) Convexity on unions of sets. *Compositio Mathematica*, **42**, 391–400. [↦ 41]

STERN, J., & LAKSHMINARAYAN, K. (1995) *Comments on Mathematical and Logical Aspects of A Model of Acquisition Behavior*. (Technical Report MBS 95-05.) University of California, Irvine: Institute for Mathematical Behavioral Sciences. [↦ 187, 191, 194, 197, 204]

SUPPES, P. (1960) *Introduction to Logic*. Princeton, N.J.: Van Norstrand. [↦ 13, 15]

SUPPES, P., KRANTZ, D.M., LUCE, R.D., & TVERSKY, A. (1989) *Foundations of Measurement*. (vol. 2.) San Diego: Academic Press. [↦ 60]

SZPILRAJN, E. (1930) Sur l'extension de l'ordre partiel. *Fundamenta Mathematica*, **16**, 386–389. [↦ 81, 271]

TAAGEPERA, M., POTTER, F., MILLER, G.E., & LAKSHMINARAYAN, K. (1997) Mapping students' thinking patterns by the use of knowledge space theory. *International Journal of Science Education*, **19**, 283–302. [↦ 146, 172, 199, 204]

TALLIS, G.M. (1962) The maximum likelihood estimation of correlation for contingency tables. *Biometrics*, **9**, 342–353. [↦ 292]

TROTTER, W.T. (1992) *Combinatorics and Partially Ordered Sets: Dimension Theory*. Baltimore, Maryland: The Johns Hopkins University Press. [↦ 60, 81, 271]

VAN DE VEL, M. (1993) *The Theory of Convex Structures*. Amsterdam: North Holland. [↦ 41, 43, 83]

VAN LEHN, K. (1988) Student Modeling. In Polson, M.C., & Richardson, J.J. (Eds.), *Foundations of Intelligent Tutoring Systems*. (pp. 55–78). Hillsdale, N.J.: Erlbaum. [↦ 8]

VILLANO, M. (1991) *Computerized Knowledge Assessment: Building the Knowledge Structure and Calibrating the Assessment Routine*. Doctoral Dissertation, New York University, New York. *Dissertation Abstracts International*, **552**, 12B. [↦ 105, 111, 145, 146, 155, 172, 200, 245, 300, 304, 308]

VILLANO, M., FALMAGNE, J.-CL., JOHANNESEN, L., & DOIGNON, J.-P. (1987) Stochastic procedures for assessing an individual's state of knowledge. *Proceedings of the International Conference on Computer-assisted Learning in Post-Secondary Education, Calgary*. (pp. 369–371). Calgary, Canada: University of Calgary Press. [↦ 172, 245]

WAINER W., & MESSICK, S. (1983) *Principles of Modern Psychological Measurement. A Festschrift for Frederic M. Lord*. Hillsdale, N. J.: Erlbaum. [↦ 17, 86]

WEISS, D.J. (1983) *New Horizons in Testing: Latent Trait Theory and Computerized Testing*. New York: Academic Press. [↦ 17, 86, 264]

WILD, M. (1994) A theory of finite closure spaces based on implications. *Advances in Mathematics*, **108**, 118–139. [↦ 117]

Index of Names

A

Aczél, J.   188, 230, *310*[1].
Adke, S.   191, *310*.
Albert, D.   viii, ix, 10, 86, 102, *310*, *318*.
Armstrong, W.   117, *310*.
Arnold, K.   288, *310*.

B

Babbage, C.   vii.
Baker, B.   ix.
Barbut, M.   129, 139, *310*.
Barr, A.   83, *310*.
Bayes, T.   231.
Bernoulli, J.   263.
Binet, A.   vii.
Birkhoff, G.   5, 28, 39, 41, 69, 105, 120, 123, 129, 130, 139, 141, *310*, *311*.
Blyth, T.   139, *311*.
Brent, R.   149, 150, *311*.
Brunk, H.   150, *311*.
Buchanan, D. G.   10, 16, *319*.
Buekenhout, F.   41, *311*.
Bush, R.   245, *311*.

C

Chamberlain, N.   37.
Ching-Fang Seu   ix.
Chubb, C.   ix.
Churchill, W.   37.
Cogis, O.   60, *311*.
Cohn, P.   43, *311*.
Cosyn, E.   ix, 105, 110, 172, 200, 275, 300–306, 308, *311*.

---

[1]Number pages in slanted characters refer to the list of references.

C (continued)

Cramér, H.   150, *311*.

D

Davey, B.   28, 83, *311*.
Degreef, E.   218, 220, *311*.
Doignon, J.-P.   ix, 3, 16, 40, 52, 53, 55, 59–61, 82, 83, 86, 102, 117, 118, 139, 146, 172, 199, 200, 218, 220, 222, 245, 271, *311*, *312*, *314*, *317*, *320*.
Dowling, C. E.   ix, 10, 29, 30, 33, 41, 117, 118, 139, 199, 308, *312* (see also Müller).
Drasgow, F.   292, *313*.
Drösler, J.   ix.
Ducamp, A.   52, 60, 218, 220, *311*, *313*.
Duda, R. O.   11, 16, *313*.
Dugundji, J.   15, 93, *313*.
Düntsch, I.   102, 118, *313*.
Duquenne, V.   118, *315*.
Durnin, J.   17, *313*.

E

Edelman, P. H.   59, 60, 83, *313*.

F

Falmagne, J.-C.   v, ix, x, 16, 40, 52, 53, 55, 59–61, 82, 83, 86, 102, 117, 118, 139, 146, 159, 172, 175, 187, 194, 196, 197, 199, 200, 203, 204, 218, 220, 222, 245, 271, 275, 288, 299, 300, 308, *311*–*314*, *316*, *320*.
Feigenbaum, E.   83, *310*.
Feinberg, S. E.   149, *314*.

Index of Terms

---

[1]Number pages in boldface characters refer to definitions.

well-formed expressions   11, 24.
well-graded   **46**–51, 53, 55, **56**–
        61, 63, 64, 73, 81, 85, 111,
        114, 115, 158–160, 173,
        177–179, 197, 198, 203,
        206, 219, 261, 272.
wellgradedness   55, 60, 83, 118,
        272.
word   **210**.
worst case number of a tree   **209**.

Z

Zorn's Lemma   **15**.

Springer
and the
environment

At Springer we firmly believe that an international science publisher has a special obligation to the environment, and our corporate policies consistently reflect this conviction.

We also expect our business partners – paper mills, printers, packaging manufacturers, etc. – to commit themselves to using materials and production processes that do not harm the environment. The paper in this book is made from low- or no-chlorine pulp and is acid free, in conformance with international standards for paper permanency.

Springer

Printing: Druckhaus Beltz, Hemsbach
Binding: Buchbinderei Schäffer, Grünstadt